The Mountain

The Mountain

A Political History from the Enlightenment to the Present

BERNARD DEBARBIEUX
AND GILLES RUDAZ

Translated by Jane Marie Todd
With a Foreword by Martin F. Price

The University of Chicago Press
Chicago and London

Bernard Debarbieux is professor of geography and regional planning at the University of Geneva, Switzerland. **Gilles Rudaz** is senior lecturer and associate researcher of geography at the University of Geneva, Switzerland, and a scientific collaborator at the Swiss Federal Office for the Environment. **Jane Marie Todd** has translated some seventy books, including Dominique Charpin's *Writing Law, and Kingship in Old Babylonian Mesopotamia*, also published by the University of Chicago Press.

The University of Chicago Press, Chicago 60637
The University of Chicago Press, Ltd., London
© 2015 by The University of Chicago
All rights reserved. Published 2015.
Printed in the United States of America

24 23 22 21 20 19 18 17 16 15   1 2 3 4 5

ISBN-13: 978-0-226-03111-8 (cloth)
ISBN-13: 978-0-226-03125-5 (e-book)
DOI: 10.7208/chicago/9780226031255.001.0001

Originally published as *Les faiseurs de montagne*. © CRNS Éditions, 2010.

This work, published as part of a program providing publication assistance, received financial support from the French Ministry of Foreign Affairs, the Cultural Services of the French Embassy in the United States and FACE (French American Cultural Exchange).

FRENCH
VOICES

French Voices Logo designed by Serge Bloch

Library of Congress Cataloging-in-Publication Data

Debarbieux, Bernard, author.
   [Faiseurs de montagne. English]
   The mountain : a political history from the Enlightenment to the present / by Bernard Debarbieux and Gilles Rudaz ; translated by Jane Marie Todd ; with a foreword by Martin F. Price.
     pages cm
   "Originally published as Les faiseurs de montagne. © CNRS Éditions, 2010"—Title page verso.
   Includes bibliographical references and index.
   ISBN 978-0-226-03111-8 (cloth : alkaline paper) — ISBN 978-0-226-03125-5 (e-book) 1. Mountain life—History. 2. Mountains—Political aspects—History. 3. Mountain people—History. 4. Human geography. 5. Mountain ecology. I. Rudaz, Gilles, author. II. Todd, Jane Marie, 1957–translator. III. Price, Martin F., writer of Foreword. IV. Title.
   GT3490.D4313 2015
   910.914′3—dc23

2014045148

CONTENTS

Over the centuries there have been many theories about the origins of mountains ranging from the theological (creationism) to the geological (plate tectonics). Yet, as Bernard Debarbieux and Gilles Rudaz's wide-ranging and fascinating book shows, mountains are also cultural constructions, defined using diverse scientific, social, and political criteria; and mountain people in different parts of the world have been perceived and (mis-)understood in many ways over the past three centuries or so. Recognizing the various understandings of mountains and their inhabitants is important not only for people living in mountains and those who are concerned with them for other reasons, but also as a basis for political action and policy making. This is a particular emphasis of the later chapters of this book, which provide a thorough analysis of the interacting processes linked to the definition of mountains and mountain people and their inclusion (or not) in policies and politics at both the global and the European levels.

A dominant theme running through the book is that of diversity and contrasts, both between mountains and other places and among and even within mountains. In topographic and ecological terms, mountains—however one defines them—are clearly different from the plains below, and this is also often true of their people. From the perspective of those living on the plains, mountain people may be seen as reckless, untrustworthy, "wild," or backward—yet also as hardworking, reliable (especially as soldiers), and spiritually pure, with many other desirable characteristics, especially those related to long-lasting cultures. Mountains are often the borders of administrative entities on all sorts of spatial scales, from districts to empires, and thus are relatively marginalized or peripheral. They are often regarded as barriers to be conquered, broken down, or diminished through the construction of transport corridors. Yet in some cases in both the past (such as the Inca empire and Savoy) and the present (such as Ethiopia and Switzerland), mountains—and sometimes individual peaks—have been central to cultural identity at national and other levels.

Given the diversity of mountains, generalizing about them is fraught with many challenges and even risks. Although individual mountains, and even whole mountain ranges, in different parts of the world may look quite similar, the ecological and cultural processes that have shaped their landscapes vary greatly. Consequently, the transfer of "scientific" models for managing mountain ecosystems from the mountains of one continent to those of another may have unexpected and often very unfortunate consequences for both mountain ecosystems and, especially, the people who depend on them for their livelihoods, as happened many times during the colonial period and has continued until very recently or perhaps is continuing even now.

This is an important book, one that goes beyond Jon Mathieu's historical analysis to bring a new and multidisciplinary understanding of the ways in which mountains in many parts of the world, and across the globe, have been and are given scientific, social, and political meanings.[1] While mountains have been a particular focus of rhetoric and policy development over the past twenty-five years or so, they have had political value for much longer; for instance, within these pages we are introduced to "oropolitics," the climbing of mountains for national or imperialistic political purposes from the nineteenth century onward.

What stands out throughout this book is the role that individuals have played in putting mountains "on the map"—sometimes literally, as with the case of explorers such as Alexander von Humboldt, Halford Mackinder, and Francis Younghusband—but also in terms of aesthetics, literature, politics, and science. Other well-known names—such as Jean-Jacques Rousseau, John Ruskin, and Carl von Clausewitz, as well as those of a number of French scientists and authors who will be less well-known to most English-speaking readers—have also had great influences on Western understandings of mountains. In recent decades a number of mountain-born politicians, scientists active in mountain areas, and certain well-informed journalists have played key roles. This may be true with regard to the understanding and development of policies for any type of environment or ecosystem. Yet although, as the authors note, mountains have never had their own Jacques Cousteau, they have had, and continue to have, many strong advocates who can also benefit from the remarkable and memorable images of mountain people and their environments presented here.

For those with academic interests in a very wide range of disciplines, this book provides an excellent set of foundations for understanding how mountains have been and still are represented in many different contexts— and for those espousing the mountain cause and developing and implementing policies to take into account the particularities of mountain areas, it offers some lessons that may be very valuable for the future.

Martin F. Price
Director, Centre for Mountain Studies, Perth College,
University of the Highlands & Islands, Scotland, and
UNESCO Chair in Sustainable Mountain Development

ACKNOWLEDGMENTS

This book is based in part on research we were able to conduct with the assistance of funding from public institutions. In particular, we would like to thank the Swiss National Science Foundation (SNSF) for its funding of several of Bernard Debarbieux's research projects (nos. 101412-103642, 100013-114004, and CR1I1-137989) and for the research grants awarded to Gilles Rudaz (PBGE-11924 and PA01-117432). We are also grateful to the Ernest Boninchi Foundation (Geneva) for the support it provided for a research project devoted to mountain populations, and to the Swiss Network for International Studies for its support of the Mountlennium project.

We would also like to thank those who read over certain chapters and suggested improvements, especially Martine Edard, Nicolas Evrard, Martin Price, Renato Scariati, and Marie-Karine Schaub. To these we add all with whom we engaged in fruitful exchanges, often over the course of many years, especially Jörg Balsiger, Denis Blamont, Said Boujrouf, Jean Bourliaud, Jo-Ann Carmin, Cristina Del Biaggio, Harald Egerer, Don Funnell, Erik Gloersen, Gregory Grenwood, Jean-Paul Guérin, Thomas Hofer, Louca Lerch, Rafael Matos-Wasem, Bruno Messerli, Alexandre Mignotte, Mathieu Petite, Guido Plassmann, Thomas Scheurer, François Walter, and Ron Witt.

Finally, several people and institutions were of valuable assistance to us in locating illustrations. In that regard we are particularly grateful to Renato Scariati for his aid throughout the book.

Our warm thanks to Mary Laur, Christie Henry, and Abby Collier, editors at the University of Chicago Press, for their assistance in preparing this book for publication. We are also grateful to Jane Marie Todd for her extraordinary translation work. Finally, we express our thanks to Marine Bertea for her commitment to this project and to the Central National du Livre for its role in seeing this book to its conclusion.

INTRODUCTION

Scientific and Political Orogenesis

Human beings' fascination with the mountains dates back to ancient times and is probably universal. Their composition, purpose, and contribution to the order or disorder of the world have always preoccupied the societies who have lived and thought in their vicinity. As a result, mountains occupy a key place in a large number of narratives, particularly cosmogonic ones— whether mythic, religious, or scientific.

One of the recurring questions these narratives have sought to answer is how mountains were formed. In 1840 the Swiss geologist Amanz Gressly proposed the term "orogenesis" to designate the formation process (*genesis* in ancient Greek) of the mountains (*oros*).[1] Although the word itself is of recent coinage, the idea is ancient. It is central to most cosmogonic narratives that give a significant place to mountains. Such narratives may invoke divine wrath, original cataclysms, or the dismemberment of a giant. The idea now prevailing among geologists and geophysicists is completely different: mountain chains are formed by dynamic contact between continental plates, which either collide on the Earth's surface or move away from each other to allow magma to force its way from the core through to the surface.

In this book we add nothing new to that scientific and naturalistic conception of orogenesis. This is not a work of natural science, though we deal in part with that field of knowledge. We favor metaphor and images and focus on a different kind of orogenesis: political. Rather than concern ourselves with the forces outside the human world that created the mountains, we set out to study the social forces at work in their identification and classification. To put it succinctly, we shall study the processes by which socie-ties construct their mountains.

That formulation may seem fallacious. Those who believe that moun-tains preexist any human action say that one can speak only of practices relating to (preexisting) mountains or of social and cultural represen-tations of (preexisting) mountains. Excellent works have adopted that posture: John Grand-Carteret's groundbreaking books on the history of mountaineering and Marjorie Hope Nicolson's in cultural history.[2] Others followed that studied the mountain imaginary in its various incarnations.[3] Still others have focused on the incredible diversity of ways of inhabiting the mountains, enjoying oneself, and producing in that environment the necessities of life or goods for the present-day economy.[4] That approach makes it possible to identify, in the history of societies and no longer in natural history, the constants and the shifts in human sensibilities vis-à-vis the mountains.

As interesting as these approaches may be, however, they should not prevent us from looking at mountains from the opposite direction. We shall argue that the mountain, far from being a given of nature on which these representations and imaginaries come to be grafted, deserves to be studied as a notion in itself, as the product of a social and political con-struction. That may sound like a provocation. Some may concede without great difficulty that the nation-state, the school, and marriage, for example, being social institutions, are constructs that presuppose beliefs and con-ventions. It is harder to imagine such a thing for notions that refer to an ex-ternal material reality such as animals, continents, or mountains. The epis-temologists John R. Searle and Ian Hacking have even used the example of the mountain or rock as an illustration of what they believe belongs to an indisputable order of reality, one that must be distinguished from purely social conventions.[5] Searle establishes a very clear distinction between a peak such as Mount Everest, which he ranks among the "brute facts"—those whose existence is in itself indisputable—and institutions such as money or property. "Institutional facts" require that the people using them believe in the value of certain documents: the bank note, the notarized deed. Common sense makes the same distinction. To those who publicly express doubts about the reality of a rock, common sense will retort that you cannot doubt something that crushes a car or a pedestrian. It thus re-minds us that mountains and rocks truly are constitutive of the material reality of the world.

We make no claim to the contrary. We will readily concede, following Searle, that the notion of the mountain refers to a brute fact—to a material reality with its own characteristics, one independent of the collective or individual imagination and of any form of cognition. But the mountain as a category of knowledge and of collective action is in fact a social construct. Its history can be written. The definition of the mountain, the entities in the real world related to it, and the qualities associated with them can be shown to be the result of conventions. This opens the way to a "constructionist" analysis of the mountain.[6]

Some authors who have adopted that point of view propose to speak of the "invention" of the mountains, such as France's Mont Blanc or Mont Meyzenc.[7] Others speak in the same terms of one or another mountain massif or chain: the Pyrenees; the Alps; the mountains of Honshu, Japan; the Caucasus; the Catskills; and many others as well.[8] They have even done so for chains that are no longer considered such—the Transhimalaya, for example—and for subcategories, such as the "Mediterranean mountains."[9] In reality, these authors do not always make explicit what was ultimately "invented." For though none of them argues that the materiality of mountains is itself invented, the invention they are considering is sometimes a category of objects, sometimes a summit or geographical region, sometimes a referent for public policies, and sometimes a collective attitude or feeling.

We will adopt a more general perspective on what are called mountains and on the prescriptive and normative practices associated with that designation. We analyze the mountain as a category of thought and as a class of objects: single mountains, "mountain regions," and "mountain environments." We will show how and why modern societies and states tend to demarcate and characterize entities described as mountainous (regions, environments, societies, landscapes) in certain ways and then act accordingly in relation to them. And we will ultimately see this labor of definition, delimitation, and characterization of the mountain as one modality among others for constructing worlds that are at once natural, social, and political. We will attempt in this roundabout way to make fully intelligible the fashioning of mountain reality.

From One Definition to Another

Because we propose to set aside a naturalistic conception in favor of a constructionist one, the definition of "mountain" has to be adapted to its object. Paradoxically, it is easier to agree on a constructionist definition than on a naturalistic one.

Many have tried to come up with a naturalistic definition but have been forced to scale back their ambitions. Raoul Blanchard, a French geographer well-known to specialists, was one of them. In the introduction to a book on the relationship between "man and the mountain" (1933), he wrote: "A definition of the mountain which would be clear and inclusive is in itself almost impossible to provide."[10] Others who contributed greatly to our knowledge of the mountains, such as Alexander von Humboldt and Carl Troll (see chapter 1), did not even try. Why such difficulty? The notion of the mountain belongs to ordinary language, to the vernacular. It usually refers to the upper portion of a marked topographical contrast as it is apprehended by the senses. In terms of prescientific or nonscientific cognition, the mountain is a thing known from afar and from a lower vantage point. For a long time the notion played only a marginal role in defining a geographical entity; it did not come to designate a geological form or ecological environment until sometime in the eighteenth century. Subsequent definitions generally sought to make that popular conception coincide with objective criteria, but these proved to be more or less arbitrary and sometimes ethnocentric.

Since then scientists have lost interest in coming up with an objective definition, either because they feared a logical impasse or because they understood that the notion could only be somewhat conventional. That is the conclusion of authors who have reflected on formal definitions of the mountain.[11] In the introduction to *Mountains of the World*, an important reference work for anyone wishing to understand the political import of the mountain's singularity, the editors, Bruno Messerli and Jack Ives, point out that "the inability of mountain scholars to produce a rigorous definition that has universal application and acceptance has often led to time-consuming debate with no satisfactory results."[12] A little further on they agree on a minimalist definition, adequate for understanding the main thesis and the illustrations that follow: "Thus, we have relied upon the juxtaposition of 'steep' slopes and 'altitude,' facets of mountain landscapes that individually, or in tandem, lead to marginality in the sense of human utilisation and adaptation."

Indeed, that is one of the great paradoxes of the notion of "mountain": although the mountain seems impossible to define by logical and systematic rules, it also seems very natural to most of us. Ask children to draw a mountain and they will do so without much hesitation. Query the passengers of a ship approaching the coast and they will easily point to the mountainous features of the landscape (if, of course, the landscape cooperates). We could no doubt apply to the mountain what Saint Augustine said about time, a notion that appears much more abstract: "If no one asks me about it, I know: but when I try to explain it on demand, I do not know!" Back in the mid-1930s the U.S. geographer Ronald Peattie proposed that we take these common conceptions of the mountain as our starting point in circumscribing it: "To a large extent, then, a mountain is a mountain because of the part it plays in popular imagination. It may be hardly more than a hill but if it has distinct individuality, or plays a more or less symbolic role to the people, it is likely to be rated a mountain by those who live at its base."[13]

When compared to a naturalistic approach, a constructionist approach to the notion of mountain defines its object in very different, and on the whole simpler, terms. If it is agreed that the notion of the mountain is intuitive but informal, it is easy to surmise that such a notion will be reinforced by conventional representations (maps or landscape paintings, for example) or by conventional methods that enlist various criteria (measurements, markers, procedures). The definition of the mountain lies in great part in these figurations and methods, which establish standards and norms. We may also surmise that this definition will be the object of controversies and disagreements, that it will be enlisted in various ways in different social, political, and regulatory contexts (to establish boundaries, to attribute a certified label, and so on). In other words, the mountain is easy to conceptualize as a conventional entity and as a political object precisely because its status within naturalistic knowledge is so uncertain.

The Mountain as Category of Knowledge and of Collective Action

If the notion of the mountain is a convention that varies by context, how is it possible to conduct an analysis and to understand its social construction and the political uses to which it has been put? What concepts and procedures can be enlisted within that perspective?

The geographical literature written in French has made abundant use of the notion of representation to account for the distinctive features by which an individual, group, or society apprehends an environment. This literature also refers, albeit less often, to "enacted representations" in order to convey how representations are translated into actions, formal plans, or public policies.[14] But the term "representation" has disadvantages as well. In French, it designates a large number of different things. Most people understand it to mean simply the interpretation of an existing reality whose intrinsic characteristics can be identified. But if the category of the mountain is purely an artifact of thought, if neither nature nor any person has ever built a mountain to conform to such a category, then the mountain cannot constitute the referent of its representations. At best, it is only the product of them.

In place of the single term "representation," we shall prefer four concepts—objectification, problematization, paradigm, and intervention—complemented by two others: figure and configuration. Together they will constitute our analytical tool kit.[15]

Objectification will designate the act of establishing an entity that one agrees to see as a component of the real world. It names a practice that consists of setting in front of oneself (*ob-jectare*, to cast before oneself) an object or family of objects because it is judged pertinent for understanding one's surroundings or the world as a whole, and, if need be, for acting on them. That is an unusual formulation, particularly when speaking of a mountain, since it is at odds with the widespread idea that objectivity designates an attitude of thought, one that refers to reality as it is. Let us simply acknowledge that there are several ways of conceiving of reality and therefore several ways of conceiving of the mountain. Let us acknowledge as well that it can be interesting to compare these conceptions not from the standpoint of their accuracy or truth, but from that of their respective purposes and advantages. Such will be our position here: to compare different projects of mountain objectification while studying the respective roles attributed to that objectified mountain. But we shall also consider the modalities of objectification: texts (exploration narratives, scientific articles, laws and regulations), data (measurements, statistics), and images (sketches, maps)—in other words, a set of material representations or inscriptions. All these artifacts, which constitute as many languages, make it possible to designate particular mountains but also to circumscribe the category as a whole—"the" mountain—by making all particular mountains comparable and commensurable. In other words, throughout this book the expression "the mountain" refers to the generic category according to which Western thought has organized its knowledge of landforms. The terms "mountains" or "the mountains" refer to the empirical objects belonging to that category.

The objectification of "mountaineers" will be analyzed in the same way. In the following pages, the term "mountaineer" refers to a generic human type (for instance, people living in traditional communities whose character is said to be determined by their mountain environments) or to a social entity (mountain climbers, for example, who form their collective identity with reference to the mountain), defined ontologically by their dependence on the natural category. When we refer to these groups in neutral terms, without essentialist connotations, we use the expressions "mountain people," "mountain dwellers," or "mountain populations" and "mountain climbers," respectively.

Problematization designates the expectations relating to or motivating that objectification. What questions can the category of the mountain contribute to answering? To what degree do these questions shape the category itself? Once again the formulation runs counter to our usual ideas. It is easy to imagine that categories of natural objects have the sole objective of describing the world as it is. But there are several ways of conceiving of that description, and each corresponds to a different project of cognition. We will have occasion to see that the notion of the mountain is not conceptualized in the same way within the cosmogonies of monotheistic religions, the natural history of the eighteenth century, human geography in the late nineteenth century, and geophysics in the twentieth. It is also not conceptualized or objectified similarly within the framework of forest policies in the late nineteenth century, colonial expansion in Africa and Asia, or present-day international organizations. To come to an understanding of these different forms of objectification, we will need to understand what motivates them and identify the problems to which they are seeking to respond. In other words, it will be necessary to analyze how problems have been constructed by scientists and philosophers, by national or colonial administrations, by mountain climbers and the inhabitants of the regions concerned, by environmentalist organizations, and by those involved in the exploitation of resources. From such analyses we will be able to deduce the mountain to which each is referring. In this book the mountain is not conceived as a thing in itself; it is a response to the formulation of problems.

Paradigm designates the ideological context within which the problematization takes place. It constitutes the exclusive or dominant prism through which one apprehends reality, from the dual perspective of understanding its components and operation and intervening to adjust reality to one's own expectations. Such paradigms include objectives of territorial control, social cohesion, sustainable development, and environmental protection. They are the backdrop for any formulation of a problem and any procedure of objectification.

Intervention designates the set of actions undertaken directly (development, the building of infrastructure) or indirectly (laws, regulations) on the materiality of geographical entities. Each of these actions stems from a particular cognitive framework that combines objects, problems, and a paradigm.

Because objectification, problematization, paradigms, and interventions are complementary and interdependent, together they compose **figures** of the mountain, through a complex operation of adjusting statements and images, of forms of action taken in situ and representations existing ex situ, · which together will be called a **configuration**. In thus individualizing a set of conceptions and actions, processes and states of being, we seek to show that they constitute a meaningful and operational totality.

In this book we present several of these figures and configurations, which may operate in succession or concurrently and have led to different ways of circumscribing the notion of the mountain and to different interventions directed at the reality thus designated. We will show how the mountains being referenced vary a great deal depending on the questions asked, the procedures for measuring or analyzing reality, and the images, statistics, and texts produced. It is from that vantage point that the mountain can be conceived—as some have proposed for the city—as a geographical category of collective action and public policies.[16]

Forms of Knowledge and Policies

The aim of this book, then, is not to analyze mountains as a component of geographical reality or to consider the "social representations of the mountain," which would assume a clear distinction between the reality of mountains and the reality of their representations. We will study the various ways that have existed for coming to terms with the idea of the mountain, so as to conceptualize and act upon the places and environments, the territories and social collectives associated with that category. It is a broad field including a large array of contexts, both historical (from the eighteenth to the early twenty-first century) and geographical. We will limit ourselves to cases where the mountain is invoked as a general, descriptive, or normative category. In fact, we will consider only the particular situations in which interventions are motivated or justified by that general conception of the mountain. We wish to show to what extent a natural category, universal in scope, has been enlisted for specific actions.

That limitation in the scope of the book should make clear the status we grant to scholarly and scientific knowledge. Such knowledge will not be analyzed for its own sake, except on rare occasions. It will, however, be omnipresent, since for the period under consideration the collective actions, public policies, and controversies studied here often enlist, explicitly or implicitly, one or another of these scholarly and scientific conceptions. Ever since the Age of Enlightenment forged the ideal of a science that would shed light on politics and of a politics that would cherish science, the concepts, procedures, and images that scientists have proposed have never been far removed from political objects. This has become increasingly true over the last few decades, now that the scientist is officially called upon as an expert within the framework of a given configuration.

The primary objective of this book is to analyze the figures by means of which the idea of the mountain has participated in the social and political life of the last three centuries. There is a second objective, subordinated to the first: that of understanding to what degree and in what form scholarly and scientific knowledge has contributed to these configurations. In reality, however, these two inquiries amount to the same thing—both focus on the nature and modalities of the connections established between knowledge and politics in modern times. For, as we have already noted, the mountains participate in modernity not only because they were configured in accordance with the modern project, but also because they have been a laboratory or (as some have said of the Pyrenees) a privileged site of "experimentation for triumphant modernity."[17]

The Mountain and Shifts in Scale

In this book we privilege several different levels of analysis. Although we will discuss a large number of local contexts in which development has occurred, conflicts have arisen, and social policies have been set in place, we will include only those cases where the normative notion of the mountain has been invoked and enlisted in the process.

We have therefore chosen to consider three scales where the category of the mountain has acquired a normative value: the modern state, colonial empires, and the global space as it now exists. We are not saying that alternative scales of knowledge making and action have never led to figures worthy of interest; but we postulate that these three political entities have been particularly decisive in the configurations through which the notion of the mountain has been carved out and put into operation. It is clear that, within national administrations and at international conferences such as the one in Rio de Janeiro in 1992, these institutional frameworks have promoted forms of rationality and coherent modalities for apprehending the physical world, forms and modalities that have provided the basis for the actions taken by the corresponding institutions. These frameworks have also subordinated local realities and local modes of apprehending that material world to a rationality conceived on a broader scale.

Let us point out here that the scales considered must not be understood as a chronological sequence, in which the local makes way for the national and imperial, and the national and imperial for the global. As a matter of fact, national modes of thought have persisted and have sometimes even been reinforced by the increasing influence of the imperial and the global. As for the local, it is constantly reinvented whenever a new scale for apprehending issues comes into being. Even in the context of empire or globalization, adjustments are always being made between local modes of thought and larger-scale notions of the mountain as object. The configurations of the mountains considered in this book are thus multiscalar, and the evolution in these shifts of scale has influenced the configurations. Finally, let us note that we are interested in the scholarly and scientific meanings of the notion of the mountain largely because science and philosophy have been powerful vectors for the construction of universalizing, and hence global, conceptions.

Our project in this book gives rise to a few simple questions:

> How were the figures of the mountain that emerged and became prominent over the past three centuries constructed, and how do they function?
> On what scale does each of them find its relevance?
> How have competing processes of configuration faced off or made adjustments to respond to one another?
> To what degree have scholarly and political conceptions of the mountain influenced each other in processes of configuration?
> What different arrangements of actors have been constituted around the various figures examined?

These questions are all of a political nature: they allow us to identify the initiatives intended to configure, and often to naturalize, a category of knowledge and territorial practices. They also allow us to point out the alliances and controversies arising from these configurations. They are political in the sense that Pierre Bourdieu said action is political: its aim is "to produce and impose representations (mental, verbal, visual, or theatrical) of the social world which may be capable of acting on this world by acting on agents' representations of it."[18]

These questions, however, do not have the same relevance in all contexts. For pedagogical reasons, we have chosen to divide this book into three sections.

Chapter 1 introduces the scholarly modalities for constructing the category of the mountain in the eighteenth and nineteenth centuries and the paradigms that gradually became associated with it. In it we will shed further light on questions already raised in the introduction, particularly the definition of the mountain, the modalities of its objectification, and the forms of knowledge and argument associated with that operation.

Chapters 2–5 (part I) are devoted to figures of the mountain within the context of nation-states at a time when those institutions were setting most of the rules. This part focuses in particular on the place granted to the mountain in the territorial claims of nations of the Western world, on so-called mountaineers in the corresponding societies, and on the public policies that have sometimes denied and sometimes promoted a specificity of the mountain.

Chapters 6–11 (part II) analyze the shift away from national frameworks as a result of colonization and the globalization of interest in the mountain. One of the principal shifts consists of diversification and increasing complexity of the actors, scales, and paradigms at work in the production of figures of the mountain. The actors that dominated in the modern configurations of the mountain—scientists and national administrations—have now been joined by others, such as international organizations, nongovernmental organizations (NGOs), and the mountain populations themselves. That increasing complexity has led to reconfigurations in which the state is still a stakeholder but the scale of reference has often changed. For example, the problematization of the mountain becomes European, Andean, regional, transboundary, or global, depending on the context. Furthermore, the shift in the scale at which the mountain is reconfigured is often accompanied by a change in paradigm and hence in problematization. Thus the emergence of mountains as a global issue in the early 1990s has gone hand in hand with the adoption of sustainable development as a model for action in intergovernmental agencies.

This book focuses on the construction of various figures of the mountain, a notion that turns out to be more flexible than is generally imagined. Beyond the particular case of the mountain, we will analyze the processes of social and spatial differentiation and the institutions and mechanisms by means of which societies produce differentiation both within themselves and in their physical surroundings. In his best seller *The World Is Flat*,[19] the American essayist Thomas Friedman defends the idea that modernity has deployed itself in the world by means of norms and techniques that make the world's places and societies commensurable, if not equivalent— even to the point to denying all specificities. We take the opposite view: although the standardization introduced by the modern project has been at work for centuries, it has also had the effect of institutionalizing a large number of differences. The notion of the mountain has played a key role in that regard.

ONE

The Mountain as Object of Knowledge

Popular Conceptions of the Mountain

The city of Reims in northern France was founded on a rolling plain on the banks of the Vesle River. To the south and west a geographical formation blocks the horizon. Contemporary geologists and geomorphologists call it a *cuesta*, an escarpment that turns quickly from a limestone plateau to a clay-marl plain. It is typical of the Paris Basin but also of many of the world's sedimentary basins. But the residents of Reims did not wait for geologists and geomorphologists, or even for eighteenth-century naturalists, to identify and name the landform. They baptized it the Mountain of Reims. That "mountain" does not display any of the attributes now associated with such a landscape: its highest elevation is very modest (about 280 meters); the difference in altitude, though apparent from Reims and the surrounding area, remains unremarkable, since the city is located only 200 meters below.

Although that Mountain of Reims may have been a mountain in the eyes of the city's residents, it is not so for specialists in the natural sciences. To be precise, it is no longer so. Even in the mid-eighteenth century the term was used at the Académie Royale des Sciences to designate geographical formations of modest scope. A report published in 1755 refers to a "Mountain in the area of Étampes."[1] We would now say that Étampes, about fifty kilometers south of Paris, lies at the foot of a plateau escarpment. In the *Encyclopédie* edited by Denis Diderot and Jean Le Rond d'Alembert, which is roughly contemporary with that report, the word "mountain" is used to designate many different things, at least by our current criteria. The *Encyclopédie* mentions the "Mountains of Rome" to refer to landforms we would now call hills. It is said that Angoulême is on a mountain; Cape Town, South Africa, has its "Table Mountain"; as for the Cordillera de los Andes, it is characterized as a group of mountains. The meaning of the term does not seem to have been fixed at the time or held in common by the many authors of the *Encyclopédie*. The only definition of "mountains" that appears in it, attributable to the baron d'Holbach, leaves the door open to a large number of different things: "large masses or irregularities in the ground, which make its surface uneven."[2]

We may therefore wonder what the word "mountain" is supposed to circumscribe, and in the name of what worldview and what form of knowledge. In popular toponymy, Reims and Rome are not isolated examples. On the contrary, "mounts" and "mountains" (the two terms were long synonymous) abound near ancient cities: Paris has its Montagne Sainte-Geneviève but also its Montmartre and Mont-Valérien; Montreal has its Mount Royal, which its residents have tended to call *la Montagne* ever since the city's origin as a colony.[3] Elevations and differences in altitude are as modest there as in Reims and Rome. In Germanic countries, many cities have their *Berg*, like Maastricht and its Maastricht Berg, though both are located in what are called the Low Countries. In English-speaking regions, Vancouver, British Columbia, for example, in a setting of large mountain slopes, has its "Little Mountain" near the city center. The Mountain of Reims is therefore not a local whim. It is one instance of a very common way of using the term, in French, English, and a number of other languages, to refer to external reality.

That practice depends on a single point of view, in the literal sense of the term: that of city dwellers constantly exposed to a contrast in the landscape. The designation exists independently of any easily circumscribable physical object: although Maastricht Berg and Mount Royal constitute forms clearly differentiated from their physical surroundings, that is not the case for the "mountains" of Paris and Reims. The terminology also exists independent of elevation—and for good reason. The concept of elevation, which emerged in the seventeenth century, became stable only in the following century,[4] after most of these "mountains" had already received their name. The mountain of popular conception is thus not a natural object in the sense that the term is now understood, namely as an entity in the external world that can be characterized by its intrinsic form or content. It is primarily one of the terms that take into account a contrast, both in the landscape as perceived from below and in modes of use: for a long time the "mountains" of Reims, Montreal, and Maastricht were wooded areas used as a source of timber for the residents of the region.

Similar ways of differentiating one's surroundings through language, on the basis of a point of view and predominant uses, can be found in many regions of the world. In Nahuatl, the principal language of the Aztec empire, communal territory was traditionally called *altepetl*. The root *alt* means "water," and *petl*, "mountain." The *petl* is a component of local territory, and in particular the counterpoint to the inhabited and cultivated places located near the *alt*.[5] In Ladakh, the term closest to "mountain" as it is used by people living in Reims and to the Aztec *petl* signifies etymologically "the other of the village."[6] Sometimes the contrast in use prevails over the contrast in form: the Basotho of present-day Lesotho, immersed in an environment with a complex and omnipresent jumble of landforms, seem to possess no generic term for a family of topographical forms. By contrast, they do have a word for the rangelands that abound at high elevations, and they use it to differentiate these lands from those used for cultivating corn and sorghum.[7] On the whole, their approach is not very different from that of the Alpine populations in the area of Mont Blanc: in Savoy, Valais, and Val d'Aosta, farmers long used the term "mountain" not to designate Mont Blanc itself or the glacial or rocky peaks surrounding it, but to refer to the high mountain pastures where their cattle grazed in the summertime. Here as well the mountain is "above," as designated from "below"—an elsewhere and an other useful for conceptualizing the complementarity of resources and locales. It seems, therefore, that mountains have repeatedly been designated in vernacular French on the basis of a phenomenal experience—that of a topographical contrast or of a contrast in use, and often both at once. People were thereby led to conceive of the highest part as a thing entirely different from the place from which it was observed.

A Class of Natural, Objectified, Purified Objects

Eighteenth-century natural history and geography proposed a far-reaching alternative to the traditional and popular conceptions of the mountain.[8] They engaged in what aspired to be a radical objectification: the mountain became a category of comparable and commensurable physical objects characterized by a set of attributes, all purportedly objective and hence independent of the particular points of view from which the inhabitants of one place or another might see and describe their surroundings.

Unlike cosmogonic notions, the concepts of modern natural history and geography were supposed to be empirical—guided by the facts. The distinction Michel Foucault proposed between prescientific and scientific knowledge certainly applies to the change in the concept of the mountain.[9]

From the eighteenth century on the natural sciences, using methods of systematic comparison, strove to understand the identity of things, the differences among them, the causal relationships that connected them, their transformation over time, and the spatial arrangements characteristic of them. The notion of the mountain became from that point on a component of a "universal science of order and measurement"[10]—referring to a class of physical objects, to evidence about the history of the earth, and to a setting in which a number of diverse phenomena interacted. The mountain, because of the magnitude of what it allowed one to think about, became one of the privileged categories of scientific knowledge, linked to a number of fields of knowledge about the natural world.

Thanks to that change in status, mountains were stripped of their intrinsically human dimensions. Prior designations, which gave priority to the criterion of economic utility (in the high mountain pastures in the area around Mont Blanc, for example), were discredited. A mountain was no longer a mountain by virtue of the use made of it, but rather by virtue of the characteristics proper to it. Discredited as well were designations valid only from a local and perceptual point of view, as in Reims, Paris, Montreal, and Maastricht. A mountain was no longer a mountain based on the criterion of its appearance in the eyes of an observer located below. It became so through instrument measurements and intermediary representations (primarily maps and statistical tables) that claimed to provide proof of its objective existence. Following Bruno Latour, we could say that the category was purified:[11] placed on the side of nature, set aside from culture, the term "mountain" was used to designate a family of supposedly objective entities, and therefore a collection of natural objects. The human dimension of mountains became a secondary characteristic, independent of their definition.

Although the modern episteme can be identified by these shared characteristics, recognizable in the great majority of scientific works from the eighteenth century on, these works did not all enlist the notion of the mountain in the same way. That notion gradually came to designate a single type of object for all moderns, but it belonged to heterogeneous projects of cognition. Although we cannot be exhaustive here, let us present a few of these variants, giving particular emphasis to the scientific conceptions to be found in the following chapters, placed in the service of the political figures under study there. In that exploration Alexander von Humboldt will be our guide, not only because of the decisive role he played in the emergence and affirmation of a modern conception of the mountain, but also because of the diversity of his own research programs.

The Casiquiare Controversy

On May 20, 1800, Alexander von Humboldt and Aimé Bonpland, both lovers of science and exploration, disembarked in the port of Cumaná, east of Caracas.[12] Their itinerary and the length of their journey were not firmly set. They had negotiated with the king of Spain for permission to explore that part of the empire, between the Caribbean and Lima.

From Cumaná they traveled up the Orinoco River and slipped into the Amazon Basin; they then continued up the Casiquiare, a tributary of the Rio Negro. From there, without setting foot on land, they rejoined the Orinoco, thus proving that the Orinoco and Amazon basins communicated via that odd waterway, an imposing natural canal leading from one basin to the other. Humboldt was actually sure of his facts before leaving Cumaná: his readings and his local informants alike had persuaded him of the existence of that communication route. Nevertheless, anxious to record the event and make later verification possible, he plotted the geographical coordinates of the site of his observation: "3°10′ of north latitude, and 68°37′ of longitude west of the meridian of Paris."[13]

In his notes and in several later publications, that unique observation is cited in support of a general reflection on the locations of mountains and the method that ought to be used to determine them. Humboldt was challenging a theory, very popular in his time, developed by Philippe Buache: that of the continuity of mountain chains. Buache was a mathematician and architect by training and since 1730 had been an "assistant geographer" at the Académie Royale des Sciences.[14] On November 15, 1752, he presented a paper with a title as colossal as his ambition:[15] to propose a new planetwide theory about the location of the seas, mountains, islands, and waterways. His thesis held that the earth was traversed by chains of "mountains" connected to one another from one end of the continent to the other; these mountain chains separated enormous "river basins," which opened onto the four great "seas," which he names the Ocean (the Atlantic), the Sea of the Indies (Indian Ocean), the Great Sea (Pacific Ocean), and the Arctic Glacial Sea. Each of these expanses is split into maritime "basins," three per "sea," which are separated from one another by "marine" mountain chains. These are merely the extension underwater of the land-based chains, invisible to the observer except when these chains approach or extend above the surface of the seas as "islands, reefs, or shoals."

That organizational principle, when applied to South America, assumed the existence of several mountain chains of major importance: those, like the Cordillera de los Andes, of which Buache already had many descriptions; and those deduced from his theory. These chains ought to have run between the basins of the Orinoco, the Amazon, and the Rio de la Plata. In reality, Buache knew that an impressive mountain chain between the Orinoco and the Amazon had been drawn on many maps from the seventeenth century on. He also knew that some doubted its existence, on the basis of information provided by the native populations and by a few explorers. But Buache did not take these eyewitness accounts seriously, since their conclusion invalidated the entire theory he was defending.

Humboldt was quite familiar with the existing theories. He wanted to end the controversy once and for all by making observations on the ground. His report is not lacking in irony: "I was fortunate enough to reconnoiter this chain on the spot. I passed with my boat in the night of the 24th of May, along that part of the Oroonoko, where Mr. Buache supposes the bed of the river to be cut by a Cordillera."[16] He goes on to mock the ways of thinking characteristic of mid-century geographers. "This bifurcation, which has so long confounded the geographers who have constructed maps of America,"[17] gives him the opportunity to denounce simplistic arguments: "The chains of mountains [in the New World] do not rise like walls on horizontal plains."[18]

He also takes the opportunity to cast aspersions on an overly theoretical and speculative form of science and to promote, as an alternative, a science of observation and empirical argument. Indeed, the account of the trip up the Casiquiare is no mere anecdote. The journey was much more than a way of deciding a controversy definitively by presenting evidence. Rather, Humboldt's narrative illustrates two different ways of conceiving of science and two contradictory ways of enlisting the notion of the mountain.

The Mountain According to Buache: An Element in a Connected System of Objects

In the eighteenth century it was customary to distinguish at least three branches of geographical knowledge: mathematical geography, sometimes also called "astronomical," which was primarily concerned with measurements, geographical coordinates, and cartography; physical or natural geography; and historical or political geography. Until the early part of that century, a specialization in political geography was often considered the most noble, the best able to enlighten kings and princes and to account for the diversity of the known world. Physical descriptions of the surface of the earth were therefore usually associated with the territory of a nation-state.

An alternative was proposed in 1726 in a little book written in Latin, in which its German author, Polycarpus Leyser, promoted a "true geographical method" (*vera geographiae methodo*).[19] Unlike the proponents of "state geography" (*Staatgeographie*) from whom he wanted to differentiate himself, Leyser believed that a knowledge of natural facts must precede any historical or political considerations and must be conceived independently of them. That knowledge of natural entities and natural boundaries, whether "dry" (mountains) or "wet" (seas and rivers), must stand on its own. For the geographical description Leyser was promoting, such boundaries, because they are fixed and tangible, had a higher value than political boundaries, which fluctuate and are perceived as being largely arbitrary.

Buache was one of the most illustrious representatives in France of that physical geography which aspired to be independent of political and historical considerations. An "armchair geographer" for the king of France, he used travel narratives, accounts of explorations, and various maps to support his general theory. In fact, the location of mountain chains across the surface of the globe was one of the major questions troubling scholars in the modern age. It was still common in the mid-eighteenth century to claim that landforms were randomly distributed over the earth's surface. The comte de Buffon, whose curiosity prompted him to seek a law on the location of mountains, reflected on the subject: "This immense globe displays in its surface heights and depths, plains, seas, marshes, rivers, caves, chasms, volcanoes, and upon our first inspection we can see no regularity, no order to it."[20] In that context, the interest of Buache's proposal lay in its great simplicity and extreme coherence. A connected system of mountains and basins was appealing because it was consistent with common sense: the waters of rivers flowed from high elevations to the sea; logically, then, the principal eminences ought to be found near streams. The theory also had the advantage of suggesting that geographical arguments could be deductive, even predictive, and hence closer to scientific methodology, while still responding to practical questions. It was possible to deduce the location of mountains from that of rivers, which were better known because they were the principal circulation routes of the time: "I thought that . . . I had to use the clues left by the rivers. We can't deny that the origins of rivers and streams naturally indicate the height of the terrains where they source their water to nourish and fertilize the lands they cross as they descend from the high places, whether it be by steeper or shallower slopes, until they empty themselves into the sea. Neither can we doubt the liaison and the relationship that mountains have with rivers."[21] In the end, the argument effectively rested on the mediation of maps. Mountain chains were in fact deduced from drawings, representations of the known hydro-

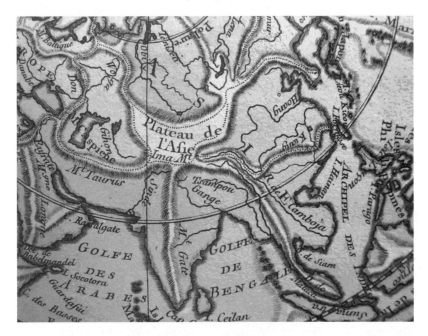

1. Philippe Buache ranked mountain chains not according to their altitude or their rocks, but according to their location. Located in the heart of a continent, between river basins that flow to one ocean or another, are what he calls "plateaus," in this case, the "Plateau of Tartary" in Asia. Between ocean basins fed by the principal rivers are "back mountains" (*montagnes de revers*), such as the one that supposedly crosses India from one end to the other. Absent from this map is a final ranking, called "coastal mountains." Buache, detail of the map of Asia published in *Cartes et tables de la géographie physique ou naturelle* (Paris: Quay de l'Horloge, 1754).

graphic networks. That method constituted a further sign of scientificity.[22] In the first place, then, Buache's system was a logical and visual arrangement of natural objects in space, objects classified within a system of fairly simple and complementary categories (fig. 1). The simplicity of Buache's argument, the deductive and supposedly predictive character of his theory, and the maps to which it gave rise together explain the great popularity of his proposal and the remarkable influence it had for many generations. Traces of it can still be found in geography textbooks and peace treaties (see chapter 2) at the start of the twentieth century.

Studying the Materiality of Mountains in Situ

From the beginning, however, observers in the field contradicted that postulated continuity of mountain chains. Travelers and explorers did not observe any mountain range on the periphery of the major river basins of central Russia, or between the Amazon and the Orinoco, or even, just outside Paris, between the Loire and the Seine. The gap between theory and observation contributed to the emergence of a completely different approach, defended by naturalists: the study in situ of the materials from which the earth's crust is made in order to understand their formation and explain the distribution of landforms.

In the article of the *Encyclopédie* devoted to physical geography, Nicolas Desmarets exhorts scholars to consult nature directly, without the mediation of their informants—to do fieldwork, as it would later be called. He formulates vigorous critiques of armchair geography and its propensity to reinforce ready-made systems: "We are now fairly convinced of the inconveniences associated with this idle presumption which leads us to guess at nature without consulting it. . . . Therefore we want facts & observers appropriate for grasping and successfully gathering them. . . . The observer must guard against any preconception, any bias that is static and dependent on a system which has already been devised."[23] He recommends that in the observation of nature particular attention be paid to the materiality of natural objects: "An intelligent observer will not restrict himself so much in his technical discussions to the external forms or structure of an object, without also taking an exact knowledge of the matter itself which in its diverse amalgams contributed to producing it; he will even exactly link one idea with the other. The matter, he will say, affects this form; he will conclude one from the other, & vice versa." The physical geography Desmarets advocated is thus the exact opposite of Buache's geography. It sent forth generations of scientists into the mountains of Europe to study the rocks they are made of, to refine the typologies by which they are classified, and to seek to understand how they are positioned. Among them in the 1760s–80s were Horace-Benedicte de Saussure, who initiated the first ascent of Mont Blanc in 1786; and Ramond de Carbonnière, one of the first to travel in the Pyrenees and write natural history accounts.

Humboldt's quest led him from the Alps to the Canary Islands and from there to the Andes and Central Asia. Primarily trained in geology by Abraham Gottlob Werner, he collected samples of rock and expounded at length on their nature. Many of his detailed maps attest to his attention to the minutiae of forms, the substance of topographical features. But, on the Casiquiare Canal, Humboldt was not the first to contest Buache's system and the conception of science underlying it. Nor was he the first to seek responses to the questions that arose about the mountains by studying their materiality. Leonardo da Vinci, who was curious about fossils and anatomy, had already argued the advantages of that approach. The eighteenth-century naturalists, however, distinguished themselves from their predecessors by the methods they adopted to organize knowledge. Here as well Descartes's principles—order and measurement—prevailed. For example, several scientists from the second half of the eighteenth century, particularly Pierre Bourguet on Chimborazo and Nevil Maskelyne on Schiehallion, observed that the presence of mountains affected their instruments and measurements.[24] From these changes they deduced the mass of each of these two mountains. Maskelyne even estimated the mass of the earth as a whole. The materiality of mountains had eluded no one: it was becoming an object of measurement.

The order that was supposed to organize knowledge relating to the materiality of the earth's crust was twofold in nature: spatial and temporal. The types of rocks that scientists learned to distinguish were recorded on maps, and their spatial arrangement, coupled with their physicochemical characteristics, became a principle of intelligibility. The location of rocky outcrops, especially in basins where layers of sediment had accumulated, was used as an indicator of their place in a chronological sequence. That working method, which is characteristic of stratigraphy, had a counterpart in tectonics, the specialty concerned with the formation of the most strongly marked contrasts in topography. In the nineteenth century the most fashionable theories in that field suggested that the overall disposition of mountains had something to do with the conditions of their formation.

That was also Humboldt's view. He wrote that the way mountains are positioned provides one of the characteristic features of "the internal makeup of our planet."[25] Convinced there was a general order, even a harmony, in the organization of natural objects, he thought that the location of mountains was part of that order, governed by a few laws that needed to be brought to light. In that respect he was a geographer. But inasmuch as he was convinced that the answer to his questions lay in the nature of mountains' rocky materials and in the history of their arrangement, he remained a geologist.[26] The order of nature that Humboldt sought to reconstruct combined the spatial with the temporal.

Central Asia as Laboratory

In the early nineteenth century relatively little was known about the spatial disposition of mountains in a number of regions of the world. Hypotheses abounded. In 1805, when Humboldt made a stop in the United States at the end of his travels in the Americas (see chapter 4), President Thomas Jefferson questioned him primarily on that subject. And Czar Nicholas I, wishing to learn the location of the mountains in his expanding empire and the mining resources they contained, asked Humboldt in 1829 to come explore the Urals and Central Asia.[27]

Central Asia was a largely unknown region at the time. Few scientists had used it as their laboratory. Peter Simon Pallas, after a visit in 1768–74 at the invitation of Catherine II, had written one of the first general studies of Asian mountain ranges (according to him, they were arranged in a star around a center) and elaborated one of the first theories on the formation of mountain chains.[28] But Humboldt viewed Pallas only as a proponent of a "dogmatic and careless geology."[29] Humboldt wanted to promote the methodical analysis of "this massif vaguely called *le plateau de Tartarie*."[30] No doubt he was once again targeting Buache, who made abundant use of that expression. For the major chains he encountered in the field and in his readings, Humboldt set about to measure the average elevation and the directions in which they ran and endeavored to link his observations to the nature of the rocks found there.

Ferdinand von Richthofen would continue Humboldt's research. Richthofen, a geologist by training, worked first on the geological structure of Tyrol and Transylvania and then on mineral deposits in California. The Shanghai chamber of commerce and the Bank of California sent him on assignment to China to search for likely coal deposits. After completing his journey, he published an atlas and several books (beginning in 1877). His first volume was devoted to the mountains of Asia.[31] Like Humboldt, but even more systematically, Richthofen identified and attempted to explain the alignment of the mountains traversing all of Central Asia by taking into account the various rocky materials he recorded.

Léonce Élie de Beaumont's theory for explaining these alignments of rocks was the most popular at the time. A geologist whose reputation rested on his knowledge of stratigraphy, Élie de Beaumont had suggested in the 1830s that mountains running in the same direction were formed at the same time. The chains running from western Europe to East Asia, generally oriented east to west, would thus belong to a different era than those in America, which are usually oriented north to south. According to Élie de Beaumont, they had been formed by the cooling and contraction of the earth's crust at key periods in its history. He also endeavored to determine the age of the earth.[32] The huge amount of energy released at the time had led to the upthrust of the mountains along the principal axes. That theory captivated nineteenth-century geologists and drew their attention to the formation of mountain ranges. It too relied on a coupling of the spatial and temporal orders.

Élie de Beaumont's theory was not invalidated until the early twentieth century. The scientist Alfred Wegener suggested that mountain chains came into being through the collision of continental plates moving over the surface of the earth's magma. His thesis, though well argued, was met with skepticism by all but a few geologists. Yet Émile Argand, for one, sought evidence to validate the theory for Central Asia.[33] With Wegener and Argand, a different conception of the order of time and space prevailed: the slow and constant time of movement over the surface of the earth and the space of continental plates, the horizontal shifting of which led to the upthrust of great masses of rock. Along the way continental plates became the privileged object of tectonics; mountains, a second and secondary form.

Instruments for Measuring Mountains

In the Americas as in Asia, Humboldt made extensive use of instruments in his observations. He dedicated much of his activity to various measurements: geometrical position, the altitude of mountain peaks, the hour of sunrise and sunset, atmospheric pressure, temperature, the intensity of the blue of the sky, and so on. Everything that the instruments of his time could measure, Humboldt measured. He carried with him a collection of instruments—sextants, theodolites, barometers, thermometers, chronometers, quadrants, compasses, eudiometers, electrometers, hygrometers—which he had accumulated before his departure and had been trained to use in Germany.[34]

One of his preferred methods of measurement was to determine elevation by means of a barometer, based on the decrease in atmospheric pressure that accompanies an increase in altitude. The technique was not new. Since the seventeenth century, in the wake of studies by Galileo, Torricelli, and Pascal, formulas had existed for linking measures of altitude to observations of the column of mercury in a barometer. Books too had attested to the efforts to measure mountains using these formulas and to compare the results obtained.[35] But it was only in the early nineteenth century, after many improvements in the manufacture of instruments and the accuracy of the formulas, that it was possible to imagine the systematic measurement of mountains throughout the world.

Humboldt thus proceeded to measure the elevation of several hundred summits and constructed countless elevation tables in which he presented his results. He did not seek to conduct a systematic inventory. Rather, he wanted to have at hand information that was strictly comparable for all the places measured. He also used elevation as a standard value by means of which to compare all his various observations. That led Humboldt to make great use of a mode of representation that had played only a marginal role before his time, the juxtaposition of stylized mountains on a single image. He could thereby indicate their respective heights, independent of the geometrical location of each one.

The value he attached to measuring altitude also led him to travel to a large number of peaks to observe its effects. In June 1802, midway through his travels in South America, he nearly reached the summit of Chimborazo, considered at the time to be the highest mountain on the planet. That ascent increased his already enormous fame. In the report he later gave of that ascent, however, he tried to minimize his achievement, suggesting that science is a more serious affair than feats of mountain climbing. Instead, he presented the observations he had recorded on site, including details (the presence of birds at very high altitudes, the nature of the rock that gashed his skin, his bloodshot eyes—apparently caused by the air pressure) when he believed they had scientific value. His curiosity led him to make connections between his own observations and those of the authors he had consulted, each time comparing altitudes. For example, M. Zumstein "showed blood at a much lower height on Mont-Rosa";[36] and on September 16, 1804, M. Gay-Lussac, "a sure and extremely exact observer," traveling in a balloon, had reached "the prodigious height of 21,600 feet, therefore between that of Chimborazo and that of Illimani," but "did not show blood."[37]

That curiosity also encouraged him to gather together, on his cross-sections of slopes, elevations measured on-site (fig. 2). As Humboldt often repeated, altitudes in themselves did not interest him. They possessed scientific value only if they could be used as an element for analyzing other natural phenomena or could be compared to average volumes across an entire chain, for example.

For Humboldt, unlike for Buache, mountains had dimensions (height at summit, average height of the chain, surface area of the massifs, and so on), which he sought to understand in relation to their constitution. His mountains were objects, whose principal intrinsic characteristics were their substance and their deployment in three dimensions. That conception of the mountain as object, substance, and volume, though it had originated before Humboldt, triumphed with him. It would be widely adopted in popular works and in the schools.

The importance Humboldt and his contemporaries granted to the mea-

2. "Journey toward the summit of Chimborazo, attempted on June 23, 1802, by Alexander von Humboldt, Aimé Bonpland, and Carlos Montufar," detail. This profile of Chimborazo combines several items, all related to Humboldt's various interests. The volcano is presented as a superimposition of zones: the lowest, cleared of snow, suggests its mineral content; the highest, snow-covered, reminds us of the geographer's curiosity about the *neiges éternelles*, the permanent snowpack. On the slope located at right, annotations refer to the episodes of his ascent in 1802. At left, the form of the volcano has been hollowed out so that he can mention the names of plant species found at different altitudes. An elevation table, not visible in this detail, appears next to the image. Humboldt, *Atlas géographique et physique du nouveau continent* (Paris: Schoell, 1814).

surement of altitude had numerous consequences. The highest summits, rarely named in traditional societies, became an object of universal attention. After the near-complete ascent of Chimborazo in 1802, scientific expeditions and triangulation campaigns moved into regions of the world where higher peaks could be found. That inquisitiveness partly explains why Westerners converged on Central Asia and competed to identify and measure the mountain chains.[38] From the nineteenth century on, mountain climbers too worshipped at the altar of altitude, choosing the most appropriate threshold for the region concerned and trying to conquer summits that surpassed it: over 4,000 meters in the Alps, over 8,000 meters in the Himalayas, over 6,000 meters in the Andes, over 14,000 feet in the Rockies.[39]

Altitude also came to be considered a useful concept for defining and circumscribing mountains. The world's mountains could now be compared in terms of a single value. That measurement also responded to the modern need to circumscribe the objects one was seeking to know. Most nineteenth-century authors proposed round numbers as a defining criterion for the principal families of geographical formations. But since conventions and units of measure varied from one scientific context to another, these numbers also differed. For the British Isles it was often proposed that eminences between 300 and 1,000 feet be called "hills," while those exceeding that threshold be called "mountains." But in the Himalayas there was a tendency to call formations up to 3,000 meters in height "hills."

In the twentieth century scientists became more reserved or skeptical about the relevance of these elevation thresholds. But apart from rare and very recent exceptions,[40] they usually embraced the imperative of delimiting mountains in order to approach them as objects. Two geographers, authors of an ambitious effort to come up with a definition, wrote in 1962: "The head of the mountain shines in sparkling light, but its foot is lost in the hazy plains, and it is man's role to provide the clarity that nature lacks."[41] Most recent studies, though aware of the limits of such attempts, have tended to use elevation thresholds or more or less sophisticated combinations of slope and altitude values (see also chapter 8).[42]

In any event, whether or not the definitions and thresholds are precisely fixed, the other mountains, those defined by ordinary language—in Reims, Paris, or Montreal—remain on the fringes of the category as rethought by the natural sciences: "Everyday language alerts us to troublesome uncertainties. . . . What do a molehill and a great peak have in common? The words 'mount' and 'mountain,' which indicate such dissimilar things, lack the most elementary precision. . . . These words arrived too early, when the fledgling French language knew nothing of real mountains and when no one felt the need to assign an order of magnitude to the mounds on the earth's surface."[43] That invalidation of popular ways of naming goes hand in hand with an interest in objectifying mountains in an indisputable manner. Popular modes of designation were based on points of view and impressions. Scientific modes strive to escape that patent subjectivity. The use of the criterion of elevation, more abstract than that of gradient, is part of that neutralization of the observing subject, as is the use of maps: thanks to them, once the terrain has been surveyed, the hypsometric curves can be drawn and the mountain chains and massifs delimited.

Vegetation, Zonation, and Plant Formations

Although Humboldt wrote important works on issues relating to the geology of the mountains, the measurement of elevation, and above all, the general conception of the mountain as object of knowledge, he tends to be remembered today for another of his specializations. He analyzed the distribution of plant species in space, which he called "the geography of plants." Naturalists had long noted that plant species occupy different zones along the main slopes. And it was long thought that that zonation correlated with average temperature and atmospheric pressure. These observations, already ancient, led scientists to adopt an argument of cause and effect. Joseph Pitton de Tournefort, the first to systematize that observation (during his trips to the Pyrenees and the Caucasus at the turn of the eighteenth century), thought that elevation could practically be deduced from the presence of a particular plant species.[44]

In fact, Humboldt continued Tournefort's work and often paid tribute to his predecessor, though he himself qualified the causal relationship. He was sensitive to the variation in the boundaries of the zones, not only as a function of latitude, which Tournefort already intuited, but also as a function of aspect and the nature of the soil and subsoil. Just as he did in observing the variations in the lower limit of the permanent snowpack everywhere he went, Humboldt made elevation and zonation the interpretive keys for comparing all the world's mountains. And because he was intent on understanding the physical disposition of plant species within plant formations, the relationships between different species, and their relationship to environmental features, he would later come to be seen as one of the founding fathers of ecology.

After Humboldt, the botanical and ecological sciences systematized geological surveys, species classification, the identification of symbiotic relationships between plants (their mutual influence on each other), and the analysis of environmental effects (primarily the complementarity between plants, the soil, and climate). That type of knowledge would lie behind a series of management practices and political initiatives, which will be presented in the following chapters: the observation of the differential between the actual location of species and the environments that seem to suit them would lead to attempts to transfer and "acclimatize" species (chapter 7); the identification of species native to a given region would lead to initiatives to protect the sites where these species are located (chapter 4); and the measurement of the overall biological diversity (biodiversity) of a region or type of environment would lead to the adoption of different global conservation strategies in particular regions of the world, based on the importance of each such region (chapter 8).

With respect to all these questions, Humboldt and his successors believed mountains to be of great interest as a type of environment. Many of the species found in the mountains are native to them, despite having evolved independently in ranges very distant from one another. The zo-

nation of plant life and the variety of aspects and soils confer on mountain environments a great diversity of species, especially in the tropics.[45] As
Humboldt himself said, with a certain effusiveness, the mountains located
near the equator afford "in the smallest space the greatest possible variety of impressions from the contemplation of nature. Among the colossal
mountains of Cundinamarea, of Quito, and of Peru, furrowed by deep ra-
vines, man is enabled to contemplate alike all the families of plants, and
all the stars of the firmament. . . . There the depths of the earth and the
vaults of heaven display all the richness of their forms and the variety of
their phenomena. There the different climates are ranged the one above
the other, stage by stage, like the vegetable zones, whose succession they
limit; and there the observer may readily trace the laws that regulate the
diminution of heat, as they stand indelibly inscribed on their rocky walls
and abrupt declivities of the Cordilleras."[46]

The Mountaineer as Naturalized Human Type

Between the late seventeenth and the late nineteenth century, modern
thought and science thus established mountains as a class of spatial objects, as evidence about the history of the physical world, and as a major
component of the causal relationships that connect all phenomena of the
physical world and of the living world.[47] After two centuries of natural and
geographical science, mountains became part of a system of forms and processes deployed on a global scale and with a universal scope. The rhetoric
of eighteenth-century science and exploration even conferred on them the
status of forms emblematic of the natural sciences.

The effort to impose a naturalistic order on the world by means of the
mountain did not focus solely on plants and animals. That ordering sub-
sumed the human species, whose natural history also began to be written
in the eighteenth century. A pervasive naturalism gave rise to a particular
type of human being destined for a great future: the mountaineer. Before
we consider the political import of that terminological and semantic innovation, we need to clarify why it was necessary to scientists in the Age
of Enlightenment and to their successors. At the time, the creation of that
human type was motivated by logical and philosophical considerations.

The motivation was logical in that, when the mountains were objectified and assigned a naturalistic meaning, many populations were included
within the entities designated as mountainous. It was of little consequence
at that point whether the populations thought of themselves as mountaineers; we have already observed that they usually thought of themselves

in other terms. Nevertheless, scientists and philosophers associated them with the mountains that constituted their daily environment. Hence the indigenous populations north of the Gulf of Saint Lawrence became the Montagnais: the first contacts that French explorers and colonists had with these peoples occurred at the base of the escarpments of the Laurentian Mountains, and the autochthonous peoples were assimilated to these landforms.

In logical terms, however, the populations did not become mountain-eers solely because their daily life unfolded in the mountains. They became so first and foremost because the naturalists striving to establish causal connections among phenomena had no reason to leave out the human populations. The idea of that connection predated the eighteenth century but truly blossomed with the rise of natural history.

Enlightenment authors were fond of giving climate a key position in the causal chain. Many imagined or observed the effects of cold, wind, and bright light on the bodies and temperaments of mountaineers. Philibert Commerson imagined that, because of regional climatic conditions, a race of white-skinned dwarfs lived in the uplands of Madagascar.[48] According to others, the mountain climate primarily affected one's health. Élie Bertrand noted a prevalence in mountain regions of "that malady called homesick-ness," which "arose in part from the difference in the weight of the air-stream in countries closer to sea level."[49] A little later the abbot Giraud-Soulavie reversed the diagnosis: "It is in the human species above all that the influence of climates is felt. . . . On high mountain peaks, man . . . en-joys the most robust health: the kinds of illnesses known there are few. It is the climate of strong constitutions and of health, whereas, in the lower climate, all illnesses have established their sovereignty."[50] Buffon included additional factors in the distinction he imagined between mountaineers and other types of human beings. He found "that the former are agile, re-freshed, well built, clever, and the women are generally pretty; by contrast, in the flatlands, where the soil is loamy, the air thick, and the water less pure, the peasants are uncouth, lumbering, misshapen, stupid, and almost all the peasant women are ugly."[51]

In the nineteenth century, as a result of progress in mineralogy and stra-tigraphy, authors were more attentive to the purported effects of geology and water on mountaineers' constitution.[52] Cretinism, which attracted the attention of both scientists and tourists, was generally explained by the in-fluence of rocks on water quality: "It seems therefore that this illness must be attributed to the water, which carries schist or limestone sediment, be-cause in infancy it obstructs the blood vessels with fine silt. Hence goiters are found in the mountain countries, in the high valleys of the Cordilleras in America, and even in Tartary, where their cause is not unknown."[53]

The specificity attributed to mountain populations was part of a twofold philosophical argument: the natural-philosophy argument, already perceptible in the previous interpretations, was paired with an argument stemming from the speculative philosophy of history. Because it was long believed that mountains constituted the most ancient forms on the earth's surface and were the birthplace of humanity, there was a tendency to see the individuals and societies that populated them as archaic. In 1767 Samuel Engel considered them survivors of the Flood who spoke a primitive language predating the biblical episode of the Tower of Babel and the confusion of tongues.[54] Even stripped of all biblical and religious references, philosophical and naturalistic discourse continued to see the mountaineer as one of the figures of the savage, a notion Western elites devised both to conceptualize the diversity of peoples and to conceive of the historical trajectory of their own civilization. Johann Gottfried Herder portrays mountaineers as "vestiges of the ancient nations" and locates "a few remnants of the most ancient humanity" in "hidden recesses, in a few rustic and almost impenetrable lands, where, however, their origins are still marked by traces in their languages, traditions, and customs."[55] The rustic character of the mountaineer was said to account for his bravery in combat and his aptitude for war.[56]

By contrast, for eighteenth-century antimoderns such as Jean-Jacques Rousseau and Jean-André Deluc, that primitive character of the mountaineer, protected from the world, is a guarantee of happiness: "For man is happy when he remains in the most natural state."[57] Much later Onésime Reclus wrote that the mountaineer's virtues can even regenerate flatland societies: "In the healthy air of the mountaintops, in the streaming water of the gorges, on the high prairies . . . far from luxury, from the lust for honors, from overreaching desires, from dashed hopes, from scattered lives, the people who will come to occupy the places vacated by corruption, exhaustion, calculation, suicide, and premature death are growing tougher and greater in number."[58] In an explicitly naturalistic metaphor, he speaks of the migration of mountaineers to the plains as so much "human alluvium." His contemporary, the U.S. naturalist Ernest Ingersoll, expresses a similar view: "Mountaineers are men of action, and from their heights have radiated the lines of human progress in all its varied aspects—political, social, and religious. As the hills themselves stand above the general level of the surface of the globe, so do the men in the hills stand prominently forth from the general plain of history."[59]

Nature, Modernity, and Civilization

Humboldt was not inclined to practice such a natural determinism or such a speculative philosophy of history and generally did not employ the label "mountaineer" to characterize human beings. When the term appears in his writings, it is used primarily to situate the groups to which he is referring, yet without placing them in the service of a preestablished causal chain. Because he was a young man during the height of German romanticism, however, he absorbed from his contemporaries the idea that the contemplation of nature and the sight of mountains in particular could have psychological effects. It was not uncommon for him to write passionate, sometimes raving descriptions that attest to a mental agitation that others before him, Saussure especially, had already displayed. Humboldt attributes the origin of that heightened state of the senses partly to nature itself and partly to the very nature of the senses and the imagination. He "considers phenomena relating to the human senses and the human spirit to be consubstantial with nature itself and believes that they demonstrate that humanity belongs to the great oneness of the universe."[60] In *Cosmos*, written at the end of his life, Humboldt sets out to present the entire body of naturalistic knowledge of his time. He devotes many pages to a purportedly objective analysis of the pleasure to be had from observing nature or landscape paintings ("the image reflected by the external world of the imagination") and to the ways that civilizations have experienced and cultivated that keen awareness of nature and the landscape.[61]

But for Humboldt the psychological effects of nature, mountains, and the landscape are not naturally determined: they do not result merely from exposure to the elements. They are primarily the result of an awakening of the senses that affects every civilization and the individuals who compose it. The argument holds for his own experiences: "In the forests of the Amazon, as on the slopes of the Andes, I felt that the surface of the Earth was alive everywhere with the same spirit, the very life that is found in the rocks, the plants, and the animals, as in the heart of humanity from one pole to the other. Everywhere I went I realized just how much the relationships I formed in Jena (where I conducted part of my academic training) had a profound influence on me, and how, inspired by Goethe's perspectives on Nature, I had gained new organs of perception."[62] This means that not every individual has the same access to the sensual and aesthetic experience of the mountains.

John Ruskin pursued a similar line of argument. The English critic and artist held, contrary to Humboldt's view, that the beauty of the mountains was the result of a divine intention. Like Humboldt, however, Ruskin thought that knowledge of that beauty could be constructed through the dual practice of scientific argument and aesthetic experience.[63] But that experience comes as much from the intrinsic beauty of the mountains as from the act of awakening the senses: it is revelation. Ordinary people— especially mountaineers—often fail to have that experience because they do not work specifically to achieve it.

For some, mountaineers are who they are because of nature's hold on them; for others, they are people who, though they live in the mountains, do not—or do not yet—possess the tools necessary for the aesthetic or spiritual appreciation of their surroundings. The mountaineer's status in the eighteenth and nineteenth centuries was thus caught between two forms of determination: a natural determination, according to which nature imposes on him physical or mental attributes; and a social and historical determination, by virtue of which the mountaineer's locale deprives him of access to modernity. In that second conception, the figure of the mountaineer is placed in the service of a general analysis of the civilization process and of modernity, an analysis more anthropological and historical than philosophical. Élisée Reclus conducted such an analysis of the societies of his time. In his *Nouvelle géographie universelle* he is mindful of the fate of the mountaineer under a variety of conditions.[64] But from one example to the next he tends to adopt the idea that mountaineers are modernity's outcasts. For instance, in the volume devoted to Mexico, he sees the mountains of that country as so many obstacles to travel and modernization. The enormous infrastructure projects undertaken in Mexico elicit his particular admiration because they allow the indigenous populations to be assimilated. Conversely, he is sorry that the mountain populations "who live apart, in remote territories," are still marginalized and ignored by modernity.[65]

A century later the historian Fernand Braudel made similar claims about the Mediterranean mountains. He begins by describing mountains as "the refuge of liberty, democracy, and peasant 'republics' [where] the life of the low countries and the cities has trouble penetrating . . . [and] seeps in one drop at a time." But he immediately qualifies that depiction: "And yet, life sees to it that there is a constant contact between the hill population and the lowlands. None of the Mediterranean ranges resembles the impenetrable mountains to be found in the Far East. . . . Since they have no communication with sea-level civilization, the communities found there are autonomous. . . . Mediterranean life is such a powerful force that when compelled by necessity it can break through the obstacles imposed by hostile terrain."[66]

The production of a class of natural objects (mountains) and of a class of human beings related to them (mountaineers) is thus part of a symmetrical ordering of the natural world and of the world of human beings. It partitions both spaces (since mountainous regions become strictly delimited objects) and societies (since mountaineers become a class of human beings apart). In addition, it identifies the qualities proper to the two classes not only as a whole but also as a cluster of causal relationships between these qualities. From one author to another these relationships take diverse, sometimes even diametrically opposed forms.

Thus scientists and philosophers contributed toward naturalizing mountain inhabitants. The work of naturalists on mountainous regions greatly predated the work of social scientists.[67] In addition, at least until the 1970s, even researchers in the social sciences and in human geography tended to link their analyses to the natural environment of these societies. As a result, natural objects continued to remain the implicit and universal referent for the production of knowledge. It is only recently that researchers have begun to work out how local populations think of nature and the mountain and how they might think of themselves as mountain people. Later in this book we will show that this shift is related to the political history of such regions.

The Ascendancy of the Alpine Model

For a long time, then, naturalists contended with the difficulty of speaking about the mountain at a level of generality beyond the singularity of particular mountains. This difficulty went hand in hand with the logical problems they encountered in simply trying to define the mountain. It was even more difficult to speak of the corresponding populations. In the first place, despite what general statements made about mountaineers over three centuries suggest, mountain populations possess a striking diversity of cultures, economic conditions, and modes of life. In the second, there is reason to doubt whether the category of the mountain, once it was reformulated to conform to a naturalistic perspective, remained pertinent for speaking in general about these populations.

That type of inquiry, highly methodological and epistemological in substance, was very common in the last century. Throughout the twentieth century many authors continued to delineate shared characteristics, problems, or destinies among the mountain populations. As recently as the beginning of this century, a U.S. anthropologist pointed out the "striking parallels in ecology, subsistence, ideology, social and cultural patterns, and marginality that reveal the common struggles and solutions of mountain people around the world."[68] But scientists are now questioning and analyzing researchers' very procedures and have gone back to the drawing board.

One of the conclusions they have reached has had major political effects: it has to do with the history of scientific production itself. For several generations of mountain studies, the Alps were the most influential model. A great deal of research in the seventeenth century was produced in the Alps and, to a lesser degree, in the Pyrenees. In addition, many scientists—such as Humboldt and Richthofen—who studied mountain chains around the world began in the Alps.

Many signs point to the Alps' status as the prototype for the organization of knowledge about mountains. Among the most obvious is toponymy. From the eighteenth century on, the word "Alps" was used to name many mountain regions of the world: one spoke of the Australian Alps, the Scandinavian Alps, the New Zealander Alps, the Japanese Alps, the Transylvanian Alps, the Canadian Alps, the Pontic Alps. In *Cosmos* Humboldt sometimes speaks of the Himalayas as the "Indian Alps" and also refers to the "maritime Alps of California" and the "Alps of Abyssinia."[69]

More implicit are the modalities for selecting the criteria by which the mountain in general is defined. In an article already cited, Paul Veyret and Germaine Veyret clearly have the Alpine model in mind: as factors specific to the mountains, they mention economic activities such as industry and tourism, which, at the time they were writing, had importance primarily for the Alps. Earlier Élisée Reclus had taken a similar approach in *History of a Mountain*.[70] Although he demonstrates his awareness of the variety of the world in his monumental *Géographie universelle*, this shorter work for a popular audience is a compendium of Alpine images, which, however, are presented as characteristic of mountains in general.

Between these two extremes, a number of authors have set out to present in comparative terms the mountains of the world, taking the Alps as a frame of reference or as a standard for comparison. In *L'homme et la montagne* (*Man and the Mountain*, 1933), one of the most interesting books on the question from the first half of the twentieth century, the geographer Jules Blache asks what all mountain populations have in common. Persuaded that the answer lies in the economic forms that best exploit the principal ecological particularity of the mountains—the zonation of plant life—he particularly studies herding and the forms of mobility (of human beings and animals) it entails.[71] He begins with the many varieties of Alpine herding, privileging one example, the "Helvetian model," to which he will compare all other forms encountered throughout the world.

At the end of the book he concludes that the Alpine model and his own prototype, the Helvetian model, have equivalents in Portugal, Morocco, and China—fidelity to the model being inversely proportionate to distance from the Alps—and he recognizes no other model anywhere in the world.

The role the Swiss Alps play in Blache's argument is not surprising. Switzerland appears very early on as a heuristic model competing with or complementary to the Alpine model. The advantage of the Swiss Alps was that they were explored earlier and more systematically than the rest of the chain. Between the Renaissance and the eighteenth century, they were close to a few of the centers of innovation in natural history (Bern, Geneva, Zurich, Neuchâtel). In addition, generations of travelers during the Age of Enlightenment were initiated into the mountain aesthetic in the Swiss Alps. Travel literature and scientific writings were thus particularly abundant. The force of the Swiss model lay above all in the resources it provided for eighteenth-century discourses on moral and political philosophy. The democratic and republican political system of the Swiss cantons, very original in ancien régime Europe, as well as a few features of their collective history piqued the curiosity of Enlightenment Europe, which was as quick to question nature as to compare the advantages and disadvantages of political forms of organization found in different places (see chapter 3). That juxtaposition of questions gradually led to the idea that the Swiss owed their historical and political particularism to the mountains that surrounded them. The image of the Swiss mountaineer took over the entire category to such an extent that, in the nineteenth century, a large number of descriptions and narratives implicitly adopted the traits that usually characterized the Swiss mountaineer to describe the mountaineer in general. More explicitly, it sometimes recognized the traits of the Swiss mountaineer in those of inhabitants of mountains very remote from the Alps (see chapters 3 and 7).

The Alps, and the Swiss Alps in particular, functioned as a model through which general knowledge of the mountain and its populations was forged. But that model was also a trap. Authors have known this for a long time, have debated the question, and have proposed alternatives.

Refraining from Excessive Generalities

In what is considered the most explicit book about the scientific project of French geography in the first half of the twentieth century, Lucien Febvre attacks the looseness of geographical concepts and the universality of the discourses his contemporaries were claiming to produce in using them: "It is of the utmost importance for our purposes to show the current imprecision in the notions on which one continues to base a whole series of historicogeographical theories and considerations with the most far-reaching ambitions but very little accuracy."[72] By way of illustration, he challenges the notion of mountain: "One commonly speaks of *the* mountain and its influence on man and of the specific character it impresses on mountain societies, which are contrasted feature for feature with societies of the plains. Mountain societies are said to be subject to the activity of a particularly oppressive and tyrannical natural environment. Nothing could be simpler. One takes a well-chosen example from a mountain society; one notes the most obvious characteristics of human existence in the region being considered; one leaves aside the particularities and erects into a general rule the observations thus gathered." He continues, in the same ironic vein, with respect to the "mountaineer": "'The mountaineer': What to say about him? The abstract, typical, universal mountaineer, the man of necessarily limited curiosity, his horizon restricted by the high barrier of the peaks. The traditionalist, a creature of habit by birth, kept apart from the major currents of civilization by his habitat, conservative in his soul, every fiber immersed in the past, superstitious guardian of the moral and material heritage of the ancestors who preceded him, because nothing has come to inspire a desire for change." He concludes: "The truth is that there is no unity of the mountain that could be consistently uncovered everywhere on the planet that mountainous landscapes are found, any more than there is a unity of the plateau, or a unity of the plain. . . . The chimera of unity is worse than a chimera: it is folly, and it is dangerous."

The charge was directed more at the simplifications of popular literature than at scientific production itself and was met with the expected reactions. Albert Demangeon, one of the leading figures in French geography at the time, replied: "We think, quite unlike [Febvre], that there are no surroundings more original and better differentiated than the mountains. No one can deny, on the pretext that physical geography has not yet provided a classification system for the mountains, that in human geography the notion of mountain, of highlands, is one of the most concrete, the most lively, and the most meaningful in existence, as clear in the minds of the common people as in those of the scientists who have analyzed it. To banish that notion from geography . . . would be to turn one's back on nature for the pleasure of a thesis to be defended."[73]

That polemic is of interest because it questions the very nature of the object of knowledge of researchers who study mountains. It is of interest as well because alternatives can be discerned behind it—whether to build a science of generalities by type of object or instead a knowledge of local arrangements. It also suggests the possibility of research projects relating to the construction of political or scientific myths about the mountain. Traces of these questions in research programs and policy recommendations, whether on a global scale (chapter 8) or on a European one (chapter 10), will be found later in this book.

Scientists have adopted several strategies to respond to the two difficulties that arise from the scientific use of the notion of the mountain—the problems raised by the Alpine model and the very possibility of constructing a general discourse on the mountain as such. The first is to geographically reframe the analysis on a scale smaller than that of the world, for example, with the categories "tropical mountains" or "Mediterranean mountains."[74] But the problem of heterogeneity is simply shifted from one level to another and never resolved. A researcher who has worked in many so-called tropical regions remarked: "The notion of tropical mountain has little meaning in itself, so diverse are the conditions provided human beings by these high-altitude mountain environments. The geographical reality varies by mountain massif, each of which is a geographical individual with its own personality."[75] The Alpine model has often been thwarted by a different type of reframing: a research specialization on the scale of a given mountain chain (such as the Pyrenees or the Andes) that adapts the terminology accordingly. In such cases scientific motivations are difficult to separate from ideological and political ones, especially since some researchers call for the "decolonization" of science. Another way they protect themselves from the excesses of models and generalizations, as well as from the very idea of naturalizing the local populations, is to abandon the classical categories of geographical and naturalistic knowledge and to adopt those of the populations themselves. But in that instance the meaning of the notion and its scientific value change radically. That posture is part of a diversification in the scientific status of the notion of the mountain, which is undoubtedly the key characteristic of contemporary research on that type of object.

The Diversity of the Scientific Status of the Mountain

In the preceding pages we have shown that the preoccupation with identifying, circumscribing, and describing mountains constituted a key motivation for scientific work in the eighteenth and nineteenth centuries. Generations of people took on the task, whether they called themselves naturalists, geographers, botanists, geologists, or philosophers. All shared the same idea: that science must circumscribe its objects in space and specify their particular content, composed of rocks, plants, glaciers, human beings, and a large number of other things. That conception of science has epistemological foundations that are linked to modernity itself. We need not elaborate them here. It also had practical advantages within the context of exploration and colonization, offering a tool for ranking observations and surveys done in one place or another and for mastering the extreme diversity of the situations encountered (see chapters 6 and 7).

But twentieth-century scientists tended to proceed differently. Geographers, including a large majority in France between the two world wars,[76] did retain the research objective of analyzing differences in the earth's surface. Mountain chains, mountain environments, and mountain regions continued to be their objects of choice and were still defined and delimited. But for other disciplines the mountain tended to become a mere descriptor. The structuring concepts were of a different nature.

That is true of geology, though it contributed a great deal to making the mountain an object of modern science, sometimes even claiming to define mountains and to characterize mountaineers. The discipline gradually abandoned the analysis of landforms and their effects on other realms of life, delving instead into the specialized fields of mineralogy, stratigraphy, and tectonics. In that pursuit there is less need to define the mountain and make it an object in itself. In geology the term "mountain" tends to be used as a matter of convenience, in accordance with what are in fact rather vague conventions, to describe a context or system of forms resulting from the processes studied. The same is true for biology and botany. Although the analysis and identification of so-called mountain species and formations played a fairly large role in structuring that field of knowledge and, later, those disciplines—before, alongside, and after Humboldt—references to the mountain have diminished in most current specializations, continuing to play an important role only in ecology. The case of climatology may be less well-known but is even more illustrative. For a time that branch of geography and natural history also made the mountain a structuring element of its body of knowledge. The typologies and maps in use in the discipline established the notion of a mountain climate in a lasting manner. But that has changed in the last two decades. Climatologist-geographers now prefer to speak of the mountain's influence on climates characterized by different criteria. As for geophysicists specializing in climate and atmosphere, they have largely abandoned the mountain in their analyses or limited themselves to studying the consequences in mountain environments of the general processes under study.[77]

A comparable observation, minus the evolution, can be made for the social sciences. Although sociologists and anthropologists have taken an interest in mountain societies as such since the mid-nineteenth century, they generally have not wished to formulate any analysis for the world's mountains as a whole, nor do they seek to define or circumscribe the mountain as such. For them as well, the mountain is primarily a context. The recent renewal of the social sciences' interest in the notion of the mountain has resulted from a radical displacement in the analysis. Many contemporary authors are curious to understand in modern or traditional societies the forms and consequences of conceptions of the environment, and for them the mountain has again become a subject of research—but this time as a shared representation, a convention of language, even an object of public policy.[78] For example, some have asked how the populations of the Himalayas described and characterized their surroundings and note that the terms employed in the vernacular languages were incompatible with the notion of mountain conveyed by modern science.[79] In Ethiopia even the adoption of local terms to designate the phases of settlement—which attests, however, to an effort by scientists to acclimatize themselves to the context of their studies—has been criticized for the simplifications it introduces.[80]

The place of the notion of the mountain in scientific activity was central for a time; it now varies by discipline and scientific paradigm. That shift can probably be attributed to one major factor. The mountain reigned supreme within the natural sciences and well beyond them in the heyday of causality and determinism. The mountain, or some characteristic element of the mountain, taken as either a cause (of the distribution of plants or the psychological makeup of mountaineers) or a consequence (of the cooling of the earth's crust or a decrease in air pressure) had become an essential and synthetic link in the causal chain that scientific activity aspired to demonstrate. With the adoption of different scientific paradigms—especially that of systematic thought—and with the "crisis of causality,"[81] the mountain, and altitude along with it, lost its scientific prestige and became an often vague object or just one context among others.

The evolution and diversification of the scientific status of the mountain is now leading to a paradoxical situation. The vast majority of scientists accept that notion and its naturalistic and objectivist sense because it has become part of a shared culture. But the notion has lost importance in many sciences. And yet it is now mobilizing a community of researchers who have never been so organized at the international level and who may have never had the ear of political decision makers to the extent they do now (see chapter 8).

Throughout the last few centuries, in any event, and thanks to the diversification or succession of scientific approaches, scientists have offered the societies of their time a very diverse set of figures of the mountain capable of motivating very different political practices and management models. The following chapters are devoted to presenting these practices and models.

PART ONE

The Mountain of States and Nations

With the advent of modernity, the mountain became a category of objects belonging to the natural world, which was conceived as being independent from the human world. The mountain was therefore no longer defined in terms of the places from which it was observed. The primary mediators in that metamorphosis were painting, cartography, and intellectual production. Taken together, they placed the notion of the mountain in the service of a generic and universal conception.

But even as the arts and sciences were forging that vision of the natural world in general, and of the mountain in particular, institutions were emerging—the modern state, then the nation-state—that restructured the social and political field at an intermediate level between the local and the global. And not only did the state and the nation-state become the framework for considering the fundamental social and political questions, they also organized the very apprehension of space and nature.

We now possesses the distance and the analyses necessary for understanding how the modern state reworked space, how it made the mastery of space a top priority. The term "mastery" should be understood first in its most basic sense, as the control of an area or expanse and as the full exercise of sovereignty. But it should also be understood in terms of cognition (mastery of the instruments of measurement, knowledge, and representation) and of politics: with the advent of modernity, "space becomes the 'project' of the state through the mediation of the administration. The administration, in becoming a relation that structures civil society and the state proper, produces territory."[1] In part I of this book we will show that the mountain was placed in the service of that operation, which, through a few privileged figures—in particular, that of "natural boundary" (see chapter 2)—converted an expanse into the territory of the modern state. When modern states became widespread in the nineteenth century, that figure actively participated in the operation of "partitioning the inhabited world" and gave the state its characteristic spatial form.[2]

The modern state, however, did not use the notion of the mountain simply to control space and to demarcate territory. The mountain also allowed the state to work out the idea of the nation. New figures naturally followed: the mountain as heart of the nation or as an expression of its cultural diversity (see chapter 3). Through that notion the modern state was also able to circumscribe a number of its resources—wood, water, snow, and so on—by placing them in the service of the national economy or of national well-being. At the time, then, the mountain was conceived primarily as a spatial and natural resource whose uses were conditioned by a rather monolithic notion of the public interest (see chapter 4). It was only later and very gradually that the idea began to take root that mountains could also be a place where men, women, and social groups live and pursue various projects, contributing to the richness of the nation by their diversity and initiative (see chapter 5).

Each of these figures is presented and illustrated in the chapters of part I. Taken together, they show the faces that the mountain assumed, successively or concurrently, from within the perspective of the project of statehood and nationhood. They also show why the mountain is "good to think about" in thinking about the modern state and nation. The prevailing scale of analysis was no longer that of a scientific knowledge with universalist aspirations; but in studying that nationalization of the mountain, we shall come to see how scientific knowledge was recycled within the political project and how, in fact, the conditions for producing scientific knowledge were politicized.

TWO

The Mountain and the
Territoriality of the Modern State

In 2009 the Swiss and Italian governments reached an agreement to modify the 750 kilometers of their shared border, particularly between the Rhone (Switzerland) and the Po (Italy) river basins. The line, fixed precisely during World War II, was based on the identification of hydrographic basins. This principle, which was sanctioned by international law, means that the lands drained by mountain streams and rivers that ultimately flowed into the Po belonged to Italy and those whose waters flowed into the Rhone belonged to Switzerland. But part of the ridge line between the Rhone and the Po was composed of glaciers, whose size and shape had changed a great deal over the previous fifty years. In view of these alterations, the two governments agreed to update their borders. Switzerland would gain a few hectares, and the municipality of Zermatt, also in Swiss territory, would acquire a ski lift and a cable car terminal.

For once, this border modification occurred in a peaceful Europe, between two countries that enjoyed an excellent relationship. Usually, however, the history of international borders and their demarcation on the ground has consisted more of conflicts and fortifications than of amicable arrangements. Over the last three centuries the objective of fixing borders has combined law and philosophy, geography and the art of war. Each of these disciplines in its own way was led to promote the "proper border," or, more exactly, the right way to draw borders. In that long history of international borders, mountains and watersheds have occupied a key place.

The idea that mountain ranges could or ought to play a role in fixing the contours of national territories goes back to ancient times. Several texts from antiquity, particularly Julius Caesar's famous *Gallic Wars*, had argued for adjusting political territories to the topography of their substratum. The idea that mountains could serve as natural limits to political territories can also be found in the Middle Ages and especially at the beginning of the modern age. As a result, certain mountain ranges, such as the Pyrenees and the wooded hills surrounding Bohemia, became lasting borderlines. The notion is still a major reference point for diplomats and political leaders.

The Policy of Natural Boundaries

Historians believe that this pairing of mountains and national territories did not have the same meaning in all periods and in every different context. Some think that the eighteenth century was a turning point.[1] Until that time, they say, the use of natural boundaries had been motivated by practicalities (to identify a convenient reference point) or by a desire for historical continuity (to remain faithful to earlier, especially ancient, principles for setting boundaries). With the Age of Enlightenment, however, a kind of spatial rationalization prevailed, at the expense of historical continuity. National territoriality now gave precedence to organizing entities that were easier to defend and easier to develop. Space was conceived as a continuous expanse, and places were understood to be contiguous and equivalent. The beginning of the eighteenth century also marked a shift in the project of the nation-state. Michel Foucault has suggested that, at the time, the modern state moved away from the principal objective of territorial control and the exercise of sovereignty, and toward the objective of "governmentality."[2] By that he means primarily the preoccupation on the part of the modern state with conceiving of its territory in terms of resources and population, and of its action in terms of the optimization of these resources.

References to the mountain as a way of conceiving and implementing territorial demarcations thus began to change in form and nature during the eighteenth century. The objectification of the mountain was subordinated to the idea that it must serve to identify a solid physical foundation and thus optimize the disposition and development of national territories. References to the mountain can thus be studied within the framework of what we will call a policy of natural boundaries.[3]

The notion that mountains constitute convenient or necessary boundaries to national territories was not the only one current at the time. In some cases—fewer, but equally interesting—modern nation-states chose to place mountains at the heart of their territorial project. They then served as a backbone of sorts to the territory. The mountain is hardly insignificant in either case: it is a major component of the territoriality of the state and of the territorial imaginary it adopts. The mountain, thus problematized, was circumscribed and developed accordingly.

Buache's Theory as Source of Inspiration

By the time the European nation-states were beginning to think that a policy of natural boundaries would allow them to optimize their territorial claims, many academic studies had already described the layout of waterways and mountain chains. The naturalistic turn that geography took at the time provided a few models that could be placed in the service of national territoriality.

The principal rational model was that of the geographer Philippe Buache (see chapter 1). In the mid-eighteenth century he proposed that the location and ranking of mountain chains could be deduced from knowledge of the waterways and of the lands they drained, which he called "river basins." Although the argument proved spurious, Buache's proposal had an undeniable interest, which lay less in its accuracy than in the maps that could be drawn as a result. Based on a knowledge of the waterways—which were often among the best-known geographical features, given their importance for exploration and the transport of merchandise—it ought to be possible to deduce a knowledge of the mountains and to represent them on a map.

Buache seems never to have suggested that his proposal be used to establish national borders. Although he elaborated the theory in royal courts and taught it to future monarchs, he always confined himself to naturalistic considerations. But others after him had no qualms about taking the next step. For example, authors of geographical descriptions of the entire globe, such as the Englishman John Pinkerton and the German Johann Christoph Gatterer,[4] referred abundantly to the French king's geographer and lauded the coherence of his system, which connected river basins and mountain chains across entire continents. Furthermore, both men invited the modern nation-states, often situated around topographical basins, major rivers, or confluences, to adopt as their political borders the mountain ranges postulated by Buache's theory. In other words, they extended Buache's spatial rationality further than even he had done: since mountains naturally border catchment basins, nations whose territories were located primarily on rivers or coasts found it more convenient to adopt mountains as their political borders.

Scientific Rationality Applied to the Argentina-Chile Border

It was along those lines and in the name of that geographical and political conception of national territoriality that the demarcation of the border between Argentina and Chile came about in the nineteenth century, at a time when simplistic theories about the spatial articulation of mountains and waterways were enjoying their heyday.

Granted, the Andes chain had roughly marked the boundary between Spanish colonies in the southern part of the continent well before the independence of South American countries. Although the boundary had never really been recognized, it constituted a convenient point of reference. But precolonial societies had not always used it in constructing their territories: the Mapuche, for example, had carved out a large realm in the eighteenth century, running from the Pacific to the Atlantic and cutting across the Andes at Concepción to the west and the Rio de la Plata to the east.[5]

When the countries of South America on both sides of the Andes became independent, their founding documents remained vague, if not silent, about the territorial boundaries of each state. The Chilean constitution did not give its first indication about national territory until 1822. Chile is said to be delimited by the Andes, the Pacific Ocean, Cape Horn, and the Atacama Desert. But the very vague reference to each of these entities did not allow for any demarcation or detailed cartography. Throughout the nineteenth century that state of affairs was a pretext for recurrent quarrels between Chile and Argentina, which were seeking to populate their borderlands and develop agriculture there.

In 1881 a treaty established the border in what appeared to be indisputable terms. Known as the Irigoyen-Echeverría Treaty, it stipulates in article 1 that "the boundary between Chile and the Argentine Republic is from north to south, as far as the 52nd parallel of latitude, the Cordillera de los Andes. The boundary line shall run in that extent over the highest summits of the said Cordilleras which divide the waters, and shall pass between the sources [of streams] flowing down to either side." The treaty, like most geographical writings and legal practices at the time, thus assumed that the principal ridge line of the cordillera corresponded to the watershed. The authors of the treaty, which was adopted before any systematic exploration of the cordillera had been done, did not know that for many segments of the Andes the principal ridge line of the range is located west of the watershed, and several valleys that open onto the Pacific extend fairly far east into the continent.[6]

To settle their disagreement, the two countries appealed to the United Kingdom for arbitration. The United Kingdom made its ruling in 1902 based on the reports Chile and Argentina had submitted and on a few statements from experts. Each of the countries had an interest in maximizing its gains by defending the conception of natural boundary that worked to its advantage: for Argentina, the main axis of the chain; for Chile, the watershed. The geographer Élisée Reclus, an observer of the controversy, remarked that the maps that accompanied the reports provided by the two countries differed in that respect: the Argentinian maps precisely report elevations and landforms; the Chilean maps emphasize river sources and runoff.[7]

In the ruling it made in 1902, which was accepted by both parties, the United Kingdom refrained from privileging one meaning of natural boundary over another. Intent on reaching a compromise, it sought to satisfy the claims of both countries equally, even if that meant making ad hoc decisions here and there along the borderline. A few details would be ironed out in the 1960s. The borderline adopted in 1969 attests to a compromise that was upheld during bilateral negotiations on the border shared by Chile and Argentina. In the north, the border runs between small catchment basins. In the south, it cuts through the middle of an Andean lake. Generally speaking, the boundary snakes back and forth from one side to the other of the principal ridge line of the Andes.

In this case as in others, the use of mountain chains to set borderlines is a matter of convenience, at least so long as there is no great concern to delimit the territories in detail, or so long as these boundaries, agreed to on paper, are believed to be easy to determine on the ground. That method, eminently geographical in its conception, cartographical in its representation, and legal in its realization, turns out to be much more complicated to implement once it becomes clear that nature did not design its objects to facilitate the work of surveyors.

The Order of Nature and Normative Naturalism

That first way of conceiving political borders in mountain regions—functional and practical, though open to contradictory interpretations—was complemented by a second in the eighteenth century, one that was more philosophical, even metaphysical. This view stipulated that the territorial order, as it was promoted by societies and nation-states, ought to be modeled on the natural order so as to be a part of a general balance of human beings and things. We shall call that conception "normative naturalism."

One way to justify that coordination between the two orders was to invoke a third, overarching order, the divine order or Providence. The Jesuits, whose reflections and activities in the field of education are well-known, were major promoters of that idea from the seventeenth century on. So too were the followers of natural theology in the eighteenth century.

A second justification for articulating a natural order with a territorial order by adjusting the borders was rooted in a rationalist conception of the progress of civilizations. This was the theory of Anne-Robert-Jacques Turgot. Before serving first as a progressive and reformist intendant to Louis XVI of France, then as his minister of finance, Turgot had worked on two book projects—one devoted to universal history, the other to political geography.[8] Apart from some short, fully realized sections, neither ever progressed beyond detailed outlines. In his *Plan d'un ouvrage sur la géographie politique* (Plan for a Book on Political Geography, 1751), he set out the objective of "understanding everything concerning the relationship between geography and politics under two headings: the diversity of production and the ease of communication. These are, in fact, the two variants by means of which the problems of political geography must be solved. We would have to add, however, the division of states, which depends in part on these two principles."[9] His description combines what he calls a "theoretical geography" with a "historical geography." Theoretical geography was conceived as "the relationship between the art of governance and physical geography";[10] historical geography concerned itself with variations in that relationship over time.

Turgot's book thus aspired to be a general reflection on the modalities for coordinating the geography of the modern state with physical geography, guided by an interest in optimizing national territoriality. That reflection leads him to see any shift in national territories in the direction of "natural boundaries" as a sign of a civilization's progress. The analytical value of his theory was therefore complemented by a normative proposal, namely, that the match between natural facts and political decisions constitutes an index of modernity and an aspect of moral character. Eighteenth-century governments were invited to found their power on strict respect for the "boundaries that nature had assigned it. Political geography has drawn the boundaries of the states; public law forms powers; but in the long run political geography prevails over public law, because in every domain, nature prevails in the long run over laws."[11] He then congratulates Spain for having renounced its remote provinces in the Netherlands and for being satisfied with a territory marked off by the Pyrenees. Along the way, the exemplarity of the Pyrenees, a commonplace among supporters of the policy of natural boundaries, was affirmed—at the cost of a revision in its meaning: the demarcation of French and of Spanish sovereignty in the seventeenth century had occurred in a completely different philosophical and political context. Viewed anew through the prism of normative naturalism, it became the model to imitate (fig. 3).

The Order of Nature and Hope for Perpetual Peace

For Turgot, the optimal achievement of a policy of natural boundaries was economic and public in nature: it was a guarantee of prosperity. As a result, and thanks to more general considerations as well, some came to recognize Turgot as one of the founding fathers of political economics. For other thinkers the policy of natural boundaries belongs to a different paradigm, one that promotes lasting peace among nations. Jean-Jacques Rousseau uses that argument in his commentary on Charles Irénée Castel de Saint-Pierre's *Projet de paix perpétuelle* (Project of Perpetual Peace, 1771). In that commentary, whose aim is to promote peace in Europe by establishing an overall balance in the ambitions and projects of nation-states, Rousseau invites nations to draw inspiration from the signs nature gives them: "The location of the mountains, seas, and rivers, which serve as limits to the nations that dwell there, seems to have determined the number and size of these nations. And it is possible to say that the political order of this part of the world is in certain respects the work of Nature."[12] Later he continues: "That is not to say that the Alps, the Rhine, the Sea, and the Pyrenees are insurmountable obstacles to ambitions; but these obstacles are shored up by others, which reinforce them or restore nations to their prior boundaries, when occasional efforts have extended them. The true support for Europe's system is in part provided by the give-and-take of negotiations, which almost always reach a mutual balance."[13] In other words, the geopolitical balance of Europe always reverts to the natural discontinuities, as if that adjustment came about by necessity.

Nicolas Buache de la Neuville pursued the same idea of a lasting peace inscribed within the natural order. A nephew of Philippe Buache who occasionally collaborated with his uncle, he too held the title of geographer to the king of France. He sought to defend his uncle's theory and to place it in the service of his own political geography. For Buache de la Neuville, a knowledge of physical geology was needed to shed light on political geography, which he considered the "most important" of the three branches, and to provide it with a few lessons: "Nature itself made the divisions of the globe from the beginning; it divided its surface into an infinite number of parts and separated them from one another by barriers that the passage of time and all human inventions will never be able to destroy. But human beings have not recognized that natural division; they have divided up the earth to suit their ambitions. Therein lies the origin of the strife among neighboring peoples and of most wars."[14] For Buache de la Neuville, the revolutionary era taking shape before his eyes represented the ideal moment to adjust the divisions of political geography to fit those of physical geography: "At a time when such an interesting political evolution is occurring around the globe, and when a new nation [the United States of America] is forming among a wise and enlightened people, let us provide men with a plan of division and apportionment that will forever fix their possessions."[15] Europeans and Americans were thus given the opportunity to adopt a system of territorial boundaries adapted to natural configurations and to achieve a state of eternal peace, an "end of history" of sorts, realized through the establishment of a perfect order: "That invariable natural division, which will last till the end of time, in being applied to the division of nations, would remove all occasions for strife and forever assure peace and tranquility among peoples."[16]

Therefore, many Enlightenment philosophers and scholars promoted a policy of natural boundaries based on the idea that it would guarantee perpetual peace and prosperity precisely because it would conform to a preexisting order. That natural order, because it belongs to the very *longue durée*, constitutes a frame of reference for escaping the accidents and tragedies of human history. The argument was used a great deal in Germany. Johann Christoph Gatterer, a promoter in that country of Buache's theory, believed that the determination of borders "in accordance with the eternal lines that nature itself drew by means of rivers, mountains, the seas" should be a top priority.[17] Somewhat later August Zeune wrote that physi-

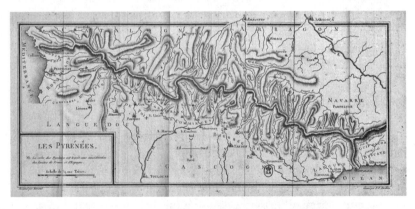

3. Map of the Pyrenees by Louis Ramond de Carbonnières, 1789. Ramond, a famous Enlightenment naturalist and a regular visitor to the Alps and the Pyrenees, here provides a very conventional map of the Pyrenees. A double line indicates the watershed, which, but for a few details, corresponds to the border between France and Spain. Thus Ramond follows the cartographical conventions that Philippe Buache popularized, even though Ramond knew very well that the chain's elevations were not distributed in that way. Bibliothèque Nationale de France.

cal geography, being "apolitical . . . has the advantage of being everlasting," and indicates the territories that nations ought to administer.[18] He also proposed that Germany be rebaptized "Hercinialand," in reference to the terminology of geologists. They linked the majority of the landforms of central Germany to the Hercynian period of the Primary Era, along with the region delimited by the Alps, the Oder, the Baltic and North Seas, and the Rhine—which thus included Denmark, the southern part of the Netherlands, and northern Switzerland.

The Age of Enlightenment, which attached such great importance to the idea of nature, thus extolled a normative naturalism. On the question of national territoriality, it recommended adopting a policy of natural boundaries, one that, because it was respectful of a preexisting order, would be the guarantee of peace and prosperity. The figures of the mountain that resulted often reduced that policy to a line on the map representing the watershed, even when the naturalists who drew it, such as Ramond de Carbonnières for the Pyrenees, were very aware of the substantiality of the mountain chain and of the multitude of foothills and secondary massifs (fig. 3).

Mountains and Peoples: Are Nations Historical or Natural?

The Age of Enlightenment added a third argument in favor of the policy of natural boundaries. From Johann Gottfried von Herder it borrowed the notion that landforms play a role in differentiating nations from one another. Herder (see chapter 1) saw nations blossoming in the heart of plains circumscribed by mountains. At a time of burgeoning nationalism and the birth of the first nation-states, it was becoming possible to present as natural and desirable the match between the populated zones of a nation, the territories of a state, and natural regions.

The theory was remarkably influential in the nineteenth century and in the first half of the twentieth. It can be found in the writings of a great many philosophers, geologists, geographers, historians, and political scientists. It is one of the ideas that recurs most often in geopolitics, in major authors such as Halford Mackinder and in lesser-known ones who, however, were destined to wield political power, such as Augusto Pinochet.[19] The theory also made inroads in popular literature, as in Rossitor Johnson's famous and premonitory *Clash of Nations*. In that book Johnson writes: "As civilization advances and population increases, the geographical rule of nationality becomes more apparent and more insistent. That rule, briefly stated, is that the tendency of mankind is to work toward union and centralization wherever these are indicated by natural boundaries."[20] Some of these authors are even more emphatic, seeing nations circumscribed by their mountain setting as the cradles of freedom and democracy, in contrast to the barbarous peoples and great empires of the Eurasian steppes. Turgot had decisively made that point back in 1751. Mackinder turned it to his own advantage in 1904: "The most remarkable contrast in the political map of modern Europe is that presented by the vast area of Russia occupying half the Continent and the group of smaller territories tenanted by the western powers. From the physical point of view, there is of course a like contrast between the unbroken lowland of the east and the rich complex of mountains and valleys, islands and peninsulas, which together form the remainder of this part of the world. At first sight, it would appear that in these familiar facts we have a correlation between natural environment and political organization so obvious as hardly to be worthy of description."[21]

Traces can also be found in nationalist rhetoric, sometimes even in decisions to modify borders, in countries best suited to such changes. In Bohemia, mountains and forests, which isolate the region on three sides from territories populated by Germanic peoples, were frequently portrayed as the very source of national singularity, especially in the mid-nineteenth century.[22] And the idea of an Italian unity guaranteed by the Alps can already be found in the literature of antiquity (Cicero), the Middle Ages (Petrarch), and the Renaissance (Machiavelli), as well as in a few peace treaties of the modern period.[23] But it was when Italy achieved its political unification (1860–71) that the barrier of the Alps really changed in status. There was a tendency to believe that the Alps had made possible the birth of a single nation, which the mid-nineteenth century political unification had merely codified.

In France, the Age of Enlightenment and the Revolution had already popularized a normative and naturalistic conception of the policy of natural boundaries. But in the nineteenth century the attention paid to the construction and evolution of the French nation led to more or less naturalizing reformulations, especially among scientists and ideologists. The geologists Dufrenoy and Élie de Beaumont, who in the first half of the nineteenth century produced a detailed map of the subsoil of France, provide a good illustration of the importance of that discourse within the natural sciences themselves. Their work would lead to the emergence of a region called "Dôme de l'Auvergne," and later "Massif Central," which grouped together topographical features (the Cévennes, the Cantal, the Monts du Lyonnais, and others) that had been considered independent in previous geographical descriptions.[24]

The book that accompanied their comprehensive map supplied commentary and a long description of geological terrain. In the introduction Dufrenoy and Élie de Beaumont present an entirely new naturalistic conception of French territory, which aspires to be a gage of legitimacy. They construct an image of France organized around two roughly circular and juxtaposed entities: a "dome," the Central Plateau; and a sedimentary basin, called the Paris Basin. On the southern fringes of the country, mountains, depicted as so many "natural borders," "separate [France] from the nations with the most natural relationships to it by virtue of the Latin or Celtic origin of their civilization and languages." The Pyrenees, the Alps, and the Jura therefore constitute the ramparts that guarantee a differentiation between the French nation and the Spanish, Italian, and Swiss nations. "Perhaps, if these barriers had not existed, the French, the Spanish, and the Italians would constitute a single nation." By contrast, in the northern part of the country, "where the natural boundaries . . . are the most vague," France "borders peoples of Germanic origin." There the contrast of languages and cultures is said to be sufficient in itself; mountain ranges are not needed to set the French nation apart.

These two geologists, then, forged an entirely new conception of the makeup of French territory, one that was widely adopted for decades in the scientific and popular literature. It combined the well-known figure of the mountain as natural rampart, at least in the southern half of France, with the figure of the mountain as heart of the nation, since the authors make the Massif Central the geological bulwark of national territory. That dual conception would structure the French imaginary throughout the nineteenth and twentieth centuries. The routes of the Tour de France, for example, have run along the country's borders since the 1920s, as if "securing the perimeter," and the mountain stages of the race have become so many acts of derring-do.[25] So too, the Tour de France illustrates the capacity of a national imaginary to use the mountain not only as a figure for the limit, but also as a figure for the nation's heart.

Theories of Mountain Wars

Various philosophical, historical, symbolic, and pseudoscientific arguments therefore favored the policy of natural boundaries. Taken together, they contributed to making a number of mountain chains into ramparts, the limit that national territoriality and the national imaginary would run up against in a number of countries. At the same time, these arguments provided for the possibility of a different status for other mountains, like those of the Massif Central in France.

Military commanders added a second set of arguments, elaborating an original discourse on the matter. They too fostered the idea that mountain chains constitute the best "natural borders" possible, but they generally did not do so in the name of philosophical or naturalistic arguments or of the perpetual peace heralded by some. Rather, they put forward practical considerations and tactical convictions shaped by the military campaigns of their time.

The most systematic statement on the subject is Carl von Clausewitz's *On War*. Yet the German general's most famous book, written in the aftermath of the Napoleonic Wars, does not contain a political theory of the proper border. Clausewitz refrains from advancing such a theory, merely proposing a strategic analysis based on the experience of the mass armies multiplying in Europe at the time. From that perspective he devotes four chapters to the specificity of mountain regions. In three of these he adopts a defensive point of view; in the fourth, an offensive one.

His conclusion is simple: large armies are always vulnerable in the mountains because of the constraints the landscape imposes on their progress. They are always open to being attacked by small groups of combatants occupying a dominant position. "The hills control the valleys," all the more so in that these hills are often in the hands of "adventurous partisans [who] always find shelter there."[26] Clausewitz therefore recommends that national armies control the high ground from a position of dominance, since whoever holds it has a considerable advantage over foreign armies attempting to force their way through. The territorial consequences of that conclusion are just as simple: a territory is easier to defend when it is surrounded here and there by fortified mountains along its ridge line—or, more precisely, "on the edge of the plateau which crowns the mountain." There, the position of dominance is optimized. "It is so easy on mountainous ground to secure a considerable tract of territory by small posts." Conversely, "when we are close to mountains, without being in actual possession of them, they are to be regarded as a constant source of disadvantage—a sort of laboratory of hostile forces."[27]

In his subsequent argument Clausewitz leaves behind practical considerations in the strict sense and positions his analysis in relation to the naturalistic studies available at the time. He demonstrates his familiarity with the uncertainties of geologists about "the origin of mountains and the laws of their formation" and with drawings of hydrographic networks that show them diverging along the ridge line. But he clearly states that military strategy needs to adopt an ad hoc conception of the mountain: "We must entirely give up the idea of a defensible line, more or less regular, and coincident with one of the geological lines, and must look upon a mountain range as merely a surface intersected and broken up by uneven terrain and obstacles strewn across it in the most various manner, of whose features we must try to make the best use permitted by the circumstances; that therefore, although a knowledge of the geological features of the soil is indispensable for a clear conception of the forms of mountain masses, it is of little value in the organization of defensive measures."[28]

Clausewitz, then, introduced a new configuration of the mountain, one that owed a great deal to topography but little to geology. This explains in part the considerable role the military would play in the detailed mapping of mountain regions, especially in the representation of cliffs, mountain passes, and escarpments. His configuration also owed a great deal to an empirical knowledge of mountain battles, which allowed Clausewitz to associate every topographical form with a potential advantage or risk. In European armies such military and strategic considerations contributed to the forging of a very particular image of the mountain. The mountain of modern armies is first and foremost a topography evaluated in terms of techniques for moving troops and equipment, and in terms of a war of position. These topographical features could be ridge lines, since the principal passes and some summits had to be controlled. Or they could be narrow passes, which are generally easier to fortify and defend than open spaces. Or, finally, they could be deep valleys, which are favorable to the movement of infantry soldiers and armored vehicles.

Alternatives to the Policy of Natural Boundaries

The idea that mountains constitute desirable borders for modern nation-states had opponents as well as proponents. In the nineteenth century a new generation of geographers set out to relativize the role of physical geography (topography in particular) in the comprehension of human territoriality. One of the first of these was Karl Ritter, who introduced geography into German universities. He was intent on bringing the *longue durée* of history into geographical reflections. Ritter noted that the Urals, for example, had not stopped Russian expansionism, nor had the Alps prevented the spread of the Latin and Germanic peoples.[29]

Somewhat later the geographer Friedrich Ratzel went even further, taking a position diametrically opposed to that of theorists of natural boundaries. Ratzel, a specialist in the state and the inventor of the concept of lebensraum,[30] disputed the notion that a state must possess fixed borders. In support of his view, he cited the fundamental mobility of peoples throughout history and the equally contingent ways that modern states are constructed. He denied that there was any rationale for mountains to serve as a natural and timeless limit of nation-states: "Mountains, by means of their barriers, combine a conservative and stimulating activity with a proud sense of self-sufficiency, which will someday allow the state to break through its protective barriers."[31]

Ratzel's analysis became widely known within the academic world, well beyond the borders of Germany. In France especially it greatly influenced a generation of geographers and historians, who refused to see the shape of a state, and its borders in particular, as a natural fact. Élisée Reclus readily admitted that modern technology diminished the barrier effect of mountain chains: "The people moving freely below the ice and cliffs can glory in having leveled the Alps."[32] Although borders in Europe still often ran along ridge lines, the reason had to do more with institutions and administrations than with constraints associated with the nature of mountains themselves: "The steepness of the slopes, the altitude of the passes, the large quantities of snow, the fatiguing climbs are minor matters when it comes to boundaries compared to the cordons of customs offices and military posts."[33] A little later Lucien Febvre, one of the founders of the Annales school of history, devoted a long chapter of his masterwork to so-called natural borders. He concludes with a formulation that perfectly sums up the book: "The state is never given: it is always forged."[34] Fifteen years later, when the borders of Central Europe had begun to shift once again, one of the first geography books wholly devoted to borders claimed: "A border does not take shape for intrinsic reasons; nature does not draw lines ready-made to oppose social action."[35] The critique of the very concept of natural border, outlined by Ratzel, thus spread rapidly to the scientific world. It also found an echo in the expansion strategies of certain European nations and in geopolitics.

In Germany Karl Haushofer, the chief theorist of geopolitics during the interwar years, took up the idea that borders are not intangible. In particular, the increase in trade associated with the Industrial Revolution meant that nothing now justified establishing borders in nature: "Absolutely nothing on the earth's surface provides protection in the form of a purely passive obstacle, not the seas or the rings of deserts or the high mountain chains or the swamp forests; everything has already been overcome by human migratory movements. By contrast, protection is assured only by a vigilant attitude toward life, an agrarian system established by living and hardworking hands behind a border drawn in the most equitable manner possible, founded on mutual consent by both sides and possessing a military aspect if need be."[36] A great admirer of Ratzel,[37] Haushofer too attributed the deployment of people across the earth's surface and the crossing of every type of physical obstacle to their efforts to satisfy their elementary needs.

The National Socialist Party echoed that argument as it was preparing to seize power. Haushofer himself had obvious affinities with Hitler's party. Although the Nazi regime made few explicit references to him and his theories,[38] in practice it promoted a similar policy of territorial annexation and control. The regime, in applying its program of uniting German-speaking peoples from both sides of natural discontinuities (the Rhine in the direction of Alsace, the Erzgebirge in the direction of Sudetenland, southern Austria on the other side of the Alps, and so on) and in constituting a vast Central European lebensraum, showed that it too, in its practice and in its ideas, had emancipated itself from any theory of natural boundaries.

In Germany, therefore, that theory of natural boundaries may have served as a foil from the apogee of romanticism to World War II. It was easy to show, moreover, that the ideology and practical implementation of that theory were rooted primarily in the cultures of neighboring peoples—the French on the one hand, the Italians on the other.

No doubt that disagreement about whether there was reason to give European nation-states lasting borders was itself of long duration. The recent declassification of the Foreign Office archives has revealed the terms in which Margaret Thatcher and François Mitterrand discussed the prospect of German reunification in 1989. According to her private secretary, the British prime minister, expressing her revulsion at the idea that the events of 1989 would relaunch the German people's centuries-old project of territorial conquest, received a retort from the French president: "Germany has never found its borders, the German people have been constantly in motion. And so they are today."[39]

Enlightenment, Nationalism, and Providentialism in North America

The British colonization of North America, followed by the territorial expansion of the United States in the nineteenth century, gave rise to an array of discourses and practices that demonstrate how natural boundaries can be invoked or rejected to justify optimal territorial configurations.

In the mid-eighteenth century the object of debate was whether to adopt the ridge line of the Appalachians as a political border. The British authority chose it as a line of demarcation between the colonies of the

Atlantic coastal plain and the Native American populations, recognizing the right of these peoples to use the lands beyond that line.[40] A 1763 proclamation fixed the boundary line on a map and established its legal status.[41] The practical value of that line took precedence: as in the case of the Argentine-Chilean Andes (but well before it), it was easy to take note of that geographical discontinuity, which could be codified even before details on the ground had been identified and mapped.

But the declaration and the line on the map, updated in 1768 and 1774, proved incapable of keeping American settlers from penetrating deeper into the territories. Their justification for ignoring British law and the boundary established on the ridge line rested on one main argument that was both geographical and nationalist: the appropriation of lands in America could not rely on the same principles as those prevailing in Europe, whose populations had already been stabilized.[42] The British defeat in the Revolutionary War removed the last impediments to the powerful push westward.

In 1802 the Louisiana Purchase shifted the question farther west. At the beginning of the nineteenth century no one knew whether the western area of the Mississippi Basin was composed of a continuous line of peaks or, as Jonathan Carver imagined,[43] of a principal massif from which the main waterways on that part of the continent flowed in both directions. Not until later in the century, when the observations of Alexander Mackensie and of William Lewis and Meriwether Clark in the north and of Zebulon Pike in the central United States could be compared to those made during the systematic explorations commissioned by John C. Frémont and the Bureau of Topographical Engineers, was it possible to deduce the existence of a relatively continuous chain, baptized the Rocky Mountains.[44]

At the time of the Louisiana Purchase, some in the United States proposed that the boundary of the new federation should follow the ridge line of the Rockies. John Quincy Adams himself used that argument in the 1820s.[45] But most authors, journalists, and politicians defended the opposite thesis. They argued that the young nation had a destiny ordained by Providence. They claimed that the Rockies too had to be crossed and that only the Pacific Ocean could serve as a territorial horizon. Thomas Gilpin, a journalist and entrepreneur and a future governor of Colorado, wrote in 1846: "The untransacted destiny of the American people is to subdue the continent—to rush over this vast field to the Pacific Ocean . . . to establish a new order in human affairs . . . to teach old nations a new civilization, to confirm the destiny of the human race."[46]

The faith that drove the architects of American expansion, however, did not lead them to deny the role of the Rockies or, more generally, of the entire physical configuration of the continent. A few recommended that the conquest of the continent, while remaining in the hands of the American people, should give rise to different nation-states whose borders would lie on the ridge line of the Rockies.[47] Most promoted the idea of a single and unified nation but propagated images that gave a certain nobility to the Rockies: the image either of the mountain range as the backbone of the United States, or of a harmonious national territory built around an enormous river basin (the Mississippi), two lateral mountain chains (the Appalachians and the Rockies), and two coastal plains beyond them (fig. 4).

4. At the summit of Pikes Peak, which was long taken to be the highest in the Rockies, a monument is dedicated to "America the Beautiful," written by Katharine Lee Bates in 1895 during a trip to the Rockies. The song extols national territory, its landscapes, and particularly its mountains: "O beautiful for spacious skies / For amber waves of grain / For purple mountain majesties / Above the fruited plain! / America! America! / God shed his grace on thee, / And crown thy good with brotherhood / From sea to shining sea!" (first stanza). Photo: Randy Rogers, www.pikespeak.us.com.

In the United States and also in Canada, the crossing of the western mountains by the first wagon trains, then by the railroads, has often been treated in a heroic mode, which captures the might and genius of two young nations that cannot be contained by the natural obstacles in their way.[48]

It was not only explorers, settlers, and politicians who produced the nationalist and expansionist rhetoric of the United States and the ad hoc discourses on the continent's mountain chains. Academics also joined in. Two of the most influential authors of their time theorized that U.S. exceptionalism: in her book on the geographical conditions of American history, Ellen Semple, who promoted Ratzel's ideas in American geography, attaches great importance to the Appalachians in the nation's destiny.[49] In her eyes the Appalachians, by blocking the horizon of British settlers for a few decades, played a role in consolidating their settlements. These mountains then allowed the colonists to move westward once they were equipped to do so. A little earlier the historian Frederick Jackson Turner had offered nationalists a useful theory of the frontier myth and its role in constituting American identity. He suggests that it was the nation's destiny to contend with the natural obstacles encountered on the westward trek. These obstacles, which had to be overcome one by one, constantly renewed the American colonists' adventurous spirit.[50]

The history of continental conquest by British colonists and American citizens provides recurrent images of the mountain as an outstanding feature in the configuration of their territory. But that image was not placed in the service of the policy of natural boundaries for any length of time. Rather, it tended to undermine that policy in the name of the people's exceptional destiny.

The Dismemberment of the Austro-Hungarian Empire and the Controversies Surrounding the Proper Borders

In the early twentieth century, then, diverse models existed for understanding the relationship between the natural order and political territoriality. Buache's theory still had its defenders and the policy of natural boundaries its promoters. But the alternatives formulated by German romanticism, late nineteenth-century geography, and geopolitics were also well-known. These various theories and the contradictory recommendations to which they gave rise were systematically enlisted when the map of Europe was redrawn—especially after World War I, when the Allied Powers oversaw the dismemberment of the Austro-Hungarian Empire. The very conditions behind that dismemberment made it possible to formulate diverse and well-argued analyses and recommendations.

The borders of the empire as it existed in 1914 had heterogeneous origins and justifications. Only a few borderlines lay along ridge lines—primarily those of Bohemia and also Hungary, where the northern and eastern borders ran along the Carpathians, separating that country from Austrian Galicia and independent Romania. After 1917 the Allied Powers, anticipating the dismemberment of the empire after the war, laid the foundations for the reshuffling of territories in Central Europe. Shortly before, the United States had entered the war only on the condition that a few principles regarding borders be respected, especially the famous "nationality principle," and, in case of disputes, the principle of self-determination by the populations concerned. Decades of academic work had yielded high-quality information about the spatial distribution of languages, which at the time were considered the principal attribute of nations. These academic studies usually concluded that no clear physical boundaries existed between nationalities and that many regions had a mix of different languages. The studies prepared for the Paris Peace Conference and peace treaties demonstrate how much attention was being paid to the issue of drawing borders along mountain chains.

South Tyrol was considered separately from the rest. In that region, located south of the Po and Danube watershed, more than 80 percent of the population had been German-speaking since the Middle Ages. In 1915 Italy agreed to enter the war alongside France and the United Kingdom only on the condition that it be allowed to annex South Tyrol. A secret accord formalized the matter. The Italian government relied on two arguments at the time. The first was geostrategic: the Italian border was easier to defend if it ran along the Alps. Otherwise it faced a threat from Austria, since South Tyrol, which a French geographer called "a formidable projecting offensive," was located close to the cities of Verona and Brescia.[51] The second argument belonged to nationalist rhetoric: the annexation of South Tyrol allowed Italy to complete its nineteenth-century project and to transform the Alps into a true national rampart. The most famous Italian poet of the late nineteenth century, Giosuè Carducci (1835–1907), winner of the 1906 Nobel Prize for Literature and official bard of the New Italy, used that image for his own purposes. In one of his works he describes the king of Italy "in the Julian Alps, riding at the head of his people and marking with his sword the natural borders of the greatest of Latin nations."[52] The campaign Mussolini undertook a few years later to lay claim to Swiss

Ticino, whose waters flowed toward the Po and the Mediterranean, relied on the same argument.

The secret accord of 1915 violated the nationality principle and one of Woodrow Wilson's Fourteen Points, which stipulated that a "readjustment of the frontiers of Italy should be effected along clearly recognizable lines of nationality." For that reason, the annexation of South Tyrol was not unanimously accepted in Italy. Irredentist militants, anxious first and foremost to unite all Italian speakers, considered the incorporation of a German-speaking population to be in contradiction with their ideals. Socialists opposed the annexation for the same reason, especially Cesare Battisti, a historian from the neighboring Italian-speaking region of Trentino, also under Austrian domination. In his prewar publications Battisti proposed that the Austro-Italian border be established at the Salurn Defile, which was both a linguistic boundary and a fortifiable site.[53]

Despite that internal opposition, the United States yielded to pressure from their allies. The modification of the border between Italy and Austria, which returned it to the ridge line and the Brenner Pass, resulted in the incorporation of more than 200,000 German-speaking Tyroleans. That annexation, accomplished in the name of the policy of natural boundaries and in defiance of the characteristics of the populations concerned, would cause lasting resentment in Austria.

By contrast, on the other edges of the Austro-Hungarian Empire the criterion of nationality had greater weight, leading to a disregard for natural boundaries or to compromises among scientists, military leaders, and diplomats. To guide their work the Allied Powers called for studies or set up commissions of experts to consider the new borders of Central Europe. These commissions were composed primarily of historians and geographers.[54] Most of the experts, including many geographers known primarily for their work in physical geography (the Frenchman Emmanuel de Martonne and the Serbian Jovan Cvijić, for example), agreed to give little weight to topographical and hydrographical factors. Several others (such as the American Isaiah Bowman, head of the delegation of American experts), though still proponents of physical and topographical determinism, allowed themselves to be persuaded. Of all the criteria they favored in their analysis, language was the most important. In fact, the American commission included in its ranks Leon Dominian, who had just published a book on linguistic frontiers in Europe.[55] The exercise was complicated by the existence of bilingual regions and by the Germanization and Magyarization policies implemented in the previous decades. The experts, especially the Serbs and Romanians, did not wish to codify the results of these policies a priori. That was especially true in Transylvania, where Hungarian tended to be spoken in the cities and Romanian in the rural areas; and in Carinthia, where German was generally spoken in the urban areas and Slovenian in the countryside. In 1919 one member of the French committee of experts, the historian Émile Haumant, a specialist in the southern Slavs and a proponent of the linguistic criterion, said of the future border between Austria and Yugoslavia: "That border will be difficult to draw. It cannot correspond to political boundaries, nor can it rely on watersheds or summits; rarely will it be able to follow a distinct demarcation between two linguistic zones; everywhere it will be necessary to compromise on the conditions commonly required of a good border."[56]

The experts, especially the geographers, conformed not only to linguistic criteria but also to theories of the state then in force. They sought to achieve coherence and economic viability in the regions concerned. Communication routes and the complementarity of economic activities between the cities and the rural regions were objects of their attention. Nor did they neglect strategic criteria: for Romania, Emmanuel de Martonne advocated a massive territory, "spherical in shape,"[57] that would be easier to defend if necessary.

With the resources provided by these criteria, experts from the different nations adopted somewhat converging proposals. They recommended that Transylvania be annexed to Romania, thereby incorporating the central part of the Carpathians into that country; and they came up with a compromise solution for Carinthia. They thus generally granted secondary importance to natural boundaries.[58]

It was when the diplomats took over the discussions, and when final negotiations opened between the governments concerned, that the diversity of conceptions of the relationship between political borders and natural boundaries really became glaring. The European diplomats indicated a clear preference for natural boundaries, which in their view were more easily objectifiable than linguistic boundaries. At one point experts even began to speak ironically about the diplomats' "geographical culture." One geographer, a French expert, made fun of diplomats who sought to discover "race watersheds."[59] The military officials too were generally supporters of natural boundaries. But they also knew that the French, Italian, and English governments wanted to make sure there were large nation-states east of Germany and Austria that could, if necessary, resist any future ambitions of the deposed powers, Russia included. That factor favored a greater Romania and a greater Yugoslavia straddling several mountain regions.

The Paris Peace Conference and then the Treaties of Saint-Germain

(September 10, 1919) and Trianon (June 4, 1920) sealed the fate of the defunct Austro-Hungarian Empire. They codified a nuanced use of the principle of natural boundaries. Romania, now straddling the Carpathians, was considerably larger than it had been. Following a referendum in which the majority of the Slovenian population voted to remain in Austria, Carinthia's Austro-Slovenian border was fixed along a secondary range in the Alps, the Karawanken. Czechoslovakia was pushed up against the ridge line of the Sudetenland in the north and west, to the great dismay of the Germans, who knew that many of their own people were located within its borders. National Socialism would not fail to remember that.

These few years of expert analysis and negotiation provided the occasion for an astonishing geographical rhetoric, first through the use of maps and then through the presentation of identity-based arguments. On the matter of maps, Bowman summed things up in a clear-sighted commentary: "Each one of the Central European nationalities had its own bagful of statistical and cartographical tricks. When statistics failed, use was made of maps in colour. It would take a huge monograph to contain an analysis of all the types of map forgeries that the war and the peace conference called forth. A new instrument was discovered—the map language."[60] The rhetoric of the cartographical image operated at full capacity, even among American and French experts, who had their own comprehensive maps.

The identity-based rhetoric was fairly literal-minded, intent on accounting for the attachment of populations and of national imaginaries to particular places and regions. On behalf of Czechoslovakia Eduard Beneš, president of the national delegation and future president of the country, invoked reasons "of a sentimental order" to justify a borderline along the mountain summit. In Romania uncertainties about the incorporation of Transylvania gave rise to a large number of books and articles vaunting the central place of the Carpathians in Romanian history and in its imaginary. In 1918, for example, Skijastikn Sekliesco published a timely book intended for French readers, knowing they supported the Romanian cause. The book, which was devoted to Romania's involvement in World War I, celebrates the Carpathians: "Romanians have always lived on both sides of the Carpathians. Nature did not intend those high mountains to be an ethnic border; rather, she intended them to form the backbone of Romanian country. As we have seen, the political boundaries before 1916 were the result of a historical iniquity. Transylvania, the wooded country on the other side of the Carpathians, has always retained its profoundly Romanian character, and its population has resisted all the assaults of aggressive Magyarism."[61] A contemporary Romanian historian has summed up the mobilization of the Romanian people in support of the incorporation of the Carpathians: "In the Romanian version [of territorial imagination], mountains unite while rivers divide."[62]A similar process of national emblematization of the mountain occurred in Poland in the last third of the nineteenth century, in the northern part of the Carpathians, shortly before the chain became the natural boundary for a reborn nation. As one historian explains it: "The stateless Poles, subjects of three different imperial powers, somehow managed to transform a remote mountain borderland along the internal Austro-Hungarian frontier into a recognizable and important national icon. While many attempts have been made to transform the *natural* environment into a *national* environment since then, this accomplishment helped to make the Carpathian Mountains, and in particular a section of them known as the Tatra Mountains, one of the most recognizable parts of 'Poland,' however it is defined."[63]

On the losing side, the nostalgia for Greater Hungary gave rise to several generations of books yearning for the ring of natural borders formed by the Carpathians to the north and east.[64] In Carinthia, finally, nationalist movements continue even today to contest the binational character of the populations located on their side of the Karawanken range since 1920. Every year these militants of *Heimatdienst*—led by Jörg Haider, the populist governor of the province, until his violent death in 2008—have assembled in the capital of Klagenfurt. They call for the definitive assimilation of the last Slovenes in the region—who now represent only 2 percent of the population of Carinthia—or their expulsion to the other side of the Karawanken Mountains.

In the early twentieth century, when nationalism was raging in many places, theories abounded of mountains as emblems for the nation, inspiring both political and military expansion campaigns and international arbitration. In particular, such emblematization made it possible for the Hungarians and Italians to exalt mountains as the territory's ramparts and for the Romanians to celebrate them as the heart of the nation-state. In other cases as well—such as in the United States and France—images of the rampart, the backbone, and the heart of national identity coexisted and did not necessarily compete with one another. That ambiguity made it all the easier to associate images of mountains with a nationalist and strategic imaginary.

Korea: The Mountain at the Center of a Nation's Construction

The nationalist exaltation of the mountain is not confined to Western nation-states. At the eastern tip of Asia, a few ancient and highly complex states have taken the same path. In Japan the Taoist religion had for centuries so valued mountains that they are omnipresent in rituals, landscape painting, and the art of the garden. In the Meiji period, when Japan first took on the principal attributes of a modern nation-state, that cultural attachment to mountains was converted into a political value. There as well the mountain was placed in the service of a nationalist rhetoric that codified in its own way Japan's entry into modernity.[65]

Korea is no doubt a less well-known but just as interesting example. The very rugged landscape of the peninsula created and popularized the image of a country where mountains are omnipresent. The founding myth of the Korean nation conferred a key role on them: the nation is said to have been constituted by gods who had come to dwell on a few peaks in the region. Mountains, especially Kŭmgangsan (known in English as the Diamond Mountains), were important and recurrent motifs in landscape painting, especially in the seventeenth century. That situation was hardly favorable to the adoption of mountain ranges as natural borders. In fact, the only land border Korea shares with its neighbors—the northern border, separating Korea from Chinese Manchuria and from Russia—follows the course of two rivers, the Yalu and the Tumen, never a ridge line. By contrast, the situation did favor the metaphorical conversion of the mountain into the nation's skeleton or underlying framework.

From that perspective a thorough analysis of the registers used to invoke the mountain casts into relief the interference between political and scientific discourses. Several analyses of the structure of the peninsula's landforms have given rise to competing political and ideological revisions.[66] Ancient descriptions of Korea stressed the existence of a principal chain, called Baekdudaegan, running from Paektusan, at 2,744 meters the highest point in the north, to the South China Sea. Both the Koreans and the Manchu consider Paektusan, which is on the border with China, the birthplace of their respective peoples.[67] Some books set out to show by their descriptions the extent to which the mountain contributes to the harmony of Korea as a whole. In the *Sansugo*, geographical records compiled by Shin Gyeong-jun (1712–1781), Korea is described as being composed of twelve mountains and twelve rivers. The perfection of its space mirrors that of time, since in Korea, as in the West, the year is divided into twelve months.

In 1903 a Japanese researcher named Koto Bunjiro, after a two-year stay in Korea, published an article on the orographic system of that country. It revolutionized scholarly representations of the region's geological structure. Trained in German science and greatly inspired by von Richthofen's theories (see chapter 1), Koto proposed that the idea of a single chain running north to south had to be abandoned. He distinguished four different chains, two to the north and two to the south. In so doing he reinforced a classic opposition, between the north and the south, that appears in descriptions of nature and society in Korea. That opposition would be used to justify the partition of Korea in 1950.

Koto's interpretation elicited several contradictory reactions that politicized the debate. Some Korean scientists saw the book as "a prelude to the birth of 'new' and scientific Korean geography" and argued for abandoning the traditional political and cosmogonic conceptions of the nature of the peninsula.[68] Others defended the classic conception of the Korean landscape, particularly in reaction to Japan's annexation of Korea in 1910. Thanks to a new upsurge of nationalism in the 1980s, the opposition between the two conceptions of the mountain, one modern and "Japanese," the other classical and Korean, again came to the fore. A newspaper article on the subject even bore the title "Japanese Imperial Power Changed Our Own Mountain Names."[69]

The Contribution of Mountain Sports to the Naturalization of National Territoriality

During the last two centuries and in various contexts, the naturalization of state territoriality did not consist of discourses alone, whether of scholars, diplomats, jurists, or bards of the national imaginary. It also took shape as actual practices in the mountain regions concerned. We have already seen the strategic notions that emerged from military practices on the ground. In the remaining chapters of part I we will see a few expressions of that territorial imaginary in the regional planning policies conducted by the state. But let us mention here another type of practice that also sought to control mountains, materially or symbolically, in the name of that imaginary: the scaling of mountains by travelers, tourists, and mountain climbers.

For a long time the ascent of peaks was understood and presented as a component of political appropriation. English-speaking authors use the term "oropolitics" to designate that practice and the state of mind that accompanies it.[70] Mountain climbing thus functioned as a metaphor (an athletic conquest stood in for a legal one) and as a metonymy (the conquest of a summit stood in for that of an entire region) for territorial appropriation itself.

Historians have shown that this practice and its political symbolism date back to the dawn of modernity.[71] Mountain climbing increased in frequency in the nineteenth and twentieth centuries, however. The peaks of the Rockies were patiently scaled one after another, first by explorers and cartographers who worked for the federal government—Pike and Frémont, for example—and then by the first travelers, who wanted to tread the high peaks of the continental backbone. The military excelled at mapping the high mountains and, to meet its own needs, climbed a large number of peaks in the Alps and Pyrenees.

In recent years hikes along the Baekdudaegan, the mythic chain in Korean territory, have become more popular. This activity has allowed South Korean hikers to celebrate explicitly the structuring power of that mountain chain in the nation's imaginary but also, when they are stopped at the North Korean border, to condemn the partition of the country. Neglecting the theories of modern geology, they argue that nothing in nature can justify that partition.[72]

Building a Legitimate Political Order

It is clear from this survey of several centuries of discourses and practices that the invocation of the mountain as a central or peripheral frame of reference for national territoriality has taken many different forms. Not all are political, or at least not in the same way. Some belong to a geographical, popular, or poetic imaginary and cannot always find effective political expression. Such invocations, when they are political, do not always refer to the same conception of the mountain or to the same forms of objectification.

Nation-states, preoccupied at the dawn of the eighteenth century with marking out their territory and exerting their full sovereignty within it, confined themselves to identifying in rough terms the mountains on the edges of their respective domains. Their subsequent concern to circumscribe the resources and populations on which they exerted their authority led them to refine their discourse. They then adopted more sophisticated conceptions of the mountain and of their territory as a whole. Their interest in protecting themselves from neighboring countries encouraged them to reflect on strategy. It led the elites in their armies to adopt the figure of the fortified mountain and to construct a detailed knowledge of that environment so as to draw a greater advantage from it. At the same time an imaginary was fashioned to encourage people to embrace the idea of the nation. This led to the promotion of narratives and practices, especially mountain sports, that made the mountain a collective frame of reference.

Such considerations play a significant role when the mountain is invoked as a natural boundary for nation-states. Kenneth Olwig has aptly shown that one of the essential issues at stake in nationalism is to make the "political body" of the nation coincide with the "geographical body" on which it is deployed.[73] The identification of central or peripheral natural reference points (whether isolated, like an emblematic summit, or vast, like mountain chains along the borderline) can contribute to the construction of a legitimate political order. The advantage of mountains, when compared to other natural forms—and especially to other emblems such as monuments—is that they belong to geological time, which in a certain sense shores up a nation's aspirations of be an ancient construction. Furthermore, mountains are part of the physical layout of a territory. If that layout lends itself to the evocation of figures of harmony (France), symmetry (the United States), or balance (Romania), it becomes possible to argue for a sort of national predestination. From that standpoint, the remark of a historian of Romania is valid for a number of other cases: "If nations are predestined, then there must be a geographical predestination, a well-defined space, marked out by clear borders, which has been reserved for them from the beginning. . . . A unitary history thus presupposes a unitary geography."[74]

But the geographical imagination has many resources and allows for the construction of a nationalist rhetoric based on very different physical configurations: "It is possible to claim almost anything about the role of geographical factors in history, as indeed about the causes of historical evolutions in general. It is not easy to say what Romania would have been like without the Carpathians—or the British Isles if they had not been isles!"[75] As a result, the invocation of the role of the mountain in the construction of national territoriality appears to be as inescapable in content as it is contingent in form.

THREE

The Mountaineer

The Other in the Heart of the Nation,
or Its Emblematic Figure?

Tempest in the Cuillins

In March 2000 John MacLeod of Macleod, twenty-ninth in the line of Mac-
Leod clan chiefs, announced that part of his property, the Cuillin Mountains, in Southern Skye, Scotland, would be put up for sale. He hoped to
make about ten million pounds sterling from the transaction and intended
to reinvest the money in reparations of the family residence, Dunvegan
Castle, located nearby. The structure, which the family has occupied almost
continuously for eight centuries, is in poor condition. The family does not
take in enough from the castle's 100,000 annual visitors to refill the coffers.

Upon the announcement of the forthcoming sale and of the existence
of a possible American buyer, the Scottish press reacted in two phases. At
first it displayed unconcealed incredulity; that quickly turned to stupefac-
tion. The Cuillin Mountains are not just any natural site. Rather, they constitute "probably the best-known mountain skyline in Britain, a symbol of
totemic significance to the islanders at home or abroad and a Mecca for
rockmen since Victorian climbers 'discovered' it in the mid-19th century."[1]
A famous major English newspaper concurred: "For the mountaineer, there
is nowhere in Britain to rival the Black Cuillin of Skye."[2]

Nor is the location of the Cuillin Mountains insignificant. The previous
year Scotland had obtained political prerogatives that conferred upon it
quasi-autonomy within the United Kingdom. In 2000 the Scots, swept up
in a wave of nationalist fervor, discovered that an age-old and emblematic
piece of land that they had believed was public could be sold, could fall
into foreign hands and become inaccessible. Journalists, their readers, hik-
ing clubs, and members of Parliament were indignant. In July 2003 John
MacLeod of Macleod backpedaled. He proposed suspending the sale and

handing over the castle property and the Cuillin Mountains to a public foundation. In return, he asked the foundation to pledge to finance the restoration work on the castle, to take charge of tourist visits to the site, and to allow his family to continue to reside there. Jack McConnell, the Scottish prime minister, was sympathetic to the proposal: "The Cuillin and Dunvegan Castle are internationally recognisable parts of Scotland's heritage. If the future of both the mountains and the castle can be secured in the way proposed, this will be good for Scotland."[3]

The passionate drama that played out in and around the Cuillin Mountains is inseparable from the fervor the Scots show for their mountains (the Highlands) and those who live there. Scottish nationalism was fueled by the narratives and vestiges of a difficult history with the English. For centuries the two armies faced each other repeatedly on the battlefield. In the eighteenth century the English army definitively gained the upper hand, and the Crown was now free to fortify strategic sites and construct the routes and bridges necessary for opening up the region.

The English victory also marked the beginning of a massive transfer of property rights in the Highlands. In two successive waves—in the mid-eighteenth century and again in the early twentieth century—many properties of clan chiefs and a large part of peasant holdings fell into the hands of English and Scots from the Lowlands, allowing the development of huge hunting grounds and sheep ranches known as estates.

It was also at the start of this movement that the image of the region and of its inhabitants changed radically. Until the mid-eighteenth century the English piled on the negative descriptions of the Highlands, which were deemed ugly and hostile, and of Highlanders, bold to be sure, especially in battle, but of an unparalleled savagery.[4] But the new feeling for nature developing at the time, then the half-admiring, half-condescending view of populations from the wild mountains, gave rise to new narratives. The poet and Cambridge professor Thomas Gray visited the Highlands in 1756 and came back with enthusiastic descriptions: "I am returned from Scotland, charmed with my expedition: it is of the Highlands I speak: the Lowlands are worth seeing once, but the mountains are ecstatic and ought to be visited in pilgrimage once a year."[5] In the following century the royal family participated in that movement: Queen Victoria acquired Balmoral, a property in the Highlands, in the 1840s. For decades she played out the elements of Scottish folklore there while at the same time displaying a real curiosity about the Highlanders and their history. That new interest in the Highlands was not unrelated to the constitution of estates between the mid-eighteenth and the early twentieth centuries—even though, paradoxically, estates contributed to the impoverishment and emigration of the Highlanders.

At the same time, nationalism seized on the dramas of history, peasant folklore, and the new aesthetics of the landscape to make the Highlands and Highlanders the emblematic images of Scotland.[6] It seemed that if the Scottish imagination were anchored in the mountains and in rural traditions, it would be possible to emancipate the country from the long-dominant cultural influence of Ireland, to celebrate a land of one's very own, and to cultivate the Scots' uniqueness vis-à-vis the English, from which even Scots from the Lowlands (between Glasgow and Edinburgh) could benefit. Nationalist and romantic authors such as Sir Walter Scott forged narratives and images that made the Highlander an inescapable icon of Scotland (fig. 5).[7] Nineteenth- and twentieth-century ideologues took on the task of describing Highlanders as martyrs of the land grab and military colonization undertaken by the English.

A recent squabble over statues shows that the debate remains heated even today. In 1994 a public campaign was launched to take down the statue of the duke of Sutherland that stands on a peak in the Golspie region.[8] The promoters of that initiative believed that the duke, a key actor in the constitution of estates in the 1830s, personified the painful metamorphosis and expulsion of the local populations. The campaign did not lead to the removal of a statue judged absurd by many, but it did result in the erection of another statue, the Emigrants Memorial, built as a tribute to the thousands of Highlanders who had to leave the country because they were not part of Highlands modernization.

Identity and Alterity of the Mountaineer

Even when the thousands of emigrants and the tens of thousands of their descendants, most of them in North America, are taken into account, Highlanders have constituted only a small minority of the Scots and a fortiori of the British. And yet they have personified a collective identity for one and a fundamental alterity for the other within the context of the military and symbolic confrontations that over the centuries have pitted not only the English against the Scots, but also the Scots against one another.

The image of the mountaineer, in the British Isles and elsewhere, has been caught up in that duality since the eighteenth century. That image functions sometimes as a foil, sometimes as a model for the idea of the nation. It has many other functions as well, in particular that of shoring up

COPE'S CIGARETTES.
36–Highlander of Montrose.

5. Highlander of Montrose, 1855.
New York Public Library.

scholarly theses about the relationship between societies and nature (see chapter 1). But when that figure enters politics, it does so first and foremost to provide an image, in one form or another, for emerging nations.

The mountaineer complements the geographical figures of the nation that we identified in the previous chapter. Rousseau, Mussolini, Mackinder, Pinochet, and many others show us that for the last three centuries the territorial foundations of modern states have been readily imagined, constructed, described, and celebrated as a function of the mountains that punctuate their topography. Herder and the generations of authors he influenced suggest that the singularity of nations is particularly strong when mountains serve as a dividing line between peoples. Within that combined imaginary of territory and people, mountains prove to be a handy reference point: they allow for the circumscription of both the nation and its territory, serving as a horizon for the low plains and basins they surround.

But the example of Scotland and the Highlander reveals the existence of another discourse and another way of pairing mountain and nation. In this discourse, mountains, now endowed with a certain substantiality and populated by so-called mountaineers who are strongly conditioned by their living environment, constitute a series of national homelands.

That idea is also pervasive in Herder, particularly since he, like many of his contemporaries, imagines that the origins of humanity are to be sought somewhere near the high plateaus of Central Asia, and that the return to the origins is a condition of virtue: "The mountains nourish the earth with water, but they also give birth to peoples. In their bosom, springs are born; in their bowels, the spirit of courage and freedom matures, while the peaceful plains remain under the yoke of the arts, money, and the vices. Even today, the high country of Asia looks like a churning mob of savage peoples. And who knows what floods, what upheavals they are predestined to cause in the coming centuries?"[9] For Herder and for those who followed in his footsteps, there are thus two ways to conceive of the pairing between mountains and nations: as barriers that constitute the natural limits of the nations they encircle; and as an environment that gives birth to specific peoples—mountaineers—who bear within them the imprint of their surroundings.

The existence of these two mental models conditioned the imaginary of the different nation-states for two centuries, once they sought to assemble representations of the men and women who collectively composed the nation, as well as representations of the territorial foundations of the state, all marked by the heterogeneity of their constituent parts. We shall attempt to show which images of mountaineers the discourses on the modern state and nation were able to bring to life. The example of Scotland—first a subject state, then an autonomous one—has already demonstrated that the mountaineer, in this case the Highlander, could either personify the national spirit or constitute its foil. The other privileged illustrations in this chapter (Switzerland, Italy, the Balkans, and others) trace a typology of political strategies whose aim is to inscribe populations characterized as mountaineers in the narratives and iconography of the nation-state or to exclude them from such constructs.

The Emblematic Individual of the Swiss Nation

Of the many ways of enlisting the mountaineer in a nation-state's imaginary, Switzerland constitutes a well-known example. Beginning in the late eighteenth century, a veritable "nationalization of the mountains" was under way there.[10] It was encouraged by the fact that the Swiss Alps, as we have already said (see chapter 1), served as a prototype for the imaginary of the mountain and of the mountaineer, by virtue of the founding texts of that imaginary. In Albrecht von Haller's *Die Alpen* (1732), in Rousseau's *La nouvelle Héloïse* (1761), and in a large number of other texts collated by many researchers,[11] we find, sometimes in an altogether new form, sometimes in a fully realized form, images of the Swiss mountaineer that have become models for positive representations of the mountaineer in general: the free peasant of the mountaintops, the cheerful and reassuring village girl, the rugged solitary shepherd, and so on.

These images have not always been associated with the image of Switzerland generally—far from it. Some texts even point out the contrast in character types and values that supposedly exists between the inhabitants of the Alps and the Jura on the one hand, and those of the cities and plains on the other, who all together constitute the Swiss people. In one of his travel narratives about the Alps (1815), for example, Philippe-Sirice Bridel, a scholar intent on constructing a naturalistic and folkloric knowledge of Switzerland, contrasts the mountains and "their virtues" to "the stormy sea of the cities."[12]

In a great many cases, however, that differentiated conception of the territory and of the inhabitants is in competition with or subordinated to another view, in which the mountain and the mountaineer represent the Swiss nation in its entirety.[13] Three types of discourse made that assimilation possible.

In the first place, *metonymic narratives* functioned in such a way as to make mountains embody Swiss territory as a whole and mountaineers the totality of the nation. These narratives proliferated in the eighteenth century, especially in works whose aim was to popularize a still-unstable national myth, even if that meant combining it with the myth, complementary in this case, of a natural rampart: "[We must] consider our Mountains in terms of their beauty and the advantages to be drawn from them, and in terms of a thousand beautiful things found there. If there were only this advantage, that they are a powerful rampart that Nature has placed around us to protect us from the insults of our Neighbors, that would already be a great deal. But that is not the only good that comes to us from them. . . . Thousands and thousands of beautiful Country Lands can be seen everywhere, with, grazing upon them, large herds of horned animals whose milk, flesh, and hide provide food, clothing, and wealth for the Inhabitants."[14] Even more explicitly, a Geneva handbook of 1911 indicates: "By living the same life, the mountain life, these people of diverse backgrounds have come to resemble one another in many respects: they are mountaineers, and that is what constitutes the Swiss people."[15] The existence of an urban Switzerland and of the regions located below the Alps is not mentioned.

Secondly, *genealogical narratives* made mountain dwellers the source of practices and values that can be beneficial when diffused to the nation as a whole. Such is the case for many texts written by visitors, particularly when they praise the political virtues of the Swiss people.[16] That representation even makes it possible to contrast the national and political destinies of different countries by comparing the influence that mountain populations were able to have on the social body as a whole. In the concluding pages of a popular work that aspires to contribute to the moral elevation and recovery of the French nation after its humiliation at the hands of Germany in 1871, Albert Dupaigne uses such images to exalt the virtues of mountains: "We cannot utter the word 'France' without turning our thoughts, finally, back to our dear country; let us rather say to our poor country, since that return is sad. We must not conceal the fact that when we leave the Swiss Alps and come home to the French Alps, and even to less rugged and less indomitable mountains such as those of our central plateau, we are painfully struck by the comparison. In Switzerland the virtues of the mountain have often saved the plain; in France, the vices of the plain have often killed the mountain. . . . The most effective means for returning life and beauty to these countries would be to give their inhabitants, through the education of the generation growing up today, the energy, the intelligence, and the patriotism that are the honor and salvation of the mountain populations."[17]

Then there were the narratives whose very *plot or modalities of production* drew a map of Switzerland, the constituent parts of which were thought to be indissociable. Such is the case for *La nouvelle Héloïse:* Rousseau, in constructing his plotline around the redemption of Saint-Preux, a melancholic city dweller, through contact with the young Julie from Valais, makes the city and the mountain complementary in their practices and values. To a certain degree it is possible to say the same of certain paintings and of many books from the eighteenth century—especially those by the naturalists Saussure and Haller, which deliver a similar message. The elites of Geneva, Bern, and Zurich produce a discourse on the mountain's virtues that weaves powerful connections between places with sharply contrasting geographical characteristics.

The political content of these narratives varies a great deal from one to another. Not all contain a naturalistic explanation for the political mores of Switzerland, and far from all are motivated by specifically nationalistic considerations. Nevertheless, an entire body of popularizing public literature, intended particularly for students, arrived on the heels of these narratives. The political instrumentalization of the federative myth of the mountaineer began "at the moment institutional Switzerland was invented," with the creation of the Helvetic Republic installed by the French in 1798 and the organization of the major Alpine festivals that took place in Unspunnen in 1805 and 1808.[18] These festivals ritualized the confrontation between peasants from the mountains and city dwellers from the region of Bern.

After decades of relative political instability, the adoption of the federal constitution in 1848 marked a turning point that allowed national cohesion to be forged as much through images and national narratives as through institutions: "The myth of the Alps made it possible to move to a higher level of identification, since it included the former state dependencies in a new form of ideological synthesis, by creating a sense of belonging that transcended the formal cantonal spaces."[19] It was in the mid-nineteenth century, therefore, that the Confederation began to make the mountaineer a true national emblem, taking advantage of literary and romantic myths, especially that of William Tell. In particular, the neutrality policy Switzerland pursued for the entire period that concerns us here—a policy that became even more pronounced during the world wars— encouraged the process of centering the nation's identity on the image of its mountaineers, who were relatively isolated in the center of Europe and jealous of their independence. That focus provided a valuable argument by which to justify international policy choices. At the same time, the image of the valorous and virtuous mountaineer provided national elites, who were primarily worried about the effects of urbanization and industrialization on social and familial values, with a useful counterpoint. The mountain was given all the more prominence in that for a long time Switzerland refused to think of itself as urban.

That institutional nationalization of the mountains also found privileged modes of expression in public iconography (such as the frescoes in the major railroad stations and federal buildings) and at national or international events designed to vaunt the merits of the modern state. The national expositions of the nineteenth and early twentieth centuries invariably placed mountains at the center of their exhibits. Switzerland similarly privileged depictions of its mountains at international expositions, first constructing and then reinforcing a stereotype that subsequently spread to the four corners of the world.

This way of staging national identity has been contested since the 1960s by urban and modernist elites, especially at the most recent national expositions, such as the one in 2002. Nevertheless, the country has continued to make use of the image of the mountain—though less and less that of the mountaineer. For example, at the World Expo in Aichi, Japan, in 2005, the Swiss pavilion once again took the form of a giant hollow mountain, inside which activities celebrating the nation's technical expertise were held. And at the London Olympic Games as well as at the Rio + 20 Conference in 2012, Switzerland once again staged its identity by means of a dense iconography of mountains.

Exalting the Nation's Internal Diversity: Italy and the Myth of the Alpini

A third model complemented the first two (the mountaineer as foil image, seen in eighteenth-century England; and as structuring image for the idea of the nation, seen in Switzerland and Scotland): that of nations whose principle of unity is clearly perceived but whose imaginary valorizes the diversity of the nation's constituent parts. Italy provides a good illustration by which to assess that view.

We noted in chapter 2 that Italian unification, centered on the idea of linguistic unity and a common civilization, came about rapidly in the 1860s, and that it used the Alps as a barrier to circumscribe a territory whose heterogeneity was everywhere apparent. Furthermore, debates on the legitimacy or illegitimacy of the annexation of South Tyrol in 1918 showed that the two logics (national-linguistic and territorial-naturalistic) could lead to recommendations that differed in their details.

The national question was not resolved in 1871, however. Once political unification was achieved in Italy, national cohesion had to be constructed. The famous words supposedly uttered in 1861 by Massimo d'Azeglio, former prime minister of the kingdom of Sardinia-Piedmont and rightist senator—"Abbiamo fatto l'Italia. Ora si tratta di fare gli Italiani" (We have made Italy. Now it's a matter of making the Italians)—mean precisely that. Although language and the prestige of history were enlisted, the contrast between the north and south, and the geographical diversity of the provinces, constituted major checks on the spread of a unitarianist myth at that time. The national imaginary as it was forged in the nineteenth century was quick to highlight the prestige of urban civilization (from ancient Rome to industrialized Turin by way of Venice, Naples, and Florence, noted for their prestige from the Renaissance to the eighteenth century). By contrast, mountain populations were sometimes lumped together in an archaic and demeaning representation. But since these mountainous regions and mountain populations existed in both the north and the south, both in the heart of the peninsula (the Apennines) and on its edges (the Alps), they did not easily fall into a single representation. As a result, the mountaineer has been invoked in national Italian discourse to explain the diversity of the nation and its various living environments and even to contribute to the very idea of the nation.

In that mise-en-scène of Italy's geographical and cultural diversity, it was the mountaineer soldier in the Alps, the *Alpino*, who had the starring role. Several factors worked in his favor following unification. The Industrial Revolution was in full swing in the cities of northwestern Italy (Turin, Milan, and Genoa). Although sustained by modernist discourse in some quarters, in Italy and elsewhere in Europe it also engendered a conservative and antiurban discourse, especially among the Catholic elites, who were worried about the effects of urbanization and industrialization on social and religious values. The Alps, which were very close to these industrialized cities, provided a model of traditional, religiously observant societies.[20] The late nineteenth century was also marked by a growing interest within the middle and upper classes in tourism, landscapes, nature sports, and mountain communities. The Alps therefore provided an important counterpart to the imagery of urbanization and industrialization in northern Italy. For denigrators of industrial civilization, they constituted the site of redemption for the burgeoning working classes. As a result of being in close contact with the Alps, "even the undisciplined workers of Turin would end up becoming good soldiers, faithful, generous, and lovable."[21]

That imaginary of the Alps and of Alpine populations, which shored up criticisms of social transformation in the industrial cities, played a clear role in all the Alpine countries and a central role in Switzerland. But in Italy it carried special weight: upon Italian unification, the Alps became the new state's only land border, the natural rampart behind which it could protect itself from continental conflicts. As a result, war, when it was not waged on the sea, necessarily became a mountain war, with mountain people at the outposts. The Alpino was vested with the role of guardian of the nation, a role that accorded well with the image of the frugal peasant, the hard worker, and the religious soul that industrial modernity attached to him at the time: "That man, who believes in and hopes for a future life, who finds peace in his soul on the word of the gospel and the priesthood . . . who . . . is not loath to obey his superiors, is the true mountain warrior, the defender par excellence of the mountain border. He will not curse the war, the militia, the nation, when he has to spend his days and nights on the mountaintops and on the edge of precipices; long fasts and bad food will not frighten him, he will be on the alert like the hunter at his post; without faltering, he will make his ten-hour marches between crests and valleys, he will traverse rocks and glaciers to leap onto the backs of his enemies; he will find no peace so long as the stranger tramps about his hills, so long as the defense is not secured."[22]

Within that context, Italy made its Alpine battalions a major component of its army. The first regiments of Alpini were constituted in Turin in 1872. Military documents, state propaganda, and popular literature draw a flattering portrait of them: "On the mountain they acquire cool and reflexive courage; . . . alone between mountains and sky, they have learned that man's iron will is sufficient to overcome danger. . . . They are soldiers of the mountain; instinctively, they preserve all the precious virtues of the ancient races. Installed in their battalions, which bear the familiar names of their villages, their hills, their valleys, they are the border guards, and when war calls on them, they surge forward, silent and strong, on the craggy summits, on the glistening glaciers, with a firm step that never wavers."[23]

Apart from the fact that this discourse invents an imaginary Alpine dweller, of which twentieth-century ethnography and human geography will be quick to provide a less idealized image, it also rests on a dual mystification. Because of the emphasis placed on the virtues of that legendary Alpino, the Italian army was praised more for being an association of remarkable personalities than for being a collective state institution. The nation's defense became solely a matter of patriotism and not of institutions and military strategy (fig. 6). Furthermore, at least until the deadly confrontations of 1917 on the border of Italy and the Austro-Hungarian Empire, the iconography of the army and of war gave more importance to mountain landscapes and scenes of ascent than to the military operations themselves. In the end, military imagery resembled the imagery of tourism and sport, with battalions often assuming the aspect of summer camps for young adults.

The myth of the mountaineer soldier reached its apogee with World War I and later with Fascism. As an Italian historian writes, "The war made

6. The cover of *La domenica del corriere*, April 30, 1916. This
Sunday magazine, very popular in Italy, consistently celebrated the
Alpini throughout the first half of the twentieth century.

the Alpine border sacred. The blood of thousands of young men sacrificed
on the peaks of the Eastern Alps . . . turned a vacation site into a *palestra*
of courage and discipline, a land of *lotta* and supreme sacrifice."[24] Once
peace was restored, Mussolini and the Italian state exalted the significance
of the militia, family, virility, and courage, for which the Alpino provided
the model to the entire nation. That myth, however, would be profoundly

shaken with the defeat of the 1940s, combined with the harsh condemnation of the war and of Fascist Italy's expansionist ambitions. In the postwar years it was often written that the Alpini were betrayed, their legendary devotion to the defense of the nation misdirected in military campaigns that sent some of its soldiers far from the peninsula.[25] The Alpine people became victims, the very embodiment of the hostage taking of the Italian people as a whole.

Between 1860 and 1945, then, the Alpino constituted an icon for the national imaginary of Italy. But unlike the Swiss mountaineer, he was not the only symbolic embodiment of the nation. He was a component of a nation that also liked to imagine itself as urban, mercantile, and coastal. He was one figure among others for the imaginary of Italian diversity, but still an emblematic figure of patriotic feeling. His status lay in the mountain surroundings associated with him and in the popularity of natural history, but especially in the specific location of the Alps, which conferred a true strategic value on him. The mountaineers of the Apennines, always identified in terms of their constitutive diversity and for whom no federative image ever emerged,[26] could never sustain a national or patriotic discourse, since the mountain chain itself had been of less geopolitical value than the Alps.

Ambivalent and Circumstantial Representations: The Balkans

One last illustration, taken from the Balkans, will shed even more light on the variable status of mountaineers in national discourse when the geopolitical context proves to be extremely changeable. In fact, southeastern Europe is made up of an imbrication of modest-sized mountains and interior plains that do not coincide with the distribution of peoples and languages. During the nineteenth and twentieth centuries the region passed through a series of contrasting geopolitical configurations that gave rise to very heterogeneous discourses on mountaineers.

The Ottoman Empire, which had been in decline since the eighteenth century, gradually gave way to the demands of nationalist movements. New independent states emerged between the 1830s (Greece) and the end of World War I. Then the borders were redrawn as a result of World War II and, after 1990, by the breakup of Yugoslavia.

Before the independence movements, political and economic power in the southeast, as in Central Europe, tended to be concentrated in the cities, on the great plains—traversed by the main communication routes—and on the coasts. Power was also in the hands of the colonial elites or of vassal populations. As a result, the mountains were often in a marginal position in terms of political and economic arrangements. Their populations were less subject than others to policies of religious and linguistic assimilation and to control by administrations. They could thus appear in the vanguard in the construction of resistance movements and then in a nationalist discourse that reconstructed their history and motivations.

One of the recurrent motifs in historical, geographical, and ethnographic descriptions in the nineteenth century and the first half of the twentieth was the individualization of these mountain populations, on a model similar to descriptions of Switzerland and Scotland written slightly earlier, through an insistence on the most specific traits: the influence of the rural economy, communal forms of social organization, attachment to Christianity, and so on.[27]

The Serbian geographer Jovan Cvijić, who was very active in the planning committees that prepared the boundary changes in that part of Europe after World War I (see chapter 2), was one of the most influential authors in that field. In a standard reference work published in 1928, he classified psychological types on the basis of natural surroundings. In particular, he praised the Serbian inhabitants of the Dinaric Alps, whom he saw as the best resisters to the Ottoman occupation and the least suited to assimilation.[28] A similar rhetoric can be found in many works devoted to Greece.[29]

The reverse image also exists. It relies on the same geographical distinction between mountaineers and herders, on the one hand, and peasants and city dwellers, on the other, but it tends rather to see in mountain societies the signs of savagery and brutality, the triumph of thieving and plunder. Found in travel narratives, this is one of the generic images of the mountaineer peddled in many parts of the world.

> Since the imagination delights in ennobling everything, these uncultivated men [the mountaineers of Greece] have been seen as patriots who, fleeing the foreign yoke, declared an eternal hatred for their oppressors and would rather tread over snow with free feet than crawl slavishly under a master's saber. We are happy to find such lofty ideas in humanity; but why must they so often be merely a brilliant sally of the imagination? It is sad to be forced to admit that the life of the Greek mountaineer is absolutely that of a bandit. He hates work that would procure him an honest comfort; but he is all the more greedy for money and, to achieve that prize, there is no means he overlooks. All his mental activity is focused on thieving. Sometimes he awaits the passing stranger to rob him; if he is bolder, he goes down to the plain

by night to pillage everything he encounters, to steal fruit, to rustle cattle, regardless of who the owner is. Such are the exploits recounted in national songs. It is said that such songs typify a people's character: were we to consider only them, thieving would be that of the Greek character, since their national songs always include it among the heroic deeds. If there is some pasha in the vicinity who rewards services lavishly, Greek patriots can be seen rushing to his court again and again to offer him theirs.[30]

The occupying forces readily borrowed these images because they take all nobility away from the mountain peoples; but the new nation-states, anxious for pacification, also employed them.[31] During the Bosnian War the disparaging image took root, proving its capacity to endure over the decades. A 1994 article in the *New York Times* made good use of it: "These mountains [the Dinaric Alps] . . . have nurtured and shaped the most extreme, combative elements of each community—the western Herzegovinian Croats, the Sandzak Muslims, and, above all, the secessionist Serbs. Like mountaineer communities around the world, these were wild, warlike, frequently lawless societies whose feuds and folklore have been passed on to the present day like the potent home-brewed plum brandy that the mountain men begin knocking back in the morning."[32] Wartime decidedly favors the resurrection of those myths of the mountaineer most deeply ingrained in history. In peacetime the images invoked are more anodyne but just as repetitive: in the 1960s Greek students continued to draw a sharp contrast between mountaineers—whom they called tough, hospitable, stubborn, happy, optimistic, and traditionalist—and the people of the plains, who according to them were quick to change, adaptable, and refined, but also sad.[33]

The Recurrence of the Imaginary of the Mountaineer

From Scotland to the Balkans and from Italy to Switzerland, the recurrence of invocations of the mountaineer within the national imaginary is striking. They are among the most stable images of the scholarly literature and of writing on travel and tourism. But in this case the mountaineer is above all an eminently political figure. Whether he is denigrated, vaunted, or even ignored when others invoke him, the mountaineer is political, inasmuch as the mention of him participates in the construction of a national imaginary, which in the nineteenth century began to be a major political issue of modern states.

Depending on the geographical and geopolitical context and on the representation of the nation within each individual, the ways of enlisting references to the mountaineer have varied. The two stereotypes at work in discourses on the nation are generally those that have structured the imaginary of the mountaineer for scholars and philosophers since the nineteenth century: on the one hand, the brutal, thieving, bellicose mountaineer; on the other, the proud, hardworking, obstinate, and courageous mountaineer.[34] They are often enlisted simultaneously in opposing arguments on how to define the national project. But official historiography and propaganda always privilege one of the two.

From the variety of cases presented here we may deduce a tendency that will hardly come as a surprise. Mountaineers tend to be valued when they represent a large portion of the national population; they are denigrated when their cultural, linguistic, and religious traits constitute threats to the unity of the nation. Conversely, in nations that have found a structuring factor of unity (language in Italy, for example), the mountaineer makes it possible to conceive of a marginal diversity that stands as a symbolic, if not a touristic, surplus value. In most cases, however, the mountaineer is truly a constituent part of the nation and has to be given a place, marginal or central, in the collective imaginary. The situation will be completely different for colonialism (see chapter 6).

FOUR

Politics of Nature

The State, Science, and Modernity

When Alexander von Humboldt arrived in Washington, D.C., in 1804 after his trip to Latin America, he already knew Thomas Jefferson through his writings. Humboldt had consulted Jefferson's *Notes on the State of Virginia* and would later call the third president of the United States an "excellent naturalist."[1] But as a scientist keenly aware of the political issues of his time, Humboldt also appreciated the role Jefferson had played in the awakening of the political conscience of America,[2] in the Revolutionary War, and in the birth of the new nation. The explorer's reputation having preceded him, Jefferson accorded him every honor. In the spring of 1804 he invited Humboldt to stay for several weeks at Monticello and in Washington; there they began discussions that would continue in their letters for years.

Their exchanges dealt with all sorts of questions: the continent's fossils, slavery in America, political revolutions. But Jefferson expected Humboldt to help him better understand the immensities that his contemporaries were discovering west of the Appalachians. The United States had just purchased Louisiana from France, and Jefferson had sent William Clark and Meriwether Lewis to the upper Missouri in search of a northwest passage to the Pacific.[3] But he lacked a general understanding of the continent and the means to assess its potential. He would not be disappointed: Humboldt explained to him his notions about the topographical layout of North America, its native populations, and the agricultural and mining potential of the Mississippi Basin and western mountains. In a letter following his first exchanges with the explorer, Jefferson soberly but emphatically attested to their import. He reported "the extreme satisfaction I have received from

Baron Humboldt's communications. The treasures of information which he possesses are inestimable."[4]

The meetings between Jefferson and Humboldt, marked by fellow feeling and mutual interest, are emblematic of an idea that had taken root in the Age of Enlightenment: the virtuous and fruitful alliance between science and politics, which was supposed to lead to a knowledge of national territories and their development. The mountains had come to be major objects of scientific knowledge; they now became major components of the practical knowledge nations were acquiring to optimize management of their territories. It was through mountains that several different development strategies and public policies, all extremely interventionist, were set in place. Forest policies were the first and initially the most ambitious.

Alexandre Surell and the Notion of the Mountain Forest

Alexandre Surell (1813–1887), a young engineer and recent graduate of the École Française des Ponts et Chaussées (French School of Bridges and Roads), was assigned to the administrative *département* of Hautes-Alpes in 1836. A native of Vosges, he had just spent a few years posted in Auvergne. He devoted most of his six years in the Alps to studying a still poorly understood phenomenon, mountain torrents and the damage they frequently caused: villages destroyed, farmlands covered with debris torn from the slopes, roads cut off. Surell visited the areas, compiled a list of the torrents, and analyzed how they functioned. In 1838 he submitted a report on the question to his central office. The very favorable response it received led him to publish the report in 1841.[5]

In the first part, his *Étude sur les torrents des Hautes-Alpes* (Study on the Torrents of Hautes-Alpes) gives a detailed and systematic description of the structure of these torrents and how they work. He attaches particular importance to the relationship between the subsoil, inclement weather, the formation of the streambed, and recurrent flooding. Surell developed terminology to describe the structure of these torrents ("channel bottom," "flow channel," "alluvial bed"), a vocabulary that later generations would adopt.

In the second part of his book Surell analyzes the deforestation of slopes by the local populations and the role this activity plays in the damage caused by flooding streams. He includes a list of proposals for reversing that process, in particular reforestation of the mountain slopes. This second part of the analysis is less original: it adopts a theory that had been formulated at length since the French Revolution and whose validity would be hotly debated in the early twentieth century.[6] Surell recommended that the administration focus its efforts upstream of the catchment basins, where the forest allows water to infiltrate the soil and impedes its flow. Invoking the existence of a natural equilibrium, he believed it desirable to return to the mountains the forests they might have lost to the woodcutters' blows: "Nature, in summoning the forests to the mountains, placed the remedy next to the ailment. She combated the active forces of water with other active forces: she countered the invasion of torrents with the gradual conquest of vegetation."[7] Unlike his predecessors, Surell attached little interest to engineering projects designed to contain the torrential floods: "How weak all our dikes appear next to the enormous means nature has at her disposal, when human beings cease to oppose her and she patiently pursues her work over the course of centuries. . . . Why, then, should humans not ask for assistance from these living forces, whose energy and efficiency are clearly revealed to them? . . . Art will then confine itself to imitating nature, to seizing on its methods and skillfully countering the forces of organic life with those of brute matter."[8] In a word, reforestation is needed if the slopes are to recover the ground cover they never should have lost. In his major reference work *Man and Nature* (1864), George Perkins Marsh, one of the pioneers of environmentalism, adopts a similar perspective. Considering human beings a disruptive agent, he proposes that the pioneer settler should "become a co-worker with nature in the reconstruction of the damaged fabric which the negligence or the wantonness of former lodgers has rendered untenantable. He must aid her in reclothing the mountain slopes with forests and vegetable mould, thereby restoring the fountains which she provided to water them."[9]

Aspects of Surell's book recall theories from the previous century about the order and harmony of nature (see chapter 1). It also reflects the ascendancy in his time of catastrophist theories, which saw the history of the mountains as a succession of tectonic and climatic crises on the one hand, and of periods of stability on the other.[10] The crisis he designated—that of deforestation brought on by generations of farmers and herders—is portrayed as an interruption in the balance, the point of harmony, among mountain, climate, and forest. A narrative of that kind can be read between the lines in a number of studies in the natural sciences, even into the early twentieth century. It was only then that geophysics radically revised its theories about the formation of mountains and that geographers and botanists began to doubt whether the Alps had ever been largely forested.

Surell's theory, cosmogonic in nature, extended far beyond the French *département* of Hautes-Alpes, the region he was studying. He often reminds his reader of how limited his observations were and notes in passing what differentiates the southern Alps from Vosges or Switzerland. He also insists that a "certain type of climate and a certain geological constitution" is required for torrents to form channels.[11] Nevertheless, the quasi-scientific method and writing style he adopts, intent on describing similarities observed in other contexts, repeatedly make use of generalizations: "There is a profound and radical distinction to be made between the flatland regions and the mountain zones. They do not resemble each other in any way, and if the danger of deforestations is far from demonstrated for the plains, it is decisively so for the mountains."[12] "When a forest disappears from a plain, it makes way for farms: one product of the soil is substituted for another, and in most cases there is nothing regrettable about that substitution. But cut down the ancient forest that covers a mountain's back and immediately everything is in turmoil. Storms and gullies rip apart the slopes: the ground cover has soon vanished, and with it, fertility and greenery. No more fields, no more farms. Left defenseless against attacks from the waters, its very bowels purged by the torrents, the mountain, finally collapsing under its own weight, descends, rolling onto the plain, which it buries under its debris and draws down with it."[13] References to the mountain generally are omnipresent; every specific observation is connected to generalities applicable to the category as a whole.

Granted, Surell was not the first to speak of the mountain forests in general. Back in the mid-eighteenth century a few scientists, such as Duhamel du Monceau in France, had sought to organize knowledge on that scale to promote adapted forestry.[14] But Surell's comparisons, being methodical, are much more convincing. In pursuing his analysis, he argues that mountain people are the ones primarily responsible for deforestation and flooding and are also the first victims of these processes. Also according to him, mountain people do not have the means to correct the situation. While sympathetic toward them, he finds them fatalistic in the face of the threat, unequipped to deal with the danger, and lacking the resources needed to implement the reforestation he calls for. He therefore advocates the intervention of the national administration to set aside highlands for reforestation.

Mountain Forest Policy in France

From the first edition of *Étude sur les torrents des Hautes-Alpes*, readers remarked on the high quality and logic of its analysis. The book, however, did not immediately give rise to the public policy it recommended. It took further flooding, in 1846, 1855, and 1856, and the considerable damage it caused in the valley and delta of the Rhone for that to happen. Surell himself, having been transferred to the Rhone dike project, personally witnessed the floods of 1846. The emotional response to these catastrophes reached as far as the National Assembly and the gates of the national palaces. The 1856 flood prompted Emperor Napoleon III to say, with his knack for the apt phrase, that "rivers, like revolution, will return to their beds and will be kept under control under my reign."[15] France was ready to enact laws.

The first law, passed in 1860, established a program for the reforestation of the mountains of France. It set up regulatory agencies and instruments for planting trees on the high slopes most exposed to erosion by mountain torrents. In carrying out that work, landowners, usually individuals or administrative districts, were encouraged to follow the advice of the forest administration. The threat of expropriation loomed over those who did not follow through. The law was thus very stringent and tended to weaken agropastoral economies and the societies themselves, which had reached their peak population. A second law, passed in 1864, was more lenient than the first. It promoted covering the high slopes with grass, which would make it possible to maintain the pastoral economy, but only with government oversight. A third law, passed in 1882 and accompanied by an updated doctrine, completed the plan of action.[16]

But even in the 1870s many observers attested to the first results. One of them was Ernest Cézanne, also from the École des Ponts et Chaussées and a National Assembly deputy from Hautes-Alpes. When a new edition of Surell's book was issued, Cézanne wrote a companion volume in which he lauded his predecessor's work. He also praised the French government's involvement in that project: "Experience has demonstrated that the *regeneration of the mountains* is a project of public interest, which the state itself can and must carry out."[17] Cézanne's analysis went further than his model, however. He included a large number of mountain regions in Europe (the Swiss and Italian Alps, the Harz in Germany, the Scottish Highlands, and others) and on other continents (the Himalayas, the mountains of Abyssinia, and the Kamchatka, for example). Mountain torrents and the link between deforestation and erosion became a theme to be studied on a global scale.

The Circulation of Analytical Models and the Reciprocal Influence of Public Policies

That change in scale also took another form at that time, namely, the circulation of ideas from one country to another. From the time of their publication, Surell's works prompted delegations to visit France to observe the phenomenon. The earliest visitors, who arrived before the enactment of the 1860s laws, took stock of the dimensions of the problem. For example, a Swiss delegation, headed by the engineer Carl Culmann, came after the 1856 floods. The analytical model began to spread, and with it reports of the powerlessness of public authorities. Culmann lamented the incapacity of Swiss municipalities and cantons to regulate what was judged to be an equivalent problem in the Swiss Alps.[18]

Visits at the end of the century evaluated the public interventions. After the terrible floods of 1882 in Carinthia and East Tyrol, a twelve-person Austrian delegation spent three months in France observing and drawing inspiration from the programs of the French administration. At the 1878 Exposition Universelle in Paris, the administration was proud enough of the results obtained to put its achievements on display. Using a collection of maps, photographs, a diorama of a catchment basin in the Gap region, and many drawings, it showed visitors from throughout the world the effectiveness of its voluntary slope-reforestation policy.[19] An observer commended "the attention of all countries to the results obtained through the application of the principles set out by Surell and the method he deduced from them."[20]

The analytical model thus became international, as did the adoption of national public policies. Several European nations had adopted forest policies in the late eighteenth century (Germany, for example) and more widely in the nineteenth century (France, 1827; Austria, 1852; Sweden, 1886; and Norway, 1893). But none of these policies targeted mountain forests in particular on a national scale. There had been corps of forest inspectors in Spain since 1835 and a federal forest corps in Switzerland since 1874.[21] Training schools had existed in France since 1824, in Spain since 1848, and in Italy since 1869. But it was not until a series of catastrophic floods (Valencia, Spain, 1864; Almería and Murcie, Spain, 1879; Valais and the Bernese Oberland, Switzerland, 1868; and Carinthia and Tyrol, Austria, 1882), along with the growth in popularity of the French model, that some projects to reroute rivers and construct dams came to be abandoned. In Spain the results of such projects had been disappointing. Now mountain reforestation was promoted everywhere. In 1877 Italy passed a national law that prohibited cutting wood located above the tree line (about 800 meters) for chestnut trees, considered an indicator species for the entire kingdom. In response to the poor results obtained, a new Italian law, adopted in 1910, reorganized the forest administration. This law proved to be more effective, and the rate of forestation in the Italian mountains rose from the 1920s on. In 1884, when a law to protect against torrents was set in place in Austria, that country established an office, constructed on the French model, for the restoration of mountain lands.[22]

Gifford Pinchot imported the French model to the United States.[23] As a student at the École Nationale des Eaux et Forêts (National School of Water Resources and Forestry) in Nancy, Pinchot had discovered Surell's work and the reforestation measures being taken in France. In 1901, after returning to the United States, he created the Society of American Foresters, which promoted large-scale scientific forestry on public and private lands. Shortly thereafter, in 1905, Pinchot became the first chief of the U.S. Forest Service. Capitalizing on President Theodore Roosevelt's interest in managing the continent's natural resources,[24] he transferred what had become a Europe-wide model to the New World: "Severe governmental regulations controlling the management of protective forests on private lands are common in Europe. There can be little doubt that similar action will be forced upon us in the United States by the results of destroying our mountain forests."[25] In the early twentieth century, then, mountain forests—and mountains along with them—became an object of public policies in many nations of Europe and in North America. These policies all had a common conception behind them, borrowed from the observations of a modest French engineer on the slopes of the Dauphiné Alps.

Convergence and Strife within National Societies

Forestry, conceived as a high-priority operation in many countries, mobilized an array of national forces. In Switzerland the early mobilization of the public authorities can be attributed in great part to the actions of the Swiss Forestry Society, founded in 1843, which sought to support foresters' goals in public debates. Although it addressed forest questions in general, it placed the emphasis on "the reforestations of our Alps and protection measures against the destruction of the Alpine forests."[26] In 1856 the society asked the Swiss Confederation to do an assessment of the mountain forests, even though the Swiss constitution of 1848 did not recognize any state prerogative in that area. The principle of forest oversight by the state became part of the federal constitution of Switzerland in 1874, the same year that the Federal Inspectorate for Forestry was created. The federal law governing the Confederation's general oversight of forest policy in upland areas was promulgated two years later.[27]

Catalonia, for its part, federated other partners, often by popular initiative. In 1898 it institutionalized a "festival of trees" designed to promote the planting of trees wherever possible in the mountains. Local and regional authorities, economic and agricultural societies, and railroad companies joined in. The initiative set a precedent for other Spanish regions. In France, shipping companies and even the Club Alpin (founded in 1874; its second president was none other than Ernest Cézanne) made their contribution to the reforestation effort. A group promoting tourism, the Touring-Club de France (TCF), also became an advocate of the forest policies. In 1909 it distributed to French schools a *Tree Manual*, written by a forest administration inspector (fig. 7). In the introduction, Abel Ballif, president of the TCF, wrote that it was fitting to "spread the reverence for trees to these generations, to draw the child's attention early on to the benefits of the forest, by means of which the climate becomes more moderate, waters flow more steadily, the grass itself grows fresher and provides more nourishment to livestock, the mountains become more picturesque and the plain more pleasant."[28]

Scientists were not to be outdone. Scientific societies, professional associations, and scientific reviews spread the message, sometimes even adopting a lyrical tone. Impatient to see Spain undertake a voluntary and effec-

7. Shepherd from the Pyrenees. The image, printed in a book that promotes forest policy for a large readership, is captioned: "The sad author of his own misfortune: the sheep seem to be complaining to him about the missing grass. Visible are the remains of the trees burned down in the last fire, which, in his obliviousness, he himself set." Manuel de l'Arbre, *Touring Club de France* (1907).

tive reforestation policy for the mountain slopes, a geographer wrote in 1908: "The huertas,[29] marked off by the dark line of the mountains, receive from them running water, the source of all richness and life, but sometimes as well the cause of ruin and death. Depending on whether the upper regions are covered with trees or denuded, the plain may be fertilized by irrigation waters or instead exposed to the ravages of torrents." When the conviction takes hold that farmers are the most affected by mountain forestation, he said, and "when every inhabitant takes care, on behalf of everyone else, to conserve his forest lands, when he is converted into a 'keeper of his mountain'—on that day, national agriculture will be saved."[30]

Mountaineers Named as the Guilty Parties

Although inhabitants of the highlands were sometimes idealized in that image of the "keeper of the mountain," nineteenth-century authors and foresters usually described them in less advantageous terms, calling them guilty and irresponsible. Surell, who undoubtedly had an understanding of mountain communities, showed restraint in that respect: "The race of people here is neither indolent nor averse to risk. It has all the qualities of the mountain: it is tough, active, persevering. None lack the real desire to defend themselves; what almost all lack is the money necessary for defenses."[31]

But many who followed after Surell let the criticisms fly, denouncing right and left communal rights, the appropriation of wood for general consumption, the habit of grazing sheep and goats in the undergrowth, and the excessive levels of transhumance. Bernard Lorentz, the first director of the École Nationale des Eaux et Forêts in Nancy, took a stand during a mountain tour in 1840: "Is it tolerable for landowners in the mountains to introduce a mode of farming and of land use that brings desolation to the valleys"?[32] Comparable accusations were made in Switzerland. In 1849 the chief of forests in the canton of Bern exclaimed: "If the misfortune affecting the deforested mountains were limited to these regions, it would be possible to say that it was punishing the guilty recklessness of their inhabitants. . . . But that misfortune affects even the valleys and the fertile plains."[33]

Harsh laws prevailed in the second half of the nineteenth century. Depriving the mountain communities of age-old rights, they supplanted mountain people as managers of their natural resources. Altercations multiplied between inhabitants and forest officials in the Swiss and Italian Alps, in Savoy, and in the French and Spanish Pyrenees. On occasion the

regional press echoed the population's distress: "People say they no longer know where to graze [their sheep] in the springtime; the foresters have
taken everything and are planting everywhere."[34]

It should be noted that those promoting interventionist public policies at the national level, potentially applicable to all mountains in the
territory, found it advantageous to lump together mountaineers from every
region in a single unflattering portrait.[35] Elected representatives from the
mountain regions needed a great deal of finesse and conviction to vaunt
France's forest policy. Cézanne, for example, convinced that forestry laws
were gradually improving, felt justified in announcing in 1872 that "the
opposition between the plains and the mountains no longer exists," even
while recognizing that centers of unrest remained in "small federations of
herders hidden in the remote recesses of the Alps."[36] Cézanne and many
others attributed the difficulties that foresters encountered on the ground
to the stereotype of the rebellious mountaineer, a stereotype that had been
patiently constructed since the Age of Enlightenment (see chapter 1).

A Political Construction of the Problem and a Technical Approach to the Solution

Within a few decades, in France, Switzerland, and then in a great number of other European countries, the mountain forest became an object of
public policy at the national level. The administration's structures and missions were reformulated within that new context, as was the relationship
between the administration and local societies and even between the administration and national or local institutions. As it took shape as a public
problem, the mountain forest conditioned in great part the apprehension
of the mountain itself: its natural characteristics, the dynamics of its evolution, its populations, and its modes of use.

However political it may have been, the definition of the mountain forest as a public problem and an object of national policy was based on a
very technical and naturalistic approach to the constitution and hydrological functions of such forests. That approach was at bottom indifferent to
the anthropological dimensions of the relationship between the populations and their forest surroundings. Many consequences followed from
that way of apprehending the mountain forest as an object of knowledge,
which tended to "simplify the decipherment" of the territory.[37] It also authorized a comparison of contexts that are manifestly different, at least
when modes of production and social uses of the forest are taken into account. It was by homogenizing the mountain forest in that way that the

state was able to argue for the legitimacy of its own intervention, based on the similar environmental degradation observed in several parts of its territory and on a complementary postulate, the inability of the populations and of local institutions to remedy that degradation on their own.[38] In itself, the technical view of the problem and of the corresponding solution played a role in disqualifying the local populations and, conversely, in institutionalizing the image of the forester as technician and engineer.

Finally, the mountain forest, far from being contained within the framework of forest planning measures, became emblematic of the new missions the state took on at the time. The discourses devoted to the mountain forest, the images diffused of it, and the scenarios adopted to deal with it were part of a reformulation of the modern state's modes of action and communication. That reformulation occurred at a time when the social, economic, and political elites were reconstituting themselves and building new alliances. At the heart of these new arrangements was the objective of promoting the state and technological modernity. The last pages of Surell's book are themselves informed by a modernist vision reminiscent of the Saint-Simonianism of some of his contemporaries. The rural populations and territories, left far behind, were hampered by a traditionalist vision that made them unqualified to participate in the modernization of the mountains in an industrial age. During those years the emergence of programs to protect the mountains for the sake of their landscape and for ecological reasons further accentuated that process.

Yosemite and the Invention of a New Paradigm

By the mid-nineteenth century the United States had almost completed its conquest of an area of continental dimensions, running from one ocean to the other. Americans had reached the Pacific Coast, seized California's Spanish missions, and built the first new cities on the coast. The control of physical space became more focused with the hunt for precious metals and the occupation of an ever-growing share of Native American lands. It was a conflict with the Native Americans, accustomed to spending part of the year in a valley they called "Yo Semite," that led Westerners to discover the place. It astonished them, and they rushed to report their amazement in stories that were reproduced in a large number of newspapers throughout the country. They had entered a deep valley formed by glaciers, with a peaceful river running below between woods and prairies. Rising along its banks were enormous granite walls. Extraordinary waterfalls cascaded down.

Everything moved quickly after that: the first tourists arrived a few years later, a California newspaper published drawings in 1856, a hotel was built the same year, and the first photos began to circulate in 1859. Before the end of the decade a few famous authors had visited to form their own idea of Yosemite. After they left, their accounts popularized throughout the country a place that few had ever seen. In a certain sense, that place gave concrete form to an image that had accompanied the conquest of the continent for decades: that of a divine Providence which had provided a chosen people with a natural world—rugged, to be sure, but also extraordinary, and here and there sublime. Yosemite was described as a miracle of nature, a gift from God. Its enormous rock faces, "created as an overflow of God's goodness," were compared to the walls of European cathedrals: "Yosemite's sacredness derived not only from its natural features but from its resemblance to the great sacred monuments of Europe."[39]

This comparison to Europe in general also applied to the principal European mountain ranges. Many authors in the second half of the nineteenth century assessed the value of Yosemite in terms of its resemblance to, or differences from, Alpine landscapes. Sometimes they merely pointed out that the United States now possessed "as lofty snow-covered peaks, and as grand panoramic views of mountain and valley, as [one] can find in Switzerland itself."[40] At other times they suggested that the site surpassed any particular Alpine locale: "Only the whole of Switzerland can surpass it; no one scene in all the Alps can match this."[41] Comparisons were rarely based on any actual knowledge of the Alps.

Often the comparisons made reference to texts that had established and propagated the Alpine model. For example, Thomas Starr King, the author of the first guide to Yosemite, invokes John Ruskin and his representation of the Matterhorn in his description of a granite peak: "An obelisk of bare granite two thousand feet high, utterly unscalable on the front, and on its backline, repeating with surprising exactness the contour of the Matterhorn on its longer side, as drawn in the fourth volume of Ruskin's 'Modern Painters' and in Hinchdiff's 'Summer among the Alps.'"[42] The scenic value of Yosemite is compared to that of other places and chains that had fixed the canons of the mountain aesthetic.

That scenic quality of Yosemite was what led to the idea of protection measures.[43] In 1864 the U.S. Congress approved the Yosemite Grant, which created a park comprising both the Yosemite Valley and a grove of giant sequoias called the Mariposa Grove. Its promoters wanted to keep loggers, miners, and farmers from encroaching on that extraordinary place. They proposed to entrust the management of Yosemite to the state of California. The protection of primal nature was not at issue at the time. It was the scenic landscape that had to be preserved from the transformation of the continent under way.[44]

The first regional planning project for the site targeted for protection was awarded to Frederick Law Olmsted, a landscape architect who lived in the area. Among his list of achievements was the design of several city parks, including New York's Central Park, and a regional plan for the area around Niagara Falls. His approach to Yosemite in the report he submitted to the state of California was to showcase the aesthetic qualities of the place.[45] He too was inspired by Ruskin's model of the sublime landscape and quoted at length from *Modern Painters* throughout his life. Like the English critic, Olmsted adopted a religious and artistic notion of the landscape and of nature. But that notion was also moral and psychological. His training as a horticulturist allowed him to compose scenic views, in Central Park and in Yosemite, that best suited the human need for beauty, rest, and moral regeneration.[46] Unlike Ruskin, however, Olmsted was anxious to put his creations into the service of the United States. He called Yosemite "a trust for the whole nation,"[47] thus anticipating its transformation from a (California) state park to a national park in 1890.

The state of California did not accept Olmsted's plan. At that date the plan was already competing with a different conception of the site and of its protection, whose best representative would be John Muir. A Scottish emigrant trained in the natural sciences in Wisconsin, Muir arrived in California in 1868 after several trips throughout the United States. He lived in the Yosemite Valley for four years and in California for nearly forty. In 1892 he founded the Sierra Club, an environmentalist group that would become the main such organization in the United States and would wage countless battles in defense of the environment. Muir's writings and initiatives attest to his particular and recurring attention to the mountains.[48]

Whereas Olmsted had become the ideal representative of the idea of civilization[49]—his writings and his praise of art and technology lean in that direction—Muir became the promoter of the idea of wilderness. The works of the Transcendental philosophers—especially Ralph Waldo Emerson and, later, Henry David Thoreau—sensitized him to the issue. In the mid-nineteenth century they had popularized the notion that the wilderness, far from being a hostile world that had to be domesticated or turned into a scenic landscape, was an environment propitious for psychological and religious experiences that elevated the soul and consciousness. Muir had also read George P. Marsh's *Man and Nature* (1864) and the works that, following in its wake, assessed the consequences of technological and economic development on the natural environment.[50] Henceforth natural scientists' attention, long fixed on the might of the natural forces and the equilibrium they established with one another over vast stretches of time, also focused on the capacity of human beings to disturb a preexisting natural balance and act as a major disruptive factor.[51] Muir was also acquainted with works by painters from the Hudson Valley to the Rockies, who between 1830 and 1860 gradually constituted a genre of landscape painting that made the wilderness the principal motif of their art.[52] These philosophers, natural scientists, and artists provided him with a set of images and concepts for understanding the wilderness and with a set of arguments for encouraging its protection. In 1912 Muir would tell how, upon his arrival in San Francisco in 1868, he had replied to a passerby who asked him where he wanted to go: "To any place that is wild."[53]

With John Muir and the concept of wilderness, the protection of the Yosemite Valley went from being a scenic notion (nature as art)—Olmsted's guiding idea—to being an ecological conception: nature as vital principle. In both conceptions the mountain was the center of attention.

The U.S. Mountain as Emblematic Category of the Wilderness

In 1890, twenty-six years after becoming a California state park, Yosemite was designated a national park. Before that time only Yellowstone, in the mountains of Montana and Wyoming, had received that status (in 1872). Sequoia and General Grant National Parks,[54] also in the Sierra Nevada, were established in 1890 as well. In both cases John Muir's influence may again have been the determining factor. Between 1872 and 1919 sixteen national parks and many state parks were created in the United States. Some are of considerable size, such as the Adirondacks; and some are ambitious, such as the series of parks established in the Colorado Rockies.[55] The vast majority are located in the mountains. They are the pride of the states where they are located and of the federal administration (fig. 8).

Granted, the idea of wilderness is not associated a priori with a mountain environment. Thoreau, for example, repeatedly attached preeminent importance to forests, wherever they were located, and even refused to associate the idea of wilderness with any one natural form: "It is in vain to dream of a wildness distant from ourselves. There is none such. It is the bog in our brain and bowels, the primitive vigor of Nature in us, that inspires that dream."[56]

Nevertheless, the mountains gradually came to be the type of environment where the nature imaginary and the wilderness imaginary would seek their principal illustrations. Several factors played a role in that conver-

8. Promotion posters published by the National Park Service
(here, an image from 1939) often depicted generic landscapes
that represented no park in particular. Mountain landscapes or
individual mountains were often used. Library of Congress.

gence. In the first place, the artistic conventions and tourism models imported from Europe provided the template for the value to be assigned to
places, as we saw for the discovery of the Yosemite Valley. Secondly, the
natural scientists who traveled the United States in the nineteenth century
focused their attention quite particularly on the mountains. Once there
was an overall conception of how the continent's landforms were arranged,
generations of botanists, geologists, and geographers, taking the advice of
Humboldt and early nineteenth-century explorers, sought to understand
the structure of the continent, the distribution of plants and animals, and

the effects of the different climates prevailing in the mountain regions. Scientists were among the most ardent promoters of protected areas, and when the ecological notion of protection was granted preeminence over the scenic notion, they became major players in defining these policies and the protection zones.

For many authors the numerous biblical references made in that context also facilitated an association between images of the mountain and the primal harmony of the Garden of Eden. It was as if that harmony had to be preserved to measure the scope of the Fall. In a guide to a region of Texas, a journalist commented on the presence of antelopes: "These graceful creatures had been shut out, by their steep hills in this enchanting recess, from any knowledge of the gloomy and bloody strife which man has been waging with himself and all God's creatures since sin and death came into the world."[57]

One last factor, more functional than natural, worked in favor of the privilege given to protecting the mountain regions. The notion of wilderness assumed its modern form in the mountains of the northeastern United States, in New England and the Hudson Valley.[58] By the mid-nineteenth century the coasts, valleys, and inner basins of these regions had already been transformed and urbanized. The mountainous areas, though often inhabited and exploited, were readily put to use for the production of wood, the provision of water, the promotion of tourism, and the protection of nature. It is for the same reason that the majority of national forests are concentrated in the mountains of the western United States. But unlike in the national parks, where the idea of wilderness conservation predominates, these forests are for the most part managed (as Gifford Pinchot wished) as water-control systems and economic resources. That differentiated management of areas and landforms was transferred to the continent as a whole as it was conquered. Landscape paintings and travel posters assumed the task of popularizing their image.

In other words, the mountain was erected as a model site for nature protection because of the convergence of an emerging paradigm—the protection of the landscape and of the wilderness—with the various ways of conceiving the mountain. Some were very old, such as the scenic archetype of the Alpine mountain that already prevailed in the mid-eighteenth century. Others were more recent, such as the idea that the mountains were complex and original ecosystems. A few extremely popular phrases (such as the slogan "thinking like a mountain," adopted by Aldo Leopold,[59] the apostle of deep ecology) and, more generally, the tendency on the part of promoters of nature protection to link their cause to the image of the mountain would anchor that association in a lasting and profound way.[60]

Variations on the U.S. Model in the
Rest of the Western World

The idea of nature protection and the concept of national parks were exported very early on to Australia (Royal National Park, 1879), New Zealand (Tongariro National Park, 1887), and Canada (Rocky Mountain National Park, 1885) and went from there to Europe in the course of the twentieth century. Europeans certainly felt no need to vaunt the sublimity of their natural world to compensate for some deficit in religious and architectural heritage. The nationalist significance of nature protection was therefore different in Europe. But many of the characteristics of such policies in the United States can also be found there.

In Europe as well, the mountain regions have been the primary objects of nature protection. The vast majority of protected sites and all the first national parks in Europe were located in the mountains of Switzerland (1914), Spain (1918), Poland (1921), Italy (1922), and Slovenia (1924). Other countries, such as the United Kingdom (1951), France (1963), Germany (1970), Portugal (1971), and Austria (1981), officially adopted the model of the national park somewhat later. But in most cases protected areas had already been created in the mountain regions.[61] In Europe as in the United States, it was the scenic quality that for a time prevailed in the defining criteria. As in the case of Yosemite, where references were constantly being made to the Matterhorn, that quality was initially measured in terms of the Alpine model, and more particularly the Swiss model. As in the United States, the ability of mountain landscapes to emblematize the national imaginary encouraged protection measures that designated them as national parks. The Triglav National Park in Slovenia (1924) and the Tatra in Poland (1954) are the most famous examples.

As they had in the United States, artists played a decisive role in awakening sensitivity to scenic views and in exerting pressure on political and administrative institutions. In France, for example, the Société de Protection des Paysages de France, created in 1901, included many artists among its members. The society played a decisive role in the design and passage of the 1906 law to protect "natural monuments." In England the poet William Wordsworth's protests against railroad projects in the Lake District in 1840 recalled those of Thomas Cole, founder of the Hudson River School of painting in the United States, against similar plans in the Catskills.[62] But of all the criticisms, the one formulated by Ruskin against railroad builders in the Alps was the most vehement. It best illustrates the process of sacralizing the mountain: "You have despised Nature; that is to say, all the deep and sacred sensations of natural scenery. The French revolutionists made stables of the cathedrals of France; you have made racecourses of the cathedrals of the earth. Your own conception of pleasure is to drive in railroad carriages round their aisles, and eat off their altars. . . . You have tunneled the cliffs of Lucerne by Tell's chapel; you have destroyed the Clarens shore of the Lake of Geneva."[63] All these artists together spread the notion that mountains ought to stand apart from industrial modernity, from the deployment of monetary practices and social transformations.[64]

Naturalist preoccupations, following in the wake of Marsh and the botanists, took root in Europe in a second phase, as they had done in the United States. By the mid-twentieth century, nature protection had begun to target what were called mountain ecosystems and biotypes, as well as the endemic or endangered species in these environments. There were also a few early cases, in Italy and Switzerland especially. The first Italian park was created in 1922 in the Alpine massif of the Gran Paradiso. It replaced a hunting reserve established by royal decree in 1821 that had sought to prevent the ibex from disappearing, as the Royal Academy of Science in Turin feared it would. The hunting reserve was decreed a royal reserve in 1856. That measure allowed the kingdom of Sardinia-Piedmont, then Italy, to save several hundred animals and later, from the 1920s on, to donate ibexes to neighboring countries such as France and Switzerland, from which they had disappeared. The ibex, even more than the chamois, is the emblematic species of the Alps in general. The fondness people feel for the ibex worked in favor of other species readily associated with the image of the mountain. Of the Alpine countries, only Slovenia had managed to keep bears in its territory, for example. These animals were later reintroduced into the French Pyrenees.

The Swiss national park in Engadine did not come into being in response to a concern for protection of an emblematic landscape. A forester, intent on setting apart a few "virgin" forest stands and on keeping them in their original state, formulated the idea for the park in the mid-nineteenth century. It was created in 1914 with a slightly broader objective: namely, to set aside a zone that would conserve all the mountain plant species and would constitute a "magnificent place of refuge for the last survivors of many Alpine animal species."[65]

The notion of the mountain that motivated nature-protection advocates and the practical initiatives undertaken were thus similar on both sides of the Atlantic. Whether scenic or environmental in nature, they usually allowed a precise place to be conceived as an emblematic instance of the category of the mountain.

A New Disenfranchisement of the
Mountaineers, and Conflicts

The idea that mountains constitute high-quality scenic areas and environments worthy of protection measures has two complementary social implications that we have already identified in forest policies: the local populations were gradually disqualified, even expropriated; and a new elite,
primarily from the bourgeoisie (artists and scientists) and government administration, was mobilized to assist the state in creating protected areas.

The disenfranchisement of local populations stemmed from the modern conception of the mountain. Since the eighteenth century it had been
the privileged place where nature manifested itself and could be experienced. The introduction of anything artificial, even when rooted in tradition, was perceived as an expression of culture of one kind or another, and
the justification for such action was therefore questioned. Antagonism was
strong, especially since the mountains and nature as a whole were seen
through the prism of an original Eden or an ideal wilderness. Once the decision was made to put Yosemite under public protection, the Native American populations who frequented the place were labeled undesirables. The
traces in the landscape of their centuries-old presence, especially the clear-
ings that resulted from occasional deforestation, were perceived and portrayed as natural formations. Unlike what had occurred in many other regions of the United States, Native American place-names were disregarded,
with the exception of the name of the valley itself. In an effort to justify
the displacement of the Native American populations, Lieutenant Philetus
Norris, the superintendent of Yellowstone National Park, constructed and
propagated the myth that the indigenous people avoided the region of the
park for superstitious reasons and out of a fear of geysers. Other examples
abound of the eradication of human traces. Mark D. Spence has written:
"Uninhabited wilderness had to be created before it was preserved."[66] Rare
were those who adopted a different point of view. One of these was George
Catlin, a romantic admirer of the Indians. Catlin believed that the original peoples were an integral part of nature and of American territory. The
national parks he desired would have allotted them a place, even if that
entailed idealizing their role within the landscape and the environment.[67]
His point of view remained marginal for a long time.

From the eighteenth century on, the purification process observed in
the United States was applied more generally in the Western world as a
whole.[68] Contrary to the aesthetic criteria that prevailed in the seventeenth
century, nature and the mountains were judged at the time to be all the

more beautiful for being unproductive, exempt from the logic of economic production and exploitation. The growing importance of this idea in the nineteenth-century sciences had comparable consequences. The dominant paradigms first of natural history, then of ecology, circumscribed notions (primal nature, climax, ecosystem) whose definitions excluded human beings and their activities.[69] As a result, these paradigms considered all humans to be disruptions of preexisting states of equilibrium. The practices of the populations who lived in the designated environments were evaluated in terms of these notions, most often negatively. The mountains had to be protected because they had their own harmony, their own operational dynamics. The best manager was one who claimed to be "thinking like a mountain," who allowed the mountains to thrive apart from any utilitarian conception of their benefits and resources.[70] In the name of that conception, "strict nature reserves" were promoted inside the protected areas themselves, both in the United States (beginning in the mid-twentieth century) and later in Europe.[71]

That conception of nature and of the mountain had two consequences for the populations living in these environments. It cast doubt on the validity of their environmental practices and on their modes of production. But it also placed the inhabitants in a state of inferiority compared to those who were producing the naturalist and protectionist discourse.

That propensity on the part of elites to construct and propagate the idea that they themselves have a noble relationship with mountains dates back a long time. It found expression early on vis-à-vis the landscape. In *Der Spaziergang* (1795), a famous meditation on the mountain landscape and the emotions it awakened in him, the German poet Friedrich Schiller suggested that the aesthetic appreciation of mountain landscapes was unique to educated city dwellers. Unlike mountain peasants, they have escaped nature's grip and the imperatives of daily labor in the fields. Authors such as Humboldt and Ruskin (see chapter 1) formulated the same propensity in theoretical and scientific terms. Ruskin wrote that the mountain landscape is to be appreciated from a distance and that those who are obliged to observe it on a daily basis are so overburdened by the mountains' beauty that they have no consciousness of it. He also said that the villages and rural dwellings he came across in his travels, especially in Savoy, distressed him for aesthetic as much as moral reasons. In his own appreciation of the landscape, he went on to distinguish explicitly between natural mountain landscapes, which evoked the loftiest aesthetic feeling, and the artifacts of rural populations, which immersed him in the greatest circumspection.

The social cleavage between an urban elite, which adopted generic templates of the mountain, and mountaineers, whose practices were perceived as utilitarian, was even more marked when it came to conflicting uses of the environment. That problem, which we have already encountered in the forest policies of the mid-nineteenth century, resurfaced in protection policies.

Examples of that kind of conflict abound. In France the first plan for a national park came into being in Oisans in the 1920s. It was supported by a coalition of foresters, mountain climbers, and promoters of mountain tourism (via the Touring-Club de France). All agreed to follow the precedent set by the administration of Eaux et Forêts in the middle of the previous century: they would restrict access to the high pasture grounds so as to relieve pressure from grazing, considered harmful to the environment.

In the hills of western Virginia, Shenandoah National Park was created in the 1920s after a few decades of observation of the Appalachian forest and timber harvesting practices there. Outdoor sports organizations and environmentalists had written damning reports on the state of the forest and of grazing practices. The creation of the national park changed the rules of the game: not only were the traditional practices banned, but a number of residents were forced to leave the park zone so that the forest could be restored to its original state.

The incompatibility between two different notions of the environment is illustrated even better by the example of large predators and how they were to be handled. From the start, the management of protected areas in the Rockies (in both the United States and Canada) granted a significant place to grizzly bears and wolves. Although never rare, these species were emblematic forms of wildlife and were monitored and protected in the region's parks. The local populations generally did not understand that practice, since their pastoral economy was threatened by the predators. A similar conflict has emerged in Western Europe with the proliferation of wolves since the 1980s. From their home territory in the Italian Apennines, the animals have regained part of the lands from which they had been absent since the Middle Ages, in the Alps, the Massif Central, and the Pyrenees.

Who Are the True Mountaineers?

A particularly revealing illustration of that tendency to disqualify local populations from managing the mountains can be found in the different names given to the inhabitants. In the Age of Enlightenment references to "mountaineers" became commonplace. The issue at hand was to naturalize the identity of the people living in mountain environments and to suggest that they all shared a common condition (see chapter 1).

In the mid-nineteenth century, mountain climbers began to use the term to refer to themselves.[72] Mountain climbing was becoming an outdoor sport, an adventure. Historians readily use the expression "heroic" to refer to such practices during that period. Mountain climbers developed skills, techniques, and know-how that made them tougher and fitter than many guides from the local populations.

Edward Whymper was the embodiment of that new generation. Passionate about mountain climbing, he made the rounds of the Alps, the Rockies, and the Andes, achieving a number of firsts and patiently building up the legend of a new hero of the mountain peaks. His writings consist of ascent narratives in which the main characters are mountain climbers, a few seasoned guides whose services he engages, and himself.[73] He had little interest in local populations, traditional ways of life, or scenic views. Whymper was one of the tourist mountain climbers who considered themselves better mountaineers than the resident populations and who laid claim to that title (fig. 9).

Alpine clubs institutionalized that evolution in social differentiation with respect to the mountain. The clubs multiplied in Europe during the second half of the nineteenth century—sometimes at the regional level, as in Catalonia, but often at the national level (the Alpine Club in the United Kingdom, 1857; the Österreichische Alpenverein in Austria, 1862; the Club Alpin Suisse in Switzerland, 1863; the Deutsche Alpenverein in Germany, 1869; the Club Alpino Italiano in Italy; and the Club Alpin Français in France, 1874).[74] In the early twentieth century the model spread to North America (American Alpine Club, 1902; and Alpine Club of Canada, 1906) and to Japan (Alpine Club of Japan, 1905).[75] Local branches multiplied in the major cities and in different mountain regions of a single country,[76] the better to make inroads in the society of their time. They recruited most of their members from the wealthiest, best-educated groups, and from those closest to the upper echelons of the government.[77]

The various alpine clubs declared common objectives: all wanted to promote mountain sports and through them a system of values, if not an ethos, that was part of the prevailing nationalism of the time.[78] All took pride in their contributions to the advancement of knowledge, encouraging scientific and cartographical studies and giving them a large place in their own publications. Interest in playing a prominent role in national public policies, however, varied a great deal from one group to another.

9. In 1865 Edward Whymper made the first ascent of the Aiguille Verte (4,127 meters) in the Mont Blanc massif. Contrary to the usual practice, he did not hire Chamonix guides (photo), preferring the Swiss guides with whom he was accustomed to working. On his way back he was attacked by residents of Chamonix, who were furious at having been bypassed. Whymper replied that, in mountain climbing, the only thing that counts is skill, and that has little to do with where one was born. Collection Yves Abraham, Musée Dauphinois.

Because there were no British mountains that reached the height of their ambitions, the members of the Alpine Club were less interested in promoting public policies than in exploring the mountains of the world and advancing knowledge and technology. Other groups sought to speak for the mountains of their respective nations and thereby to strengthen the mountain as an object of public action. Such was the case for the Club Alpin Français (CAF). In 1874, in the first volume of the *Annuaire du CAF*, Ernest Cézanne, a founding member of the club, published an article titled "The Question of the Mountains." This time he did not confine himself to the question of forests and reforestation but noted in more general terms the marginalization of the mountains in that age of French industrialization: "The plains are advancing, the mountains retreating." But there was a danger in neglecting the mountains, if only because they are "the ramparts of France; . . . how would we hold onto them if, little by little, the population deserts them?"[79] The promotion of tourism and mountain sports was said to go hand in hand with the regeneration of forests in the mountains, their economic development, and an increase in their strategic importance.[80] The meaning of the slogan adopted by the Club Alpin Français, "For the Homeland through the Mountain," is thus clear: it promoted a national and political conception of the mountain and established itself as its chief representative.

The situation was different in the Germanic world. The Austrian Alpine Club was originally very Viennese and aristocratic and concerned primarily with science. It soon merged with the German Alpine Club, which had a different composition and different objectives. The industrial bourgeoisie was a greater presence in the German club, which was anxious to include residents of the mountain regions and to contribute to the development of their economy and of tourism. That social activism contributed to the enormous success of the Austro-German club, measurable both by the number of members (more than 100,000 in 1918, which is to say, on average ten times more than either its Swiss, Italian, or French counterparts) and by the number of tourist facilities built at its initiative. Unlike the French club, it became the champion of the mountaineer, and in that respect it mirrored the celebration of the figure in popular literature.[81]

During the interwar years that Austro-German club would even make the mountaineer the archetype of the hardworking and patriotic Aryan. In so doing it positioned itself as the representative of these mountain populations, whose values and interests it claimed to be defending. That did not prevent the club from going along with national policies in forest management and nature protection: Hohe Tauern National Park in Austria came into being after the Austro-German Club purchased the lands surrounding the Grossglockner in 1913.

The Mountain as National Asset

The mountain of modern times thus has its heroes (mountain climbers and engineers) and its self-proclaimed mouthpieces (administrations and associations that may boast of representing the state, directly or indirectly). Until the interwar years, sports associations located outside Germany had few mountain residents in their ranks. Similarly, the national administrations that oversaw forest-protection policies rarely solicited the services of the inhabitants of the affected regions. In its early days Gran Paradiso National Park in Italy was an exception. But the administration itself quickly recognized that it could no longer privilege the recruitment of residents from the surrounding villages, so great were the difficulties in reconciling national interests and local conceptions.[82]

Ultimately it was the shift in scale from the local to the national level that conditioned the adoption of new concepts of the mountain and new modes of intervention. Within a few decades, nation-states readily made the mountains an emblematic place, a remarkable scenic landscape, a wild aspect of their territory, an ecological asset, even a reserve for unusual species.

That national and multifaceted interest in the mountain as a part of nature explains the variety of public policies that targeted mountains in the mid-nineteenth century, whether to reforest them or to protect existing scenic areas and ecosystems. It also explains the strong propensity of the Western nations to try to control uses of the soil or even to acquire ownership of the land itself. The modern state, continuing to respect a tradition deeply rooted in Europe—that of collective ownership of high pasture lands—early on laid claim to the highlands, for reasons of either security (when they were borderlands) or prestige. They seized considerable forest acreage, sometimes by taking it away from local populations to assure their mission of reforestation, and sometimes by taking it off the market, as in the case of the national forests in the United States.

When they did not appropriate private lands in the highland areas, Western nations placed more limitations on rights of use there than in other places. Restricted access to protected forests often came about in that way. Many national parks were built on private or communal property, but they imposed considerable restraints on possible uses.

In Europe, North America, and even Japan, the mountains of the industrial period, more than any other type of environment, were the object of close scrutiny from the nation-state. Artistic and scientific discourse on the one hand, and differentiated management practices of national territories on the other, came to confine the mountains to the role of a natural space whose management was to be conceived primarily as a function of ecological objectives and recreation. Its status as a shared national asset, which the emblematic value of some of its peaks and many international borders had previously granted it, was thereby reinforced.

FIVE

The Mountain as Living Environment

The aggressive policies conducted by the modern state in forest management and nature protection have played a large role in shaping the mountains of the Western world over the last two centuries (see chapter 4). Another set of policies adopted an alternative model of mountain knowledge and practice. In several cases, in fact, without neglecting the national interest (their primary motivation), they declared objectives relating to the people most affected: the local populations. This second group of public policies was the result of an expansion in the range of objectives taken on by the modern state in the early twentieth century, which included education, health, and an improved standard of living. They targeted the populations themselves and no longer merely a territory to be controlled or resources to be exploited. With the advent of the welfare state, the mountain was conceived as a collective living environment. The mountains became a territory.

Two factors lay behind that shift in focus: migration and differences in the standard of living. Migration away from the mountains was not new. Whether temporary or permanent, it was often even seen as typical of mountain societies (see chapter 1). For several decades even permanent migration was not perceived as a problem. When the forest policies were set in place, some were pleased by the prospect of outmigration they raised, since the reduction of demographic pressure sometimes appeared to be a condition for success. In Scotland, the outmigration of the Highlanders was even a desired and accepted consequence of the redistribution of land and the establishment of stock breeding on a large scale (see chapter 3).

Under certain conditions, however, migration was perceived as a problem. For example, a country might have a collective attachment to inhabited mountains as the keepers of tradition. In Austria since World War I, and even more in Switzerland since the mid-nineteenth century, references to the Alps and to Alpine traditions had played a decisive role in the national imaginary (see chapter 3). In addition, an antimodern or antiurban ideology sometimes raised fears that the psychological makeup or moral lives of the mountaineers would degenerate once they had moved to the industrial cities, or even that they would adopt the rebellious spirit of the "dangerous classes." That fear was behind the charges repeatedly lodged in Switzerland against the city,[1] from the early nineteenth to the late twentieth century. The conservative, Catholic-leaning literature of northern Italy sometimes formulated the same fear during the interwar years: "Mountain men have the mountain's pure and sincere force. They were born in the deep valleys, between the ranges of the majestic Alps, and they bear in their souls that solemn and austere spectacle of nature, which is like a constant and silent education. The restless passions of the big cities are unknown to them; they do not have the same needs as the people of the plains. They are accustomed to the meagerness of their rugged land, to the modesty of their rustic dwellings, to the familiar quiet of the long and monotonous snowy winters."[2]

More recently, migration and the reversion of mountain areas to fallow land have been perceived as threats to biodiversity and to scenic areas—in other words, to the characteristics of the mountain on which the tourist and environmentalist imaginary gradually conferred value. After World War II many countries became alarmed by that situation: Spain, for the mountain chains that border the Meseta; France, for the southern Alps, the southeastern part of the Massif Central, and some Pyrenean valleys; and Italy, for the southern Apennines and the Piedmontese Alps.

Above all, migration became a problem when it revealed inequalities in wealth and access to public services, which welfare states—generally from the 1950s on—wanted to fight in the name of a certain social equity. That third diagnosis is not very different from the one made at the same time for what were termed "underdeveloped" regions of the world. But except in the critical literature, many hesitated to use that term for mountains in the Western world. In the name of national cohesion, a country's mountains, perceived explicitly or implicitly as "underdeveloped," attracted renewed attention from the state and from society as a whole. With a new paradigm came a new problematization: the mountain was reconfigured as a place to live and as a territory.

Easing Forest Laws and Promoting the Alpine Economy

This shift first became apparent in the forest policies of Europe, which had tended to be authoritarian in the nineteenth century. They had held the mountain inhabitants responsible for flooding and had directed them away from traditional herding practices (see chapter 4). But that attitude sometimes proved counterproductive. Because of the hostility of local populations and local authorities, the implementation of some land re-forestation policies in the mountains was obstructed and their effects reduced. In some places, such as Italy, local resistance won out, draining the policies of their effectiveness until the 1910s. In response to that situation, a few initiatives in the late nineteenth century relied on the local populations' ability to adapt. The authoritarian approach was replaced by a more pedagogical approach, focused on the objective of agropastoral modernization.

Auguste Calvet was no doubt one of the first to move in that direction. A forest engineer, he arrived in Bigorre in the central Pyrenees of France in 1866. The French law that required the planting of grass had been in effect for two years. According to Calvet, though the law was designed to take into account the pastoral economy in forest management practices, it did not mitigate the tree worship prevailing among foresters or the administration's hostility toward the local populations.[3]

Calvet took an approach that was at odds with forestry orthodoxy. He strove first to understand the local populations, then to persuade them to modernize their herding practices to make them compatible with the objectives of forest- and water-management regulations. To do so he studied the traditional organization of pastoralism in the Pyrenees. He drew inspiration from Frédéric Le Play's monographs on that massif. Like Calvet, Le Play was an engineer, and his writings are now considered founding texts of modern sociology.[4] Calvet also studied pastoralism in the Swiss Alps and the French Jura. He was convinced that the Pyrenees were just as propitious, if not more so, for modernized livestock breeding. He advocated substituting cattle, which had little detrimental impact on the environment, for goats and sheep, which could cause considerable damage to the forests. He enlisted the services of several Jurassian cheese makers and, calling on prominent local citizens, encouraged the creation of nearly a dozen dairies, from one end of the mountain range to the other. In the Pyrenees Calvet spearheaded the conversion of traditional stock breeding, with all its local particularities, into a modern mountain economy based on the most prosperous models.

The initiative was a failure. The new approach did not suit local modes of social organization. The dairies went bankrupt one after another or were converted into state-owned or private companies. Calvet's superiors reproached him for paying too much attention to the dairies and too little to forestry proper. Calvet left the Pyrenees after that failure in 1879.

The future would vindicate him, however. His experiments were known and had been observed in many mountain regions. They gave rise to comparable undertakings, headed by Félix Briot in the Alps and by Lucien-Albert Favre in the Cévennes,[5] who in their turn promoted the milk-and-cheese economy and collective or cooperative local operations.[6] It was only much later, however, that the outlook of these atypical foresters, sometimes called "social foresters," really spread to the various echelons and offices of their administrations. The practice caught on after World War II. One sign among others: the official *Revue forestière française* devoted a special issue to the "agro-sylvo-pastoral balance." On the basis of several regional reports, the issue promoted the coordinated development of the different activities across all the mountains of France.

Even before the initiatives of the "social foresters" were adopted by the forest administration as a whole, these practices had spread to civil society, particularly through private organizations. The principal French group in that field, the Association Centrale pour l'Aménagement des Montagnes (Central Association for Mountain Planning), wanted to pursue the foresters' initiatives and to be a place where a new conception of the mountain economy would develop. Its members were diverse, some concerned with river navigation, some with the transformation of the landscape, others with the provision of wood, still others with mountain people's living conditions. When the association was founded in Bordeaux in 1904, its primary interest was the Pyrenees. But it encouraged similar initiatives in the southern Alps and in Dauphiné. Its slogan attests to its ambitious program, which extended to France as a whole: "Save the nation's land."

Academic geography, which was becoming institutionalized at the time, was not to be outdone. It produced many detailed descriptions of rural societies and rural economies, thus rehabilitating a world that the champions of industrial progress had often condemned. Two generations of local and regional monographs, published as dissertations or as articles for a broader audience, familiarized readers with the ways mountain societies were organized. Many researchers in the first half of the twentieth century gave detailed analyses of the forms of transhumance. They showed that these migrations, which took advantage of the diversity of ecological zones and varied from valley to valley, attested to the richness and sophistication of the relationship that local societies had established with their natural environment.

Other academics known for their mountain research also participated in public debate. Charles Flahaut, professor of botany in Montpellier, wrote in 1908: "The public authorities are losing interest in the mountain populations . . . [who] are being abandoned, or nearly so, to the bureaucracy to which they fall victim."[7] The same thought was repeated in travel magazines: "The forest must not drive away the mountaineer; it must protect him."[8]

Between the 1860s and 1900, then, there was a reversal in the way mountain populations were viewed. Accused of every evil in the mid-nineteenth century, a few decades later they had acquired no end of defenders. French society had in the meantime developed a curiosity about the rural world and the values it was said to safeguard. France had also become aware of the risks of a large-scale rural exodus. French society had anxieties about modernization as well, but the reasons differed by territory and population.

In that context the invocation of the mountain in general, and of the mountaineer in particular, would never be more systematic. The increasing importance of the categories, obvious in modern science (see chapter 1), could also be discerned in the applied sciences, foremost among them forest science. The transferability of models took hold as an idea, even at the cost of overestimating, as Calvet had done, the capacity of local systems to adapt to forms of organization imported from other mountain regions.

The mountain of foresters, including social foresters, was an abstract mountain, an "average" mountain in the arithmetical sense of the term, identified primarily by naturalistic criteria. Its social and cultural characteristics were never conceived of as anything but a secondary feature, local color. These could always be modified as a function of political, social, or moral imperatives, so long as these imperatives conformed to the laws of nature. The category of the mountain, thus instrumentalized, was preeminently normative.

Supporting the Farmers

Whereas the first agropastoral measures, adopted in the late nineteenth century, had the primary objective of supporting forest policies, a second set of initiatives, taken during the interwar years, sought to modernize the agricultural sector independent of forest concerns. The primary issue at stake was to guarantee producers sufficient revenue, out of concern for equity and a fear of excessive migration.

In some countries the measures adopted did not target the mountain regions in particular. In the United Kingdom, for example, the rural exodus had been going on for quite some time, and the conversion of highland sectors into moors, nature reserves, and cattle rangelands was undertaken in the early twentieth century. In the United States, the western mountains continued to be developed into the first half of the twentieth century, both as national forests (conceived as a source of wood products or as a natural resource) and as cattle rangelands. The federal government did not feel the need to design rural development measures or provide specific financial compensation. In many other countries, public policies supporting agriculture explicitly targeted the mountains and in fact played a role in calling attention to them. Two rationales were cited for these measures.

The agricultural policies of the twentieth century sought to promote a functional and rational specialization of production regions. They encouraged the products thought to be best suited to the region's natural conditions and most complementary to what other agricultural regions were supplying. The construction of railroad systems resulted in nationwide agricultural markets, which gave a comparative advantage to the mountain regions in the production of milk, cheese, and meat, since these areas were generally blessed with a wealth of meadows and grazing lands. Most of the measures adopted throughout the twentieth century were thus aimed in one way or another at livestock breeding. Switzerland was one of the few countries that, to ensure the food supply, supported grain production, generally considered obsolete in the mountains.

The second rationale that led to the adoption of specific measures to support mountain farmers lay in the difficult production conditions reportedly found there and in the wish to compensate for the resulting handicap. Some of these constraints could be counterbalanced by technical support: farms were small and the parcels split up, especially in Latin Europe, where the system for bequeathing property tended to divide it among the heirs. The public authorities would sometimes have particularly detailed cadasters drawn up, encourage the reunification of parcels of land, or promote cooperatives, particularly in the case of dairies (fig. 10).[9] Other measures included financial compensation. The growing season was shorter in the mountains than on the plains and the yields lower; the slopes were steeper, which made mechanization difficult or expensive. Financial allotments were designed to underwrite farmers' costs.

When the idea prevailed of a specificity of mountain agriculture based on specific conditions of production and specific threats, it led to the adoption of policies, whether basic or sophisticated, dedicated explicitly to the mountain regions. Such was the case for Switzerland from 1929 on, for Austria in 1960, for France in 1972, and for Spain in 1982.

The adoption of policies aimed specifically at supporting mountain agriculture required a definition of that concept and the identification of zones eligible to receive benefits. In other words, it required the objectification of mountain farming. The demarcation of zones was based on the major factors identified as handicaps to agricultural production: elevation and slope. The criterion of elevation addressed the problem of the shorter growing season. In most cases the cutoff was set at between 600 and 1,000 meters, but sometimes it was decreed to be slightly lower. In Poland, for example, the most recent law sets the threshold at 250 meters. The criterion of slope addressed the question of production techniques, compensating in particular for the difficulty or high cost of mechanization. The indicators were usually average slope in the farm zones or districts or, in Spain for example, the differential between the highest and lowest points of the territory concerned. Most countries combined the two criteria, and some added additional ones. In Switzerland, for instance, when the cadaster of agricultural production was drawn up between 1944 and 1949, the criteria of elevation and slope were complemented by others based on the "most important production factors": "length of the growing season, precipitation rates, prevalence of sunshine, accessibility, etc."[10]

Since a plot of land or a farm classified as being "in a mountain zone" was eligible for subsidies, the financial stakes for farmers were high. Various pressures thus came to bear, and the defining criteria were not always applied in an optimal or objective manner. It has been shown, for example, that the number of subsidies granted in France since 1973 has been, "curiously, inversely proportionate to the gravity of the geographical handicaps."[11]

Beginning in the 1980s, within a context of agricultural overproduction and the increasing importance of environmental questions in Europe, agricultural policies underwent a shift. They gradually abandoned direct payments to keep market prices low and increased payments to farmers and assistance to improve the production chain. Farmers were primarily encouraged to maintain scenic areas and later to promote biodiversity.

Diversifying Resources and Bringing Modernity to the Village

The task of assisting and modernizing mountain economies and societies also took the form of sectoral policies.

The installation of hydroelectric facilities was one of the sectoral policies that contributed most to forging a modern conception of the mountain. The principal innovations occurred in the Alps, particularly in France in the 1860s–80s. It was there that the first water turbines converted the kinetic energy of mountain streams into electrical energy. Then came the first pressure pipes and the first electrochemical and electrometallurgical plants, which consume on site the electrical current produced. In nearby cities, such as Grenoble and Turin, a new body of knowledge and expertise developed rapidly in the last third of the nineteenth century, especially in the fields of hydraulics and electrotechnology.

Thanks to these localized innovations, a new technological and political discourse about the mountain in general and its modernization emerged. In fact, the initial experiments and successes led to the spread of innovations to the rest of the Alps—to Slovenia, for example[12]—and to many mountain regions with comparable topographical and climatic characteristics. Hydroelectric plants and factories on the valley floor multiplied in the Pyrenees, the Massif Central, the Carpathians, and lower-elevation mountains of eastern Central Europe.[13]

When modern forms of energy and industry swept into these areas, many saw them as the vehicle for modernizing the mountain regions and their populations in general. Industries, plants, and pressure pipes, followed by dams, brought investments and infrastructure, especially roads, to regions that until then had been oriented toward an agropastoral economy. The local populations hoped to receive income from industry to complement their agricultural activities, which were in relative decline. From the mid-twentieth century on, many districts took advantage of the taxes that the electricity producers paid them to undertake their own development projects. Hydroelectricity thus contributed to bringing "modernity to the village."[14]

Simultaneously, that innovation played a role in imposing a new and therefore more modern image of the mountain, far removed from the traditional or archaic representations that had held sway for decades. The industrialists and the public authorities were key figures in the promotion of that image (fig. 10), as were academics and local scholars. One historian vaunted the modernity of Swiss Valais: "In Valais the mountains are particularly threatened by a new form of erosion, which may bring on the desertion of the high valleys. But human predictions have only so much value, and the concrete dams will soon revive life at high altitudes. A few years from now the water running through the galleries will restore circulation in the Alpine body, threatened with anemia."[15]

The early twentieth century, however, brought a competing notion of

10. Cover of the professional journal of the Association des Producteurs des Alpes Françaises (APAF) in 1931. A concern with modernizing Alpine economies and societies led to
the birth of organizations such as this, where business owners, academics, and high officials
rub shoulders. These organizations seek to promote hydroelectric facilities and the industrialization of the Alps and to spread the benefits of modern modes of production throughout
rural societies. The cover of the APAF journal illustrates metaphorically, through the parallel
established between the electric lines and the row of grapevines, the organization's interest in promoting a type of modern development that is respectful of rural traditions.

hydroelectric plants to the mountain regions. It was the result of a further technical innovation of the late nineteenth century: the long-distance transportation of electricity over cables. It was now possible to produce electrical current in the mountains that would be consumed far from the site of production. Industrialists and public authorities encouraged that alternative early on, which gave rise to major facilities without any industrial consumption site nearby.

In 1916 the Austrian state decided to promote hydroelectric systems in the Alps to supply the national railroad network. After the *Anschluss* in 1938, Germany undertook a huge hydroelectric program in the Austrian Alps, especially in Tyrol, Vorarlberg, and the Salzburg region, to fuel the Third Reich's industries.[16]

The nationalization of electric companies after World War II furthered that process: the production of hydroelectricity in the mountains tended to become one of the contributions of these regions to state energy policy. Here again the conception of the mountain at issue had already prevailed in forest- and water-management policies in the nineteenth century and in policies to protect nature and scenic areas in the twentieth: the mountains were seen as an aggregation of resources that ought first to benefit the national economy and society as a whole before providing for the well-being and development of the local populations. The exploitation of mining resources in mountain regions was often conceived on that model.

Tourism Practices and Policies

Tourism policies were marked by the same ambiguity throughout the twentieth century. Until the late nineteenth century the mountain tourist economy was modest in scope and confined to a small number of places. Overall it was not government policies promoting tourism that sparked the curiosity of travelers; rather, that curiosity arose in large part from a growing appreciation among Western elites for mountain scenery, first in the Alps and Pyrenees and later in the Highlands, the Lake District, the Rockies, and the ranges of the Pacific Coast of the United States. The Alps had long served as a frame of reference for identifying landscapes worthy of interest (see chapter 1). At the same time tourist services were generally provided by individual initiative: farm families offered lodging and guides to travelers passing through, a few family dynasties (such as the Ritz, natives of the Upper Valais) made a fortune in luxury hotels, a few capitalists invested in mountain railroads, and so forth. The demand was similar in different regions, and supply often followed suit, turning mountain tourism into a uniform tourism practice. The railroads and grand hotels built in the mountains of Europe and North America ultimately came to resemble one another.

In the twentieth century many nations confined themselves to policies that encouraged or regulated these individual or private initiatives. In Switzerland, for example, tourist development had always been decentralized: it gradually spread to Alpine villages and created a number of small, often specialized businesses, which allowed the Alpine population to remain relatively constant. In Austria as well, the development of tourism remained primarily in local hands, and a large population was able to continue living in the Alps.[17] In those two countries public policies simply facilitated control of the land by local actors or encouraged (in Austria even more than in Switzerland) the complementarity of agropastoralism and tourism. It is therefore clear that the Swiss and Austrian mountains were not uniquely conceived as a recreational area but also, and sometimes especially, as a living space rich in traditions that the state was anxious to cultivate and valorize.

The situation was completely different in other countries, including France and the United States. The French government was interventionist from an early date, in this realm as in others. It accelerated the development of tourism in the most promising mountain areas. Tourism development in the French mountains was centrally planned from the 1940s on, peaking in the 1960s–70s. It has been viewed in hindsight as a coherent infrastructure and investment program, christened "Plan Neige" (Snow Plan) in 1964.[18] Less than a true plan, however, it actually consisted of a dozen operations with common characteristics,operations that reflected a political, administrative, and technical culture. The sites chosen and the resorts that emerged have a number of things in common: a high elevation with the guarantee of snowfall, a topography favorable to the construction and management of vast ski slopes, a location far from the villages (to avoid major real estate constraints and to promote modernist architectural models and urban development projects), and the concentration of different tourist services (ski lifts, accommodations, tours, and so forth) in the hands of a small number of companies. Although that so-called plan affected only a small number of sites (Tignes, Avoriaz, Les Menuires [fig. 11], Superdévoluy, Flaine, Les Arcs, and others), it is a telling indication of a political culture in France that differed greatly from the one prevailing in Switzerland and Austria at the same moment. Strong government initiatives developed the tourism industry, regional planning policies were guided by a notion of the mountain as a rich source of revenue, little con-

11. The architectural modernity of the winter sports resort in
Les Menuires (Savoy, France). Photo: Bernard Debarbieux.

cern was shown for local characteristics, and development took a sectoral
approach. For that reason, the plan has been said to conform to a colonial
management model.[19] In the mid-1970s the French state abandoned this
model, which was judged too modernist and too burdensome for society
and the environment. Interestingly, it then adopted a "mountain directive"
(1976), which allowed it to limit and control urbanization and to keep
ski areas from extending beyond a certain elevation (1,600 meters). The
French state's reversal of its tourism development policies applied only to
the models it privileged, not to the preeminent role it recognized for itself
in deciding what ought to be done.

The model used to encourage and develop tourism in the United States
and Canada is rather similar to the French model. Extensive public ownership of the western mountains has made the federal government in both
countries an important player in that realm, as well as in the forest products industry and nature protection (see chapter 4). In the United States,
the U.S. Forest Service leases or grants zones to private companies for use
as ski areas. That authority, made official in 1986 with the passage of the
National Forest Ski Area Permit Act, provides the government with a fairly

significant source of revenue.[20] Because of the absence of long-standing societies in these zones and an indifference toward the Native American populations, the development of tourism between the Rockies and the Pacific was not conceived as a way to meet objectives of integrated economic development, social cohesion, or checks on migration. Complementing the national parks system, a series of fairly modern and functional tourist resorts has developed, their establishment and expansion often negotiated with the federal administration. Their operators have had little contact with the other economic actors in the American west, which strongly suggests that many U.S. resorts were inspired by the urban planning model of the French resorts in Plan Neige.

The Regional Construction of Problems: The Appalachians

In the eastern United States, by contrast, Appalachia was apprehended and shaped in a completely different manner. It was believed that the modernization of that area and the articulation of the social and economic dimensions of its development had to be conceived regionwide. For that to happen, however, the Appalachians had to be understood as a coherent regional entity that could be apprehended as such within the context of a federal interventionist paradigm.

This did not happen automatically. At the beginning of the nineteenth century, America's independence from England, followed by the Louisiana Purchase, sparked the idea that the Appalachians constituted a boundary separating the American settlers from the Native Americans, and that, in crossing that boundary, the people of the United States had taken their destiny in hand (see chapter 2). Behind that mythic unity, however, the idea of diversity prevailed. The ridge lines and hollows that divided the Appalachian region were primarily conceived as so many different and inhospitable hinterlands. Furthermore, the dividing line between the southern states and the northern states that would face off in the Civil War (1861–65) cut through the middle of the mountain range.

As of the late nineteenth century, however, the Appalachian range gradually came to be apprehended as a totality. Arnold Henry Guyot (1807–1884), a Swiss naturalist who had immigrated to the United States, published a scientific article and map in 1861 that presented the Appalachians as an essential component of the geological structure and geographical organization of the continent.[21] In the same years, the idea took hold that the Atlantic states possessed a common hinterland. In 1894 William G. Frost, president of Berea College in Kentucky, proposed that the "Mountain

Region of the South" should be seen as "the mountainous back yards of nine states." Frost added that he saw that region as "one of God's grand divisions, and in default of any other name we shall call it Appalachian America."[22]

Articles in the press and popular works published from the late nineteenth century on tended to speak of Appalachia as a backward region populated by a rather peculiar group of Americans.[23] The recurrence of the theme of backwardness, of a people "behind the times," "seemed sufficient basis for the construction of a generalization concerning the coherence of mountain culture and the homogeneity of mountain people. And it was this which made possible the redefinition of Appalachia as a discrete region of the nation and the redefinition of the mountaineers as a distinct population, and thereby made possible the legitimation of Appalachian otherness."[24] At the turn of the century, the image of the "hillbilly" and of "mountain whites," a degenerate version of that population, began to take hold. The hillbilly—poor, uneducated, and with a large brood of children—was caught up in feuds between rival clans, had little respect for the law, and loved his moonshine.[25] The industrialization of the northwest, combined with the Great Depression, which significantly affected agriculture and coal production, fueled the migration of large numbers of residents from the region. That migration would spread the image of the hillbilly to urban areas, which liked to style themselves temples of modernity.[26]

That dual image of the backward hillbilly and of Appalachia as a geographical backyard was fostered by the first statistical measures of wealth and economic production. They revealed that entire counties lagged behind the national average.[27] Since the late nineteenth century that identification of an enormous zone of poverty and backwardness has given rise to several charitable and political initiatives. Protestant congregations located in the northern states sent missionaries to ensure that the region had basic social institutions, especially churches and schools, which they believed would help the region make up for its "delay." An entire people had to be saved.[28] These organizations produced reports and statistical studies that together drew the perimeters of a region that was behind the times and justified their actions on the ground.[29] Photo reports documenting the situation also legitimized political action (fig. 12), and the general public was informed about the poor living conditions in the Appalachians through the publications of these photos in magazines.

The Great Depression and the various reports on the economic and social state of the nation culminated in a study called *Economic and Social Problems and Conditions of the Southern Appalachians* (1935).[30] It described

12. Pumping water by hand from the sole water supply in this section of Wilder, Tennessee, in Fentress County (Tennessee Valley Authority, 1942). Franklin D. Roosevelt Presidential Library and Museum.

the crisis the region was experiencing in even more detail. As part of the New Deal, the federal government multiplied sectoral initiatives to boost the regional economy: hydraulic engineering projects and hydroelectric plants in the Tennessee Valley, the establishment of national and state parks, the construction of scenic highways, and so on.

It was not until the 1960s, however, that an integrated regional development program targeted the area as a whole, thereby reinforcing the image of its unity. The years of President John F. Kennedy's administration, which made the war on poverty a top priority, favored that outlook and the establishment of public welfare programs to complement the systems already in place at the federal level.[31]

In 1960 state political leaders in the area founded the Conference of Appalachian Governors with the aim of taking a regional approach to development problems. Whereas federal measures at the time tended to encourage economic restructuring of the industrial and urban centers in crisis (the Area Redevelopment Act), the CAG cited Appalachia's "underdevelopment" and "depression."[32]

Coordination among the Appalachian states culminated in a highway construction project, the Appalachian Development Highway System (ADHS). The program built three thousand miles of highways, designed both to make traveling within the region easier and to improve access from the outside.[33] That coordination resulted in the federal government's adoption of the Appalachian Regional Development Act on March 9, 1965. It ordered the creation of the Appalachian Regional Commission (ARC), whose objective was to reduce the effects of poverty in the region by diversifying the economy while improving the provision of social services (health and education) to its residents. Therein lay the principal rationale for identifying Appalachia as a special region: to set up a regional development program built on the idea that this mountain zone was to be apprehended as a totality and managed in accordance with an overall plan. The zone of intervention targeted by the Appalachian Redevelopment Act occupies a large area, extending from the southern part of New York to eastern Mississippi. This area does not correspond to the entities usually recognized by geologists and geographers. The zone covers the Appalachian Mountains, but it also includes counties not considered mountainous, in the eastern part of Mississippi, for example. That region does display similar problems of poverty and unemployment, however. Indicators show that the geographical center of the region covered by the ARC (West Virginia, eastern Kentucky, and western Virginia) has the largest concentration of socioeconomic problems.

Once the Appalachians were recognized as a special region at the federal level, defining that region became the object of long and delicate negotiations. Each state was responsible for determining which of its counties belonged to the zone and were therefore eligible for ARC-funded projects. The criteria, which varied from state to state, usually involved economic indicators, especially per-capita income and the unemployment rate. Here and there political leaders sought either to join the ARC zone or to remain outside its jurisdiction, depending on whether they agreed with its philosophy.[34] But throughout the process the federal government sought to ensure that the circumscribed zone constituted a compact whole.

The aim of the measures implemented was thus not so much "to change the mountains" as "to change the mountain personality."[35] These measures sought to update the region's infrastructure and to attract jobs by funding several areas of growth.[36]

Integrated development in Appalachia was originally conceived as a temporary operation: "As the region obtains the needed physical and transportation facilities and develops its human resources, Congress expects that the region will generate a diversified industry and that the region will then be able to support itself through the workings of a strengthened free enterprise economy."[37] But in the 1980s it was conceded that this objective had not been achieved and that the image of the region's marginality continued to prevail. Analyses published since the mid-1990s have criticized the plan of action adopted for overlooking the root of the problem. According to the authors of these analyses, the Appalachian economy rests primarily on the exploitation of natural resources by entities based outside the region, which have no concern for its development.[38]

Decentralizing and Promoting Self-Reliance in the Mountains of Europe

European countries also took an approach focused on particular mountain massifs. The Massif Central, which has been more affected by outmigration and economic decline than other French massifs, received lasting attention from the French government in the second half of the twentieth century. The public policies adopted in Europe between the mid-nineteenth century and the 1970s, despite their great diversity, all tended to apprehend the mountains as a totality from a national perspective. These policies, whether authoritarian or paternalistic, whether motivated by the objectives of growth and resource development or by the paradigm of the welfare state, always expressed the desire to make the mountain a state concern. It was the state, aided in that project by scientific expertise, that decided the location of the mountain regions, the criteria by which they were to be apprehended, and the modes of action to be taken with respect to them. In short, it was national governments and the political class that configured the mountain. Although the populations may have benefited in terms of jobs, infrastructure, and living conditions, they lost some of their autonomy. Their influence over their environment and their place within society were largely determined by the conceptions that society as a whole and the modern state adopted toward them.

The principal reversal in that situation occurred in the 1970s. With the general tendency toward government decentralization and the promotion of self-governance and citizen participation in Western countries, policies were set in place to encourage local development and collective initiatives. The phenomenon was widespread and applied to every type of geographical region. It might have led to the dissolution of the very notion of the mountain in public policies. Public debates in France, Switzerland, and Italy showed that this option had its defenders. Instead, a new way of conceiving the specificity of the mountain and of giving it a political form came into being.

France, after decades of redistribution policies favoring mountain regions and their populations, gradually moved away from financial assistance. The Law on the Development and Protection of Mountain Regions (known as the Mountain Law), passed in 1985, was enacted in the wake of the decentralization laws (1982–83), which applied to all territorial collectivities in the country. The Mountain Law sought to take into account a few postulated specificities of the mountain regions. In particular, its many provisions set out the conditions under which tourism development and environmental protection ought to be pursued. In moving away from the financial aid policies that had dominated in the previous period, that new law encouraged above all the "self-reliance" of the mountain regions, based on programs designed and executed by local agents.[39] As several political scientists have pointed out,[40] the law was a real novelty in France's legislative arsenal. There the Jacobin tradition tends to promote the passage of laws of general application—and yet, the French Parliament had passed a bill dedicated to a single type of region within national territory. It was noted at the time that the mobilization of an object as symbolic and apparently apolitical as the mountain made it possible to sidestep a great deal of political opposition: "the majority [leftist] parties . . . did not necessarily approve of the intellectual postulates on which it rested or on the new management practices that it entailed, but they accepted the geographical argument of the mountain's singularity."[41]

That shift in the French public authorities' approach to the mountains therefore indicated the adoption of a new paradigm. National priorities, from infrastructure to social welfare, would never again take precedence over local dynamics. These dynamics, identified more clearly than ever before in the national legislative arsenal, were essentially taken into account in the name of diversity—of local territories, societies, and economies—which public policies were responsible for promoting in the interest of development.

That new way of apprehending the mountain in French public policies was thus part of a more general tendency to recognize the role of local resources—material resources but also, and especially, cultural resources—in economic and social development.[42] Overall, development was no longer conceived linearly, as a single objective. "Backward" or even "underdeveloped" areas, foremost among them the mountain regions, were no longer encouraged to catch up with the more advanced ones. Development became differentiated, more focused on drawing an advantage from local specificities. Any characteristic specific to the mountain regions became an asset for building their future.

It was in that context that the mountains came to be promoted as a laboratory for local development. In particular, by banking on the image of environmental quality often associated with them (thanks in part to decades of protection policies targeting scenic areas and nature), the mountains acquired an exemplary value in the promotion of sustainable development. For the last thirty years the creation of a large number of regional natural parks in the mountains of France has demonstrated the success of that idea and its institutionalization.

Italy, in its decentralization of policies related to the mountains, in large part anticipated the French initiatives. In 1971, nineteen years after the promulgation of the Law on Measures in Support of Mountain Regions, it was revised to promote greater initiative and decision-making autonomy at the local and regional levels.[43] The law transferred legislative power—especially the power to delimit the mountain regions—to the regional level. The principal innovation of the law, which broke with a policy by which the central government dispensed financial aid, lay in the establishment of a new political and administrative level: the *comunità montane* (mountain community). Several districts come together under that rubric in a concerted valorization of their territory. To achieve that objective, they define and implement a joint plan for socioeconomic development. This new institutional framework seeks to rationalize through joint projects the tasks assigned to municipalities, sometimes favoring mergers between districts.

A comparable evolution occurred in Switzerland.[44] In the 1970s the Confederation began to encourage socioeconomic actors in the Alps and the Jura to construct programs that the government would support. In 1974 the federal parliament passed the Federal Law on Investment Aid for the Mountain Regions (known as the LIM). That law encouraged the design of programs at the regional level, which the Confederation would finance with assistance from the cantons. A region is defined as "a group of communes [municipalities] closely associated geographically and economically, which propose to execute some of their tasks jointly." Most of the financing continues to go to infrastructure programs that allow the mountain regions to make up for delays in that regard. But these projects are no longer designed on a case-by-case basis; they are negotiated in an integrated manner with local actors, in order to guarantee efficient follow-through at the local level. The geographical zoning instituted by the law has given rise to debates that are of interest to anyone wishing to understand the political stakes involved in definitions of the mountain. The 1974 law stipulated that each of the zones be centered in a small city, which was to play the role of regional development center.[45] That condition was met with incomprehension by many local residents and elected officials, especially in Valais. In effect, the law placed cities at the center of the system, even though residents considered cities privileged entities: they are not in the mountains, and they have little interest in mountain problems. The mountain people's mountain was no longer the mountain of public policies.

A new phase began in the early 2000s. The New Regional Policy expects even more initiatives from local actors and substitutes local development objectives for infrastructure objectives. The policy was expanded at that time to include all of Switzerland's rural regions. But the specificity of the mountain has remained present in the design of local programs and in complementary sectoral policies that generally broaden their scope. The New Agricultural Policy, adopted around the same time, encourages the communes to work together on rural development projects wherever agricultural activity can be better coordinated with environmental management and tourism.

In France, Italy, and Switzerland, then, the mountains have functioned as laboratories for decentralization and regional development policies. That tendency is also apparent in Austria. In 1979 the Federal Chancellery of Austria launched its Bergebiets-Sonderaktion (Special Initiative for the Mountain Regions), a step toward decentralization and the promotion of endogenous regional development, particularly through the establishment of cooperative enterprises.[46]

The Consumption of Images and the Image of Consumption: The Politics of Branding

In Austria as in many other European countries, the shift in public policy toward decentralization and self-reliance in the mountain regions—toward a territorial and integrated approach—has gone hand in hand with the promotion of agricultural products bearing a certified label. That approach represents a specific aspect of public policies and the one best able to show how the political construction of the mountain rests as much on regulatory initiatives as on the social imaginary.

In 2002 the mountain farmers of Austrian Tyrol who practice organic farming created the certified label "Bio vom Berg." The program's supporters claim that the mountains ought to be the privileged environment for organic farming, its "native habitat" (*Heimat*). And in fact, 81 percent of organic farms in Austria are located in the mountains. According to the promoters of Bio vom Berg, that mountain location offers a guarantee of the superior quality of the organic product. That pursuit of a surplus value for products vaunted for their quality—because they are associated with a mountain imaginary of tradition and purity—is complemented by the valorization of the cultural characteristics of the production regions. Tyrol is thus presented by promoters of that approach as a "unique cultural landscape" in which agriculture plays a determining role.[47] The consumption of certified organic products supports that agriculture, which makes the region what it is.[48]

Within a more institutionalized context, the Appellations d'Origine Contrôlée (AOC; Controlled Designation of Origin) and the Indications Géographiques Protégées (IGP; Protected Geographical Indications) are privileged modes for implementing such strategies to valorize the overall image of the mountain and promote quality. Of course, these designations do not apply solely to mountain regions, but they have found fertile ground there.[49]

At the same time, many economic actors in Europe (professional organizations, food distribution outlets, and so on) also capitalize on the surplus value provided by the strong image of mountain products, by creating their own "mountain" label (see chapter 10). The circle of actors intent on drawing an advantage from the image of the mountain grows ever wider, attesting to the power that image exerts on consumers, even though the overuse of labels and the opacity of the defining criteria may create confusion.

That consumer fad for mountain products, and the fad among producers and distributors for the corresponding labels, can lead to abuses. To prevent them, some governments have sought to control the use of the designation "mountain." France led the way in 2000 with a decree specifying how the umbrella agricultural law passed in 1999 applies to the use of the term "mountain" on agricultural products.[50] The decree states that "all operations of production" and manufacture must take place in a mountain zone in France, as defined by the Mountain Law of 1985.[51] In 2008 Switzerland established the geographical conditions required for an agricultural product to be granted the mountain label. Ultimately, governmental authority remains the guarantor of the image and legitimacy of references to the mountain.

Organizing the Mountain Populations by Means of New Political Identities

The considerable interest that Western societies and nation-states have shown in their mountains and "mountaineers" has radically changed the

local populations' image of themselves. For a long time people characterized as mountaineers did not identify with that characterization (see
chapter 1).[52] It was mountain climbers who first designated themselves as
mountaineers, wishing thereby to indicate their intimate knowledge of the
mountains as well as their legitimate authority to say what ought to be
done there (see chapter 4). The local populations identified themselves as
mountaineers only somewhat later, and they meant something entirely different by that term.

In fact, though no specific studies exist on this matter, it seems that
these populations began to embrace the designation only in the late nineteenth century. In Switzerland the populations most affected seem to have
adopted the term *montagnard* early on[53] because they realized it was in use
in the administration, among tourists, and in Swiss society as a whole. For
example, the Swiss national exposition of 1896 made use of the term to
give these populations pride of place (see chapter 3). Since its connotations are fairly positive in that country, the adoption of the term can be
considered a symbolic advantage. By means of a process of internalization
well-known in the social sciences, the "mountaineers" appropriated an
identity that was imposed on them from the outside.

That form of self-designation, however, also occurs in different contexts, where, by contrast, the local populations feel the need to impose
a certain representation on themselves. Also in Switzerland, it has been
shown that the older uses of that self-designation were designed to counter the indifference that some travelers manifested toward the mountain
populations.[54] In Europe more generally over the last thirty years, the
adoption of the term *montagnard* or its equivalents has come as a response,
tinged with pride, to the derogatory images that have circulated about the
mountaineer with the advent of urban and industrial modernity and its
glorification.

The adoption of public policies that target the mountain regions nationwide has tended to institutionalize such questioning and repositioning
of identities. This did not happen immediately: historians do not note any
use of the term *montagnard* for political ends by those so designated in the
early days, even though the conflicts these policies engendered, especially
in the nineteenth century, had a great deal to do with identity issues. But
from the mid-twentieth century on, when public policies took on explicit
socioeconomic objectives, certain political, linguistic, and identity-based
practices evolved appreciably. Parallel to the elaboration of public policies, groups formed to defend the interests of those designated as mountain populations. They declared themselves *montagnards* and sought to

provide—or, more precisely, to impose—a legitimate interlocutor to engage with the legislature and the government administration, which were defining public policies that concerned mountain populations and territories. At first these groups, at both the local and national levels, were composed primarily of elected officials, who could defend the cause of mountain populations in regional and national parliamentary chambers.

In Italy the National Union of Mountain Municipalities, Communities, and Authorities (UNCEM) was founded in 1952, only a few months after the passage the National Law on Mountain Areas, the first of its kind in that country.[55] The Swiss Center for Mountain Regions (SAB) was created in 1943,[56] when many legislative measures were passed to keep the labor force in the mountains stable. In France the National Association of Elected Representatives from Mountain Areas (ANEM) was established just after the passage of the Mountain Law (1985). Advocates of the law even encouraged the creation of the group, believing that such an entity was a condition for success in promoting the mountain as an object of integrated public policies.[57]

The creation of these lobbies, which support a strong political and territorial identity, illustrates the importance of the national framework in the approach taken to mountain questions in the twentieth century. In centralized countries the mountains, when they became a political object as such, gave rise to representatives whose discourse could not be uniquely local or regional. The nationalization of the mountains in public policy was accompanied by a nationalization of the very institutions through which the debate took shape.

These self-proclaimed representatives of the mountain populations, taking their cue from the dominant characteristics of public policies at the time, tended to make two arguments. The first invoked the principle of national cohesion or *solidarité* (roughly, social welfare) to allow mountain dwellers to benefit from the state's redistribution policies.[58] These associations had no difficulty reminding people of the handicaps faced by the regions they represented. They demanded that policies, whether sectoral (agriculture, energy, or tourism) or territorial (planning, development, regional policy, public services), take these handicaps into account. That rhetoric triumphed with the welfare state, then became a target of criticism when liberal ideology regained ground in the 1980s.

The second argument invoked the decision-making autonomy of the mountain regions themselves. The modern state's tradition of interventionism in regional planning, from mid-nineteenth-century forest policies to the promotion of natural protected areas throughout the following century, was a recurrent source of discontent among the local populations. In promoting autonomous development, political organizations sought to exempt themselves from mandates said to come from outsiders (tourists, "ecologists," the government), on the grounds of their vast knowledge of the mountains and their long experience with the environment.

The case of Switzerland demonstrates the evolution in the problematization of the mountain, as attested by the successive names borne by the country's principal pressure group on behalf of mountains.[59] The Swiss Union of Mountain Farmers was founded in 1943, within the difficult economic context of World War II. The organization was created when it was noted that, though mountain farmers were the standard-bearers for the social demands of Swiss agriculture generally, they drew fewer benefits than others from agricultural policy.[60] The Swiss Union of Mountain Farmers, having arisen from a concern for differentiation within public debate, quickly found its legitimacy thwarted by its professional focus. Although its by-laws declared its intention to defend the interests of mountain populations as a whole, some criticized it for being primarily concerned with agricultural questions. A dissident organization therefore formed in 1972. The Swiss Union of Mountain Farmers took note of these criticisms and in 1973 became the Swiss Union for the Mountain Population. The following year the passage of the Law on Investment Assistance for Mountain Regions (1974) clearly demonstrated that the public authorities wanted to broaden their political approach to the mountain well beyond agricultural questions and to garner support from regional entities. Then, in 1988, the Swiss Union for the Mountain Population became the Swiss Union for Mountain Regions, a name it still bears.[61]

The reconfiguration of political identities in European countries thus created lobbies organized at the same administrative level as that which was determining the principal mountain policies. They have been able to turn to their own advantage a designation (*montagnard* and its equivalents) and certain themes (heritage and tradition, local experience and know-how, specificity and *solidarité*) that had been used to construct the stereotype of the mountaineer in previous centuries. As a result, these lobbies mark the political institutionalization of the identities and representations by which the mountain question had earlier been addressed.[62]

That slow emergence of the *montagnard* as a political figure is therefore inseparable from the emergence of the mountain as a political object. Their destinies are joined. Riccardo Maderloni, a leading figure in the

National Union for Municipalities, Communities, and Authorities in Italy, made this point emphatically when addressing a colloquium a short time ago: "When man abandons the mountain, the mountain follows man."[63] The shared destiny that is assumed to exist between the mountain and the *montagnard* ultimately rests on their consubstantiality.

PART TWO

The Mountain on a Global Scale

Is there at present a globalization of the mountain? That apparently simple question calls for different responses depending on what is meant by the term "globalization." If the question is whether mountainous regions are currently marked by the globalization of economic and financial fluctuations, by the increase in the circulation of goods, people, and information, then the answer is assuredly yes, as attested by a large number of the academic references that appear in this book. If the question is whether supranational and global authorities—especially intergovernmental organizations and NGOs—concern themselves with the mountain in the interest of global governance, the answer is also yes. Chapter 8 is in fact wholly devoted to that aspect.

But it would not be altogether consistent with the project of this book to speak of globalization as an external fact, a state of the world that applies to mountains and to a large number of other regional landforms. In the previous chapters we endeavored to show how the mountain as a category of objects was conceptualized to motivate, support, or enable political initiatives on a national scale. In part II we are obliged, in the interest of symmetry, to ask what the notion of the mountain makes it possible to think, say, and do once nation-states no longer constitute the principal—if not the only—level at which political questions are framed.

In other words, the globalization of mountain issues as an external fact interests us here only insofar as it attests to an evolution in the scientific and operational representations of the mountain and its uses, hence to changes in configurations. In fact, globalization is not only a transformation of the ways that socioeconomic and political systems are organized but also a propensity to view numerous phenomena as being intelligible only on a global scale.[1] On that last point we should emphasize that, in the spirit of this book, scale "is not simply an external fact awaiting discovery but a way of framing conceptions."[2] In part II as a whole, therefore, we will study the extent to which globalization has invoked the mountain as a category of intelligibility, and we will consider the specific problematizations to which that category has given rise and the forms of political action it has produced or promoted.

To that end we will consider several phases and scales of analysis. In chapter 6 (phase 1) we will study how, under colonization but also immediately after the colonies achieved independence, the models of cognition and territorial control that the Western nation-states had adopted in the eighteenth century were transferred and adapted to new contexts. In chapter 7 (phase 2) we will show how the planning and management of mountain spaces and environments functioned within that context of colonization and decolonization. In chapter 8 (phase 3) we will trace the appearance and proliferation of references to the mountain at international conferences and in international institutions, as well as the consequences of that phenomenon for the actors and objects involved. In chapter 9 (phase 4) we will show that the increasing importance of global diagnoses and recommendations regarding mountains has had particular effects on the formulation of the social and collective identities of the populations most concerned. Finally, in the last two chapters we will take into consideration the multiplicity of scales at which mountain issues are defined in the wake of globalization. Chapter 10 is devoted to the European Union, a supranational—albeit multifaceted—authority that for a long time kept its distance from the political mobilization of the notion of the mountain. In chapter 11, which is devoted to initiatives (often transboundary in nature) carried out across an entire mountain massif, we will show that in many cases territorial recompositions of the mountain have occurred not on a global scale but at an intermediate level.

If globalization means that knowledge, political transformations, and initiatives relating to mountain issues occur within a global framework, then that process has been at work since the earliest days of European colonization. It has accelerated in the last twenty years with the proliferation of informal and institutional initiatives. But globalization, far from rendering obsolete the other scales of analysis and action, is accompanied by a regionalization and a repositioning of initiatives on the part of the nation-state, and even by the redesignation of particular places as mountainous. Each scale repositions itself in relation to the others by virtue of a process that some have proposed to call "glocalization."[3]

SIX

The Mountain and Colonial and Postcolonial Territoriality

For five years, between 1889 and 1893, Colonel Algernon Durand lived in the Gilgit region of Kashmir, on the northern border of the Indian Empire, at the foot of the Karakoram.[1] At the time he was the principal British military agent on the job. His missions were many. First, he oversaw the border with the Russian Empire, which had gained a great deal of ground in Central Asia over the previous decades. Secondly, he was supposed to train local soldiers. Although they were under the orders of the princely state of Kashmir and Jammu (actually a British protectorate since 1846), they were charged with the mission of defending the interests of the empire and ensuring that its authority was recognized. These duties led him to coordinate military operations in valleys that were still in revolt but were decisive for controlling the roads that led to the Pamirs. Finally, he was in charge of civil and hydraulic engineering projects and was encouraged by the viceroyalty of India to improve travel, development, and irrigation in the region. In short, Algernon Durand had to be everywhere at once.

In 1899 he published a book about his time in India, his travels, and the operations he undertook in northwestern Kashmir.[2] The book, dedicated to "the officers and men British and Native who served at Gilgit," depicts a man who regards his surroundings with very British eyes. In the manner of a traveler, he celebrates the high mountain scenery and the agricultural valleys. Looking northward from the ridge lines, he catches himself dreaming of the greatness of the empires past that controlled Central Asia and stands as the anxious witness to the triumph of Russia over the peoples of the region.

We reached the top of the Dorah, fourteen thousand eight hundred feet in height, without difficulty, riding practically to the top of the pass. We had a lovely day, the air was perfect, and the view from the top is fine. Immediately below you the ground drops very suddenly to the Hauz-i-Dorah, Lake Dufferin as it was christened by Lockhart's party, but the lake itself is out of sight. We stayed a short time at the top, looking out over the Badakshan mountains towards that mysterious Central Asia which attracts by the glamour of its past history, by the veil which shrouds its future. Balkh, Bokhara, Samarkhand, what visions come trooping as their names arise! The armies of Alexander, the hordes of Genghis Khan and Timur go glittering by; dynasties and civilisations rise and fall like the waves of the sea; peace and prosperity again and again go down under the iron hoof of the conqueror; for centuries past death and decay have ruled in the silent heart of Asia. Are we now looking on the re-awakening? Shall the land again blossom like the rose, and proud cities rise on the ruins of the old under semi-European sway, and greater emperors than the Great Khan rule through centuries of generally lasting peace a greater and united empire? Let us hope so. But who that knows her methods can dare to prophesy that Russian rule will necessarily prove better for the people than Mohamedan.[3]

Durand also manifests an almost ethnographic curiosity about the indigenous populations and a barely veiled anxiety about dissensions between the families and clans that control the region. He reports on the military organization he has set in place with Kashmiri troops. In addition, he proudly details a few of the modernization projects he has begun in the region.

In short, he styles himself the guardian of the empire and the promoter of its interests, while taking care to report the specifics of the territory that is the object of his mission: "For four years Warden of the Marches on the northernmost point of our Indian frontier, it was my good fortune, in peace and war, to deal with the most primitive races, to penetrate mountain vastnesses where the foot of a European had never trod, and to wander through the most magnificent scenery that the eye of man has ever looked upon. I trust, therefore, that the following pages may give some idea of what life on the frontier really means."[4] His geographical curiosity, aesthetic sensibility, strategic preoccupations, and technical expertise—all imported from England—are skillfully combined and placed in the service of a multifaceted colonial project.

The same year that Algernon Durand moved to Gilgit, another British soldier arrived in the region. In 1889 Francis Edward Younghusband was not a newcomer to the mountains of Central Asia. In 1886–87 he had organized an expedition that took him from Manchuria through the Gobi

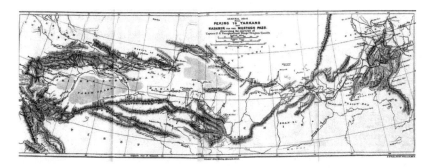

13. Map of Colonel Younghusband's expedition through Central Asia (1886–87). Francis Younghusband, *The Heart of a Continent* (1896).

Desert to Kashmir.[5] He was the first Westerner to take that route, which is bordered by mountains on either side (fig. 13). Younghusband was also the first European to cross the Mustagh Pass, at an elevation of 5,422 meters in the heart of the Karakoram range. He arrived in the Hunza Valley in 1889 in the company of a troop of Gurkhas—Nepalese recruits—with the mission of mapping the valley. He then came up against a Russian expedition, which went on to reconnoiter the upper Kashmir region at a time when the czarist empire still had ambitions of seizing northern India. Younghusband was under Durand's authority at the time, and the two made every effort to keep the Russians from crossing the mountain range. Lord Curzon, viceroy of India, named Younghusband British commissioner to Tibet (1902–4), once again to contend with Russian ambitions. The viceroy later sent him to Tibet to open up the country to the empire's merchants and cartographers. The expedition Younghusband headed in 1904 achieved its ends at the cost of bloody battles.

The Colonial Project and the Standardization of Knowledge

Durand and Younghusband were two figures among others of British influence in the highlands of Central Asia. Both were remarkable personalities, representative of the leaders who shaped the colonial enterprise and made it a reality on the ground. Their lives and their writings show that they shared the fascination of British elites for exploring the world, the high mountains in particular, and a sincere curiosity about the populations of these regions. Military men of the highest order, they looked at the world through the eyes of strategists, endeavoring to control the access roads and the terrain itself. In that sense they were the embodiment of colonial efficiency in that part of Asia. More generally, however, they carried with them a culture, at once geographical, cartographical, and ethnographic, that produced a large volume of empirical knowledge and worked to place it within a structured and standardized general plan.

Scholarly societies were grateful: Younghusband was a member of the Royal Geographical Society from 1888 on, becoming its president in 1919, a few years after Lord Curzon himself. At the time, the Royal Geographical Society was the most meticulous and benevolent observer possible of British expansion and of the growth of scientific knowledge, which according to that group went hand in hand. At the turn of the century Durand and Younghusband were also among the small number of founders of the Central Asian Society. The aim of the society, created in 1901, was to gather, discuss, and diffuse the scientific knowledge acquired by explorers, cartographers, and administrators active in the region.

That objective alliance between scientists, especially geographers and cartographers, and colonizers was apparent in many contexts in the eighteenth century, and even more in the nineteenth, in every part of the British Empire and in competing empires, both in the mountain regions and elsewhere.[6] Gone was the generation of explorers, such as Humboldt in the Andes, who claimed to stand apart from any form of imperial enterprise, or Captains Cook, Bougainville, and Surville in the Pacific, who approached new lands without really anticipating any occupation of them.[7] Soldiers and sailors adopted systematic, scientific methods of collecting observations, now with the mission of implementing colonization.

Henceforth the colonizers evaluated the characteristics of the explored regions from the standpoint not only of their capacity to increase general knowledge, as in Cook's and Humboldt's time, but also of the possibilities for development, in order to meet the needs of the metropolis.[8] At the time, colonization meant projecting onto the coveted lands a type of spatiality (characteristic of the modern map), a type of geographical rationality, and a finality that capitalized on that comparability of knowledge acquired on a global scale. "The love of geography stirs colonial ambitions," wrote Leroy-Beaulieu in 1888.[9] The spatial order imposed by the production of knowledge and by administrative standardization in the colonies, the "colonial enframing" Timothy Mitchell speaks of with reference to Egypt, made it possible to present the conquered regions as "a book to be read" by those who had mastered the language.[10] Once again knowledge and power were intertwined, this time in an imperial version. The conception of the mountain conveyed by geographers, cartographers, the military, and the administration was part of the organization of knowledge and of the imperial enterprise.

That does not mean, however, that every form of knowledge and every form of action in colonized regions was the result pure and simple of duplicating knowledge and know-how acquired in Europe. Exploration, cartography, and scientific analysis of distant lands brought its own share of knowledge and sometimes gave rise to modifications in scientific arguments about the mountain and other objects of knowledge. Indeed, the production of scientific knowledge was largely influenced by the places where it emerged and by the adaptive necessities of the colonial enterprise, in whose service it was readily placed.[11] In the situations we present here, therefore, we need to distinguish between standardization on the basis of Western models and contextual adjustments.

Across India as a whole, the first modern maps of that part of the world played a defining role in the construction of representations. Maps attached much more importance to the geometrical position of the cities and coasts, the principal rivers, and the large mountain ranges than to the complex political and ethnic arrangements, still little known at that time. They therefore suggested a homogenized image of the desired space. In placing the Deccan Plateau in the center of the image, locating the two gulfs of the Indian Ocean in the bottom corners of the map, and surrounding the northern border with sketchily drawn mountains, these maps frame the imperial project. The principal mountain ranges running northwest to northeast (Balochistan, Hindu Kush, Karakoram, Himalaya, and Arakan, to borrow the names that would become current in the nineteenth century) gradually acquired substantiality on maps in the following decades. They thus served as psychological barriers to the imperial project.[12]

The Survey of India had taken on the task of producing detailed maps. By the 1840s the coasts, the Ganges and Indus valleys, and the Deccan Plateau were all mapped. The task at hand was to map the mountain regions in order to confer substantiality on them. At the end of the century work focused on the high mountains in northern India. On that front as well the Royal Geographical Society was both vigilant and passionate. In 1896 its president, Clements Markham, wrote, referring to the Karakoram, Himalaya, and Tibet: "It is to the desirability of completing the exploration of this mighty range, that I am anxious to turn the attention of geographers."[13] The exercise would keep the Survey of India occupied until the beginning of the following century. It enlisted the assistance of the indigenous people, the *pundits*, to whom the British provided training in triangulation. The Indians could thus work in Tibetan territory, where the British were not allowed before Younghusband's forcible takeover. The training of *pundits* in Western cartographical techniques is one of the most interesting examples of modes of acculturation of the indigenous populations, and the knowledge they acquired became a weapon in the conquest of minds and lands.[14]

The Czarist Empire and the Status
Conferred on the Mountains

The czars were not to be outdone. When the Russia of Peter the Great (1672–1725) decided to shore up and further expand its empire in the direction of Persia, India, and China, it equipped itself with cartographical knowledge, a geographical vision, and a political project in which references to the mountain again played a major role. The czar dispatched many explorers to the east and south, ordered geographical and naturalistic descriptions, and had a large number of maps and atlases drawn up. The philosophes and scientists behind the *Encyclopédie*, who bore witness to the czar's modernity, were grateful.[15] At the same time Peter turned to his own advantage ideas borrowed from many geographers and soldiers, in particular Vasily Nikititch Tatishchev,[16] who believed that Europe ended at the Urals. The czar's objective in adopting these ideas was twofold: first, to place historic Russia solidly within the sphere of European civilization, for which the Urals would constitute the eastern limit; and second, to demarcate a sphere of influence—marked with the seal of difference—located beyond the mountain range and destined to be part of the empire.[17] Siberia emerged from that representation as the principal Asian component of an empire being constituted.[18] The Urals came to be not only a continental frontier but also a transitional zone within a fledgling empire.

For a long time the czars vacillated, seeing the high mountain ranges to the south and southeast (the Caucasus, the Pamir, and the Hindu Kush) sometimes as the natural boundaries of their empire, sometimes as so many spans to be crossed on the way to Persia, Turkey, and India.[19] Russian soldiers traversed the Caucasus in the hope of acquiring access to the "southern seas," which the empire had long coveted. In the Hindu Kush, by contrast, their many close encounters with soldiers of the British Empire, especially with Younghusband's and Durand's men on the borders of Kashmir in the 1890s, along with the determination of the viceroys of India, dissuaded the Russians from moving beyond the main ridges.

Desirous of spreading out across an entire continent, the Russians came up against a world of mountains whose like they had never known in their homeland. Russia thus found itself obliged to construct a geopolitical vision that would give these mountains their due. Between the 1830s and 1910 the czarist regime had at its disposal a collection of scientific works from which it could draw inspiration. The czar had sent Humboldt on a mission extending from the Urals to Central Asia in the hope that he would find the metal ore necessary for the country's industrialization. He also wanted Humboldt to provide him with a relief map of the southern and southeastern part of the empire (see chapter 1).[20] Humboldt did not merely meet the czar's expectations: based on his observations, he deduced an overall conception of the entire region. He was the first to use the term "Eurasia," indicating thereby that the distinction between Europe and Asia had no basis in nature—even though that claim contradicted the geographical vision of Peter the Great's descendants. He also promoted the idea that the high mountains of "Middle Asia" constituted a region all its own, in terms of which Eurasia as a whole could be conceived and understood. The idea of the centrality of the highlands of Asia had a bright future ahead of it: for Humboldt, that centrality was merely spatial, but for the German geologist Richthofen it was structural. In his view, every part of the continent branched off from that highland bulwark. For Halford Mackinder, the creator of modern geopolitics, the centrality of the Asian highlands was geopolitical. He wrote a famous article in 1904, while a Russophobe campaign was raging in England,[21] in which he suggested that to conquer the continent one would have to gain control of these highlands and the plateaus that bordered them to the north. He called these areas, taken together, the "pivot of history."[22] Shortly before, Younghusband himself had chosen the title *The Heart of the Continent* for the book in which he recounts his journey through Asia.[23]

Between Humboldt's journey and Mackinder's article, Russia had conquered the entire region, reaching the doorstep of Kashmir and Tibet. The empire adopted only a few terms from these comprehensive visions. The concept of Eurasia displeased the czars, since it contradicted the idea that their empire was the only transcontinental nation-state on earth. By contrast, the identification of an Asia made up of highlands, and later of a "pivot of history," suited their imperial vision.[24] It seemed to justify extending Russian power to the roof of the world. But the czars refuted the distinction between highlands and lowlands. For the empire's administrators, the mountains and the steppes were all of a piece, and the czar's hold on them had to be so as well. The czarist empire had every intention of rightfully occupying the mountains located on its fringes and until the early 1900s held on to the hope of making Tibet a vassal state.

The French Empire and the Topography of Indochina

France took a similar approach in Indochina. The French, contained to the east and south by the sea and to the west and north by two countries (Siam and China) that managed to preserve their independence, quickly set about taking over the plains of Annam and the deltas of Cochinchina and Tonkin to the east. They then headed back through the Mekong Valley and subdued Cambodia and Laos to the west.[25] The hills the French discovered between the two axes of their expansion were modest in elevation. Topographically, however, they contrasted sharply with the surrounding plains and were real constraints on movement, thus hampering the efforts of the French to establish a link between the Mekong Valley and the China Sea.

Auguste Pavie headed a mission that explored the entire region over some fifteen years (1879–95), collecting a great many observations and performing a number of topographical surveys. The mission then produced a game plan, refined year after year, that gradually provided an understanding of the topography and hydrography, necessary to establish a settlement. That plan was the first to accompany its maps with a legend whose terms ("mountain," for example) were translated into the principal regional languages (Thai, Khmer, Chinese, and Assamese) but never into the languages used in the highlands.

The three principal colonial powers in Asia thus adopted similar methods with regard to the mountain regions in their zones of expansion—systematic exploration, detailed cartography, and the construction of a general representation of the mountainous terrain—enabling them better to conceptualize what role should be assigned to the mountains in the conquered areas. But though the techniques and types of knowledge enlisted were quite similar, all having been borrowed from European expertise in the matter, the colonial vision projected onto these mountain regions differed from one power to another.

States and Buffer Zones on the Mountainous Fringes of the Indian Empire

The British Empire chose to confer a special status on the mountain regions and populations, which placed them in a subordinate but autonomous position within the empire. The East India Company, followed by the colonial administration, discovered a mosaic of kingdoms more or less subject to the rapidly declining Moghul Empire.[26] All possessed a complex territorial organization: the authorities were secure in their hold over the centers, but the fringes, especially in the mountain regions, were poorly integrated. Anthropologists have invented several concepts to characterize that relationship of subjection of one prince to another and of that territorial configuration, including the vivid term "mandala state" (or simply "mandala").[27] It refers to a state "in the shape of a pyramid, based on a hierarchy of increasingly inclusive political units, one inside the other. . . . Power is strong, nearly absolute, in the central region, and gradually diminishes toward the periphery, becoming in the outermost ring mere hegemony founded on ritual."[28]

In the late eighteenth century the British replaced this model with a colonial territoriality that aspired to be both standardized and differentiated: standardized, in that the colonizers sought to grasp that entire heterogeneous entity with the same instruments, maps especially, and with the same military and administrative systems; and differentiated, in that they also treated the regions characterized as central differently from those called peripheral. The heart of British India was Bengal and the Ganges Valley. It was from that center that the gradual incorporation of the major regions of the peninsula was organized. The mountain regions running northwest to northeast were administered in such as way as to reinforce their imperial function as a barrier. There the British preferred to maintain or institute protectorates, such as Kashmir, which Durand controlled, or buffer states, such as Nepal. These constituted a vast security zone in contact with the competing empires: Russia and also China, whose territorial ambitions regarding the "mountains of the west" increased under the Qing Dynasty and then, after 1911, under the Republic. Territorial arrangements varied in their details from one sector to another.

In the northwest, diplomats and military personnel discussed several options. The diplomats were responsible for constituting the large region between Russian Central Asia, Persia, and the Indus Valley, while the military was assigned to control the political borders adopted. In the late nineteenth century the failure of the British to take control of Afghanistan (during the wars of 1839–42 and 1878–80) led them to adopt a system consisting of three boundary lines with complementary functions.[29] The first line ran to the base of the mountains occupied by the Pashtun and Baloch tribes: up to that limit the colonial authority exercised full power and pursued a policy to develop systematically the colony's resources. The second, called the Durand Line after Mortimer Durand, Foreign Secretary of colonial India between 1884 and 1894 and Algernon Durand's brother, constituted the official boundary of the Indian Empire. The line ran through Pashtun and Baloch territories, following the ridges and passes that the military was able to control. The third line, called the external line, corresponded to the border between Afghanistan and the countries of Persia, Russia, and Tibet. That was the limit of what the British considered to be their sphere of influence. The system made it possible to demarcate a first buffer state, Afghanistan.[30] In 1896 an eccentric choice of borders completed the desired neutralization of the mountain regions: an odd appendix, the Wakhan Corridor, was attached to Afghan territory, guaranteeing that there would be no geographical contact between the Russian and British empires.

As for the Russians, they abandoned the model of buffer states. Their rapid progress on the plains and steppes of Central Asia convinced them that they could fully occupy and Russify these regions. Intent on exercising complete control of the highlands in the center of the continent, they wanted to be in direct contact with Tibet and Afghanistan, their aim being to make inroads into those areas later on. The Wakhan Corridor was not their doing; nor was the buffer zone desired by China, between Chinese Xinjiang and the neighboring republics of Central Asia.[31]

On the northern border east of Kashmir, where Algernon Durand was in control, the British crown pursued its strategy of mountainous buffer states. It guaranteed autonomy to Nepal, Sikkim, and Bhutan, striving thereby to keep the growing Chinese influence at bay. Within that region Tibet was the most uninhabited and the most isolated zone but also the most coveted. Britain's interest in the region can be explained by the position Tibet occupied as a go-between for the three regional powers and also by the manifest ambitions of Russia and China. Thanks to Lord Curzon, viceroy of India, the expedition headed by Francis Younghusband in 1904 put Tibet in a subordinate position, comparable to that of the buffer states located farther to the south. In 1914 a border called the MacMahon Line, named after one of Mortimer Durand's successors, was established between the British Empire and Tibet. It ran from the Karakoram to the border of Burma, passing north of Nepal, Sikkim, and Bhutan and also following the principal ridge line of the Himalayas. But that border was not ratified by China, because of the suzerainty that country claimed over Tibet and all the Tibetan populations, including those located south of the MacMahon Line.

To the northeast, beyond Bhutan, the empire established somewhat vague boundaries for mountain zones (the North-East Frontier Agency, the Bengal Eastern Frontier Regulation, the Chin Hills Regulation, and others) and granted them a special status. As in the northwest, however, the British adopted a multifaceted system. In 1873 they drew an inner line at the point of contact between the Assam Plains and the hills in an attempt to circumscribe the rice fields cultivated by the people of the plains, whom the administration had encouraged to settle there.[32] Beyond it, an outer line marked the limit of the empire's ambitions, between India and Tibet. Farther to the south, all along the Rakhine (Arakan) Mountains, this line fixed the border between India and Burma, both British colonies. Between the inner and outer lines, the hill tribes (the Naga and the Chin, for example), though officially subject peoples, possessed a de facto autonomy that seemed to suit them. The lowland farmers were not permitted to enter that intermediate zone, and the hill tribes were dissuaded from going down to the plains. In 1910 an administrator of the viceroyalty defined these two lines: "Up to the Inner Line we administer in the ordinary way. Between the Inner Line and the Outer Line we only administer politically."[33]

Throughout the nineteenth century the government in Calcutta, unlike czarist Russia, systematized that policy of double or even triple borders based on a dual conception of the mountain. From the outside, the empire took the highest mountain chains as the horizon of its ambitions. There was no dearth of references to the theory of "natural borders" to justify that choice.[34] Inside, one or several lines of internal demarcation and military control (the Durand Line, for example) ran along the base of slopes, a river, or a relatively low ridge line. Between the two, buffer states, protectorates, or de facto autonomous border zones, located in the dense areas of mountains and hills, constituted the protective shield behind which the "pearl of the British Empire" constructed its uniqueness.

Constructing an Adequate Image of the Indigenous Populations

The mountainous fringes of India therefore constituted an area of tribal settlement that the British wished to control but not to administer directly. The distance taken from the mountains can be explained by the images the British had formed of the local populations. Despite the efforts of a few military officers and administrators such as Durand, the populations were little known, and a number of legends circulated about them. Most often the indigenous people were seen as threats. The Pashtuns' raids to the west and those of the Nagas to the east reinforced that belief. Usually, however, the populations seem to have been feared because they were unknown. Once they were studied—or, better, subjugated—the British view of them sometimes changed radically.

So it was with the invasion of Tibet. Many articles and reports from the nineteenth century describe the Tibetans as savage and cruel. And yet, as soon as Tibet was conquered, a proliferation of flattering descriptions evoked a peaceful, devout, and cheerful people. Younghusband contributed to that change in image. After the conquest of Tibet, his greater familiarity with its inhabitants led him to spiritual and mystical experiences, which he spoke of in esoteric works published near the end of his life. The man who had brought Tibet to heel by force of arms was also one of the first to promote the myth of a happy and spiritual people, a myth with a great future ahead of it.[35]

More generally, the colonizers constructed naturalized representations of the indigenous populations, associating them with their mountain surroundings. These representations were undistinguished in both senses of the term: despite a great diversity of populations, the colonial power confined itself to a few uniform images, denying the ethnographic complexity of the sites.[36] It therefore imposed an order on the world that corresponded to the imperialist enterprise, based on a distinction between modernity and premodernity. A preconstructed grid, somewhat like that used for cartography, was applied to the indigenous peoples to suit the colonizers' objectives of subordination.

Despite a few traits in common, however, colonial histories gave rise to representations of the indigenous populations that varied by context. A detour through Indochina, the Caucasus, and the Maghreb will provide ample illustrations.

The Moïs and Mountaineers of Southeast Asia

The French attitude toward the populations of Indochina was noticeably different from that of the English in India. In the first place, French colonization got its foothold in preexisting agrarian states, which were in the hands of Siamese or Indochinese populations living on the outlying deltas and on plains traversed by the major rivers. As in India, the mandala states of the region had little control over the mountainous areas, though these were in close proximity to the center.[37]

In these mountains and high plateaus, explorers and administrators encountered an extraordinary diversity of populations, who had intermarried a great deal and were perceived as very different from the Siamese and Annamite peoples. Jules Harmand, a French physician and explorer, even spoke of the "atomization of nations," since the character of the populations often varied from one village to another.[38] To simplify references to these highlanders, the colonial administration generally borrowed the generic term by which the peoples of the plains and coasts designated mountain dwellers in their own languages. But since the seat of the colonial administration was in Cochinchina, the Assamese designation "Moï" quickly came to be preferred to other equally contemptuous and simplistic terms: "Khas," employed by the Laotians,[39] and "Penongs," used by the Cambodians. The term "Moï" thus appears in the titles of a great number of books and articles written in the late nineteenth and early twentieth centuries.

Yet the originality of French colonization lay in the importance granted to analyses of the region's ethnic diversity, thanks especially to the work of the Pavie Mission. Auguste Pavie himself criticized the simplifications that the Assamese and the Cambodians had introduced and declared his interest in breaking free of them. He wrote that when the Thais, the Assamese, and the Cambodians invaded, each of them "pushed back toward the ridge lines many peoples, either autochthonous or stemming from ancient migrations, whom they considered to be in a state of savagery. Some of them do live at a very inferior level, and they include types indicating Negrito, Malay, and Tibetan origins. These peoples are generally lumped together under the name 'Moïs' by the Annamites, 'Stiengs' or 'Penongs' by the Cambodians, and 'Khas' by the Thais. These designations apply solely to the populations closest to each of them. We translate them all by the term—rarely justified, however—'savages.'"[40] In the maps produced within the context of the Pavie Mission, every effort is made to take these differences into account. The maps mention many different peoples without by any means establishing a strict typology.

The majority of colonizers did not share Pavie's curiosity, however. Instead, they adopted a generic representation of the mountain populations. In the travel literature and in administrative documents these populations are portrayed as either noble savages or degenerate races. Often they are represented as communities to be civilized and as populations resistant to the policies conducted from the lowlands. But whether the images are sympathetic or unsympathetic, they are always used to devise a strategy of control. As a member of the Pavie Mission, himself very curious about "Moï civilization," wrote: "The only means for preventing future insurrections is to hold the mountains."[41] Initially neglected, administrative and military control of the highlands became a top priority for the colonial administration armed with that conception.

Thanks to general overviews, usually written in France, the discourse on the highland populations took a more academic form at the beginning of the twentieth century. Geographers and ethnologists from the metropolis strove to portray the regions and their peoples in a manner applicable to Indochina as a whole or even to all of Southeast Asia. It was then that the

Moï were most readily characterized as "mountaineers." Academics were intent on identifying characteristics shared by places and populations planetwide, and geographers especially wanted to link populations to the different types of topography and climate groups that could be distinguished. The Moï became one mountaineer people among others.

French geography in the interwar years reformulated its analyses within the framework of these categories. In a vast survey of "the Asia of the Monsoons," Jules Sion repeatedly uses the distinction between inhabitants of the plains and mountaineers to describe the settlement of the region and the relationship between groups, now and then adapting a few stereotypes to serve his purpose. He describes the Lolos, located between China and Indochina, as "a vigorous race of mountaineers. Energetic, agile, and well proportioned, they hunt wild boars and bears, and graze their flocks of sheep."[42] Referring to Southeast Asia as a whole, Pierre Gourou speaks of a "geographical divorce between mountains and plains" and contrasts the region's mountains to the "Alpine mountains," where he observed an interpenetration of mountains and plains, in the form of commerce and population movements.[43]

German geographers, by contrast, tended to emphasize the match between ethnolinguistic groups and ecological environments. For the mountains in proximity to Thailand, both Wilhelm Credner and Harald Uhlig observe that the zonation of agrarian systems seems to correspond to that of vegetation: the Thai populations, organized into states, practice rice farming in the lowlands; the Tibetan-Burmese and Hmong communities, who came from the north in the nineteenth century, practice slash-and-burn farming in the highlands (over 1,000 meters); and many different groups practice mixed systems in intermediate zones (300–1,000 meters).[44] European science thus worked to standardize ethnic distinctions through a naturalistic reading.

The Legacy of Indochinese Mountaineer Images

That human type, the mountaineer, which the colonial administration and European science adopted, persisted in several forms in the following decades. English-speaking scientists readily adopted the term "mountain people," even if that meant abandoning earlier designations ("highlanders," "highland-minorities," and "hill-tribes").[45] The U.S. Army even began to use the term when it moved into Indochina in the 1960s (fig. 14). Like France in 1947–54, the United States believed at the time that, in its fight

14. Facing Viet Cong guerilla warfare in the interior of Vietnam, the U.S. Army
(Robert McNamara, U.S. Secretary of Defense, pointing to the refuge zones in
the highlands of Laos) forged a dual conception of the Indochinese mountain:
as a refuge for belligerents, who had to be dislodged, and as a land populated by
ethnic groups likely to be enlisted in the cause. Photo: Marion S. Trikosko.

against the Viet Cong, it could count on the support of the people living in
the mountains.

It was not until the 1950s that a different analysis of the nature and
specificity of the mountain populations emerged among scientists working in the region. Research in political anthropology showed that, over the
long term, the hierarchical societies of rice-growing populations and the
agrarian states in the lowlands had influenced the highland populations
more than previously thought.[46] Studies by historians pointed to the existence of ancient upward-migration movements of populations anxious
to escape nation-states that were quick to subordinate (whether militar-
ily, economically, or by means of religion) social groups of an inferior
status.[47] Some suggested at the time that it was Western science that had
constructed, or even essentialized, the cleavage between mountain populations (highlanders, mountain people, or hill tribes) and populations on
the plains and coasts.[48]

If we are to believe these recent studies, many so-called ethnic groups
originally had the same characteristics as the peoples of the plains and cen-

tral cities and emerged in response to a desire for emancipation: "Hill people [in Southeast Asia] are best understood as runaway, fugitive, maroon
communities who have, over the course of two millennia, been fleeing the
oppressions of state-making projects in the valleys."[49] According to this
view, these communities then rebuilt themselves and individualized themselves on the basis of specific cultural and ecological practices. The distinctions existing between mountain populations and those of the plains may
thus have arisen from a recurrent concern for political differentiation. The
different modes of life can be explained less by natural facts than by political attitudes. That "postcolonialist" turn in historical and ethnographic
analysis invites us to take a second look at these mountaineers of Southeast
Asia. Taking the long view, such research points to the simplifications of
the colonial administrations and of the scientific analyses contemporary
with them. Both were primarily concerned with connecting the ethnolinguistic map and the otherness of the regions' mountain populations to
natural characteristics.

The Russian Empire and the Peoples of the Caucasus

When the Russians arrived in the Caucasus, they came into contact with
a population clearly perceived as original, speaking languages very different from their own and in most cases practicing Islam. The conquest of
the mountain range over several decades, from the late eighteenth to the
mid-nineteenth century, entailed a takeover of physical space but also the
submission of these populations. Under the Ottoman and Persian empires,
in fact, the people of the Caucasus had grown accustomed to a great deal
of political autonomy. The Russian advance southward therefore met with
strong resistance to annexation and Russification.

Even before the war of conquest (1816–64), the czarist regime had
dispatched scientists and soldiers to the location, and they disseminated
the image of savage and bellicose Caucasian populations. During the war
that image was shored up by the fierce resistance of several of the North
Caucasian groups, especially the Chechens. The writings of the explorer Si-
mione Bronevski, very popular in their time, explained the temperament
of these populations in terms of the natural facts, in the tradition of Mon-
tesquieu and Malte-Brun: "The subjugated peoples live in the hot zone, the
conquerors in the high cold countries. Although Caucasus is part of the
moderate zone, because of its mountain landscape it has more cold than
hot climates, from which stemmed the bellicose spirit of the Caucasian
populations."[50] The military leader Valery Zubov also used the nature of

the mountain to explain the "particular propensity" of the Chechens "for thieving and oppression, their zeal for looting and murder, their perfidy, their bellicose spirit, their daring, their stubbornness, their savagery, their boldness, and their unbridled insolence."[51] These determinist theories, forged a century earlier in the Alps and Scotland, endured for a long time in Russian literature and science, both in ethnology and "culturology" and in geography.[52]

Soon after the military conquest of the Caucasus, the czarist empire's strategy consisted of imposing administrative control and Russification on the local populations. But this strategy had little success. In the early years of the Soviet Union, the Caucasian peoples attempted to achieve independence as a short-lived "Mountainous Republic." Soon afterward, the populations of the Caucasus and Transcaucasia were accorded limited recognition through the establishment of the federal system (1924), and some ethnic groups in the USSR were granted a certain autonomy. But the policy of subordination and assimilation quickly resumed. It involved mass displacements of populations, when deemed necessary, to regions north of the Caucasus Mountains and to Kazakhstan, particularly during World War II. These measures resembled the exile of many communities resistant to the Ottoman Empire at the end of the war of conquest or their displacement to piedmont regions, where their populations declined.

The czarist empire and the USSR thus manifested a somewhat similar interest in the Caucasus. Their strategies toward populations from the Caucasus shifted back and forth between two extremes, however: assimilation through acculturation and forced migration on the one hand, and, on the other, recognition of ethnic specificities, sometimes even in the form of political autonomy. Nevertheless, it should be noted that the second of these two strategies was more timid, and usually more erratic, than the first, which served as a guiding principle for nearly two and a half centuries. In Russia's takeover of the Caucasus, the major bifurcations in that state's political history would ultimately have little influence.

The Berbers and the Colonial Project in the Maghreb

France attached just as much importance to the mountains in organizing its colonization of the Maghreb and to the image of the "Berber" in conceiving of the differentiation among the local populations. The term "Berber" had been used since antiquity to designate the people living in the Maghreb before the Arab Muslim invasion. Generations of authors contemporary with the apogee of the major Arab dynasties used it specifically to designate the pre-Arab populations. But these authors, especially the famous historian Ibn Khaldun, author of *History of the Berbers* (1375–79), employed the term more to construct a historical narrative than to shore up a geographical, topographical, and ethnographic reading of the region.

It was the foreign powers, first the Turks and then the French, who tended to characterize the Berbers in terms of the mountains, which Arabization had barely reached at the time. Not all Berberophones were inhabitants of the Maghrebi mountains: some of these communities occupied the Saharan oases and even the middle of the Sahara Desert. But French colonizers, more familiar with the northern regions than with the desert, made a convenient association to structure their reading of the region they were occupying. In any case, since the borders of the French colonies did not lie in the mountain regions, the strategic concerns specific to such locations did not arise.

In Algeria in the 1830s, and slightly later in Morocco, the French state began to encourage scientific expeditions, such as the one Bory de Saint-Vincent headed in Algeria (1839–42). France also invited isolated studies, such as those of Captain Devaux.[53] These missions had the dual objective of reconnoitering the terrain and studying the local societies.[54] They showed particular interest in these mountain populations called Berbers, even before research on them was organized into a quasi-institutional framework known as "Berber Studies."[55]

The analyses conducted by the military, then by geographers and ethnologists, linked the history and mode of life of the Berber populations to their mountain surroundings in two major ways. First, many geographers tried to naturalize the Berbers' characteristics. Some authors took a determinist point of view consistent with the dominant model of the nineteenth century. Henri Busson, for example, endeavored to explain Berber singularity in the region of the Aurès (Algeria) in terms of the geological substratum and the supposed mountain climate of that region. He also sought to explain the diversity of peoples and ways of life in terms of nuances linked to soil quality.[56] Secondly, some authors linked Berber singularity to the mountains by foregrounding the difficulty of reaching these regions, which supposedly explained why the vernacular languages had survived and why the precepts and rules of Islam had not penetrated very deeply. That historicist reading had a few widely known spokesmen, such as Émile Masqueray (1843–1894), who had considerable influence.

Between these two extremes, a few authors, such as Jean Célérier in the case of Morocco, attempted to draw a distinction between nature and history. To explain the Arabization of the peoples of the plains, he differentiated between the psychological influence of the topography and its physiological influence, contrasting, on the one hand, "the groups that allowed themselves to grow soft as a result of the plain's influence, submission to the [royal] Makhzen order, and hence Arabization," and, on the other, the "rebel tribes," Berbers "trained in harsh mountain life," "born warriors" who "lack imagination."[57]

Colonial science thus produced a set of fairly convergent discourses on the mountain character of the Berber peoples. Colonial expositions depicted them in their human and geographical exhibits. But scientists disagreed about the consequences. Colonial institutions, taking their cue from the scientists, adopted various strategies for taking that singularity into account.

In Algeria the proponents of total occupation faced off against those advocating a different colonial strategy: an occupation limited to the plains and the coast, leaving aside the "Berber mountains." Another line of cleavage divided advocates of Berber assimilation from those who favored maintaining a strong autonomy, the regions concerned being less desirable than the so-called Arab lands. The debate arose within the colonial administration, especially the Arab Bureaus created in Algeria in 1833 for the administration of the "natives." But the question was also debated among French Catholic missionaries, in universities, and in political circles within the metropolis.

In Morocco the French state adopted in succession two very different policies. In the early days of the protectorate, Louis Hubert Lyautey, its chief representative, was leery of the "Berber myth," which showed too much consideration for the populations in question. He focused his attention on so-called pacification operations—military missions aimed at imposing French authority in the Atlas and Rif mountains. It was within this context that cartographers in Morocco began to provide detailed representations of the topography. They made widespread mention of the "Berber" tribes and used the term "Djebel" (or "Jebel," or "Jbel") to refer to the most impressive peaks, which constituted points of reference for the colonial army. In 1930 the colonial authorities took a different attitude and recognized Berber specificity. The government at the time promulgated a decree, the Berber Dahir, designed to exempt those of the Berber culture from Islamic law (Sharia) and to modernize the customary law in force.

Nevertheless, neither in Morocco nor in Algeria did the French state encourage the adoption of an administrative division that would separate the two populations. In the effort to optimize colonial control, a pragmatic policy of power dispersal and destructuration generally took precedence.

Conceiving of Otherness within the Colonial Project

In the British Indies and in French Indochina, in the Caucasus and in Russified Central Asia, the mountain was therefore a category widely used to conceptualize and implement the colonial project, which, according to Edward Said, was first and foremost a geographical project.[58] Other studies have shown that the same was true in Mexico, the Andes, and even Indonesia (see chapter 3), though each time the regional context gave rise to specificities of approach.

The mountain is a recurrent figure for conceptualizing two types of configurations simultaneously: the overall topographical layout and the natural surroundings of the colonized region on the one hand, and the diversity of the indigenous peoples on the other. The first configuration was guided by a recurrent curiosity on the part of geographical science; it was also motivated by an interest in establishing the boundaries of the colony and its internal subdivisions and in controlling movement within it. The second arose in response to an ethnographic curiosity, to the need to organize control over the indigenous populations, and to an interest in promoting the best-adapted policies. That process has sometimes been called the "territorialization of political identities."[59]

Although the policies conducted in the metropolis and those conducted in the colonies had many characteristics in common, it was on that last point that they most differed. Mountaineers in Europe and North America, however differently they might be seen, were always thought of as being an integral part of their respective nations. The specificities attributed to them certainly singled them out, but they never resulted in the mountaineers' being exiled from the nation. Public policies were in place to assure national unity if need be. The situation was different in the colonies. In general, the construction of the otherness of the "native" was necessary for the colonial project.[60] It often found a balance between two extremes: a radical difference, which made coexistence and any hope of integration difficult; and too great a proximity, which made maintaining the distance between races and peoples impossible. Missionaries were often fond of using euphemisms to signify differences. The peoples to be Christianized had to be viewed as nonbelievers or pagans, hence different; but it was also believed that they were capable of being educated in the Christian faith.[61] What was true for the "natives" in general also held for those called "mountaineers" in particular. The imperative of acculturation, assimilation, and modernization—part of the project of nationhood in Europe especially—

played less of a role in the colonial project, which simply maintained differences in status and condition or even promoted them.

The Mountain and Independence for the Colonies

When the former colonies became independent, the rules of the game radically changed. The new nation-states, often spurred on by staunch nationalists, had to rethink political territoriality and ethnic distinctions. At the time, the young nations hesitated between perpetuating colonial paradigms that had proven themselves in the past and adopting different ones in the name of the radical change in political context. The negotiations that followed generally led to the adoption of Western conceptions of state territoriality.[62]

In the Maghreb the independent nations of Morocco (1954) and Algeria (1962) denied Berber specificity. Moroccan nationalists had strongly protested the Berber Dahir and, more generally, the "Berber myth," which they saw as an attempt on the colonizer's part to divide Moroccan society. In Algeria the presence in Kabylie of Christian missionaries, who were convinced that the Berbers were less susceptible to Islamic doctrines than the Arabs, was perceived as divisive of Algerian society. Upon independence, therefore, the Berberophones fell under suspicion from nationalist elites, the majority of whom were Arabic speakers. Except in a few intellectual and academic circles, the theory of Berber singularity disappeared from national discourses and from the national imaginary. States promoted a national history that emphasized the millenarian melting pot of peoples. For example, Ahmed Taleb Ibrahimi, a minister in the first independent Algerian government, wrote in 1973: "When you read everything that has been written on the Arabs and the Berbers in Algeria, you realize that a real effort was made to undermine the Algerian people, to divide them. The claim, for example, that the Algerian population is composed of Arabs and Berbers is historically false. The first Arabs who settled Algeria in the seventh century married autochthonous women and produced offspring of mixed blood. . . . As a result, Algeria is not a juxtaposition of Arabs and Berbers but an Arabo-Berber mix of people who, embracing the same faith and adhering to the same system of values, are inspired by their love of the same land."[63] At the time, the Moroccan constitution and the urban elites promoted a conception of the modern nation (*watan*) to replace that of the community of faithful (*ummah*). In the eyes of the elites, the notion of *watan* was better able to foster the idea of national unity.[64]

The countries of Southeast Asia went through the same process, nationalizing representations of the territory and of the populations. But this process took different forms because of the divergent colonial trajectories that had gone before. One country, Thailand, had even managed to avoid being colonized altogether. From the mid-twentieth century on, however, the policies relating to mountain regions and mountain populations converged in several countries. All continued the colonial policy of controlling the mountainous fringes and border zones. But their success varied. Thailand and Vietnam largely achieved that control. Before colonization Vietnam had had a state with wide-ranging cartographical, military, and administrative expertise. Once it became an independent nation, it organized an early and systematic colonization of the central high plateaus and of the mountains in the northern part of the country.[65] It favored the settlement there of members of the Kihn ethnic group, at the expense of the populations in place. And in 1993 it established an institution that paired ethnic issues and development issues: the Committee of Ethnic Minority and Mountain Development.

Burma, Cambodia, and Laos obtained fewer results in these areas, sometimes for lack of means, sometimes for lack of political will. For a time Laos even promoted a policy that ran counter to the territorial strategies of its neighbors. In the 1950s it adopted an integration policy for mountain minorities that entailed identifying them by their agroecological zone. The Lao Lum (Lao of the Valleys), the Lao Thaung (Lao of the Slopes), and the Lao Sung (Lao of the Peaks) together came to compose an ethnic triptych, representations of which appeared in national iconography.[66] That system of organization lasted until 2002, at which time the regime replaced it with classifications that identified about forty different ethnic groups.

That movement to integrate the margins and to assimilate the mountain populations seems to have accelerated in the last twenty years, but with variations from country to country. International borders have opened up, regional economies have become more integrated, and trade has increased. The populations long settled in the mountains have reacted in various ways to growing state pressure in the region: integration or resignation in some cases; armed resistance in others; and migration or emigration for those who, like the Hmong, have been anxious to avoid both assimilation and confrontation.[67]

Finally, the independence of India and Pakistan (1947) and the Communist revolution in China (1949) led to an exacerbation of political tensions in the mountain regions. There are several reasons for this. First, each of these countries has sought to exert full sovereignty over its territory; and in many cases the armies, where nationalist sentiment was often most keen, displayed an astonishing zeal.[68] The inherited borders dating to colonization had to be made explicit, even though they were sometimes fixed arbitrarily or, like the MacMahon Line in the case of China, were never recognized. In Kashmir (between India and Pakistan) and in Aksai Chin (north Ladakh) and the border regions of Assam between India and China, that state of affairs set off a period of conflicts and tensions that have yet to be resolved.[69] As strife has increased, international borders have grown more rigid. And shifts in effective borders and military tactics often entail moving international borders closer to the watersheds of the principal mountain chains. The policy of natural boundaries still has many adherents.

Within their territories, the nation-states of the region have tended to integrate the populations of the mountainous fringes, even though their constitutions (especially when they are federal, as in India) or their cultural policies sometimes promote policies favorable to ethnic diversity. In northeastern India the initial attempts to integrate the margins and to force assimilation of the hill populations failed.[70] Hill people such as the Nagas,[71] intent on defending the autonomy they had acquired under the empire between the Inner Line and the Outer Line, organized a resistance movement that has to some extent paid off. All around Assam, on the border of China and Burma, India is now hemmed in by small federated mountain states—Arunachal Pradesh, Nagaland, Manipur, and Mizoram—that seem to have reinvented the buffer zones of the imperial state.

To the northwest the Durand Line, which in 1947 became an international border between Pakistan and Afghanistan, has been the scene of recurrent confrontations between Pakistan and the Pashtun tribes, which do not recognize the line as it stands. The border problem was reactivated when the Taliban came to power in Afghanistan and constituted an Islamist homeland, before the attacks of September 11, 2001. In December of that year the hunt for Osama bin Laden in the very mountainous region of Tora Bora gave the press an opportunity to dust off the stereotype of the rebellious and cruel mountaineer. In the preceding cases, the nation-states, whether they had always been independent (Thailand) or became independent after colonization (all the others), adopted or preserved a territorial arrangement in which the mountains retained their peripheral and usually marginal status within political territoriality. In a few countries, however, the symbolic and functional centrality of the mountain serves as a recurrent motif of territoriality. Ethiopia is one example. The political heart of that country, which managed to avoid colonization, lies in the center and highlands of the territory, the settlement zone of primarily monotheistic ethnic groups. The populations of the savannas below, around the central core, are majority Muslim and occupy a marginal position in the nation's society and institutions: "The [official] history of Ethiopia could essentially be summed up as the constant struggle of the populations on the high plateaus, possessors of an original civilization, against the disinherited tribes of the surrounding regions, who are eager to invade that paradise. That is why, from the torrid deserts to the temperate summits, the zonation by degrees of increasingly elevated levels of civilization can be observed."[72]

Propaganda and Oropolitics

The colonial powers and the nation-states that emerged from decolonization thus enlisted the mountains in the project of territorial appropriation. By way of conclusion, one mountain practice that proved itself in the West in the nineteenth century merits consideration: the contribution of mountain climbing to colonial techniques and nationalist rhetoric. We have seen that this sport was placed in the service of the idea of the nation at that time, especially in Europe and North America (see chapter 4). Four examples will serve to demonstrate the oropolitics at work in the rest of the world, thanks to the exploration of the high mountains of the globe or to the colonial project itself.

Tingri, 1905: Taking advantage of the forced opening of Tibet by the Younghusband expedition, Captain Cecil Godfrey Rawling took on the mission of reconnoitering and mapping the upper Brahmaputra Basin. This was the first time Western travelers were able move freely on the northern slope of the Himalayas. With his team, Rawling approached within eighty kilometers of Mount Everest and confirmed the measurements taken half a century earlier: yes, Everest truly was the highest peak on earth. It was impossible to approach the summit from Nepal, which was closed to foreigners. From Tibet Rawling began to dream of an ascent of Everest, but his mission did not allow for that detour. He continued on his way and later reported his observations. The information managed to reach Lord Curzon, viceroy of India. In a 1905 letter to Douglas William Freshfield, one of the leading figures in the Royal Geographical Society and the Alpine Club, the viceroy wrote: "It has always seemed to me a reproach that with the second highest mountain in the world for the most part in British territory and with the highest in a neighbouring and friendly state, we, the mountaineers and pioneers par excellence of the universe, make no sustained and scientific attempt to climb to the top of either of them."[73] Lord

Curzon, who stepped down from his post soon afterward, would not succeed in prompting such ascents. In an irony of history, after several British attempts from the 1920s on, it was a New Zealander, Edmund Hillary, and a Nepalese, Tenzing Norgay, who reached the summit first, half a century after the viceroy of India's appeals.

Nanga Parbat, the 1930s: Nanga Parbat is one of the principal summits of the Karakoram, in Kashmir territory. It has a reputation for being one of the most difficult summits to scale of all the high Asian mountains. Many German expeditions, strongly encouraged by the National Socialist government, tried their luck from 1933 on. All failed, leaving behind them the greatest accumulation of mountain-climbing victims in history. The propaganda press called the peak, which defied modern Germany, "the mountain of German destiny" (*Schicksalsberg der Deutschen*). In the 1930s Heinrich Himmler, head of the SS, believed he had located the origins of the Aryan race in Tibet.[74] All of Nazi Germany cultivated the myth of the superman along with him. The elite of Germanic mountain climbers carried that myth to the summit of Nanga Parbat. In another irony of history, however, it was an Austrian, Hermann Buhl, who first reached the summit in 1953. The nationalist spirit that lay behind mountain climbing in the late nineteenth and early twentieth centuries did not find its sole expression in national territory and in the colonies. It was deployed independently of traditional geopolitics, apart from conventional forms of territorial sovereignty, wherever the greatest feats could still be achieved. Annapurna played a comparable role in the French national imaginary, minus the accumulation of victims. The first summit of more than 8,000 meters to be scaled, in 1950, it elevated a few French mountain climbers to the heights of national glory.

Siachen, November 2008: India organized for tourists the Siachen Glacier Trek to one of the largest glaciers in the world.[75] For twenty-four years the Indian and Pakistani armies have been encamped there, somewhere between 5,000 and 7,000 meters in altitude, fighting to control the glacier and the surrounding ridges. The entire region of Kashmir has been the object of dispute between the two countries since 1947. Both rival powers have a particular interest in holding the ridges of the Karakoram, which would allow them to control the Karakoram Pass, one of the major passages leading to Pamir, and would assure contact with one of the disputed border zones between China and India. It is obvious, however, that nationalist rhetoric has taken precedence over strategic interests: "Siachen has acquired a sanctity of its own, which is part folklore, part military legend, part mythology, and a substantial measure of national pride."[76] That is not an isolated case in Central Asia. Peaks located on the border of Arunachal Pradesh—the Nyegyi Kangsang (7,047 meters) and the Kangto (7,090 meters), to use their Indian names—were scaled in the 1990s by expeditions whose particular objective was to remind people of the Indian claims on the region. In mountain climbing as well, the symbolics of the mountain in the West filtered down, even in its details, to the territoriality of what had become independent nation-states.

Huascarán, 1908: The principal summit of the Cordillera Blanca was identified long ago and immediately found a place on Spanish maps. But the cordillera was not systematically explored and mapped in detail until the 1820s. At the time, Peru was independent and welcomed these expeditions. But the Army Geographical Service of Peru turned away from the high elevations of the cordillera, preferring to focus on mapping the low mountain zones because of their economic, social, and political interest. A genuine competition, in sports and in technical expertise, took hold at the time among Western (primarily German, English, American, and Italian) mountain climbers, naturalists, and cartographers.[77] Once again national rhetoric played a role. In yet another irony of history, however, the winner of the competition was Annie Smith Peck. An American, she said she had climbed the Huascarán not in the name of her nation but as a woman: "I decided in my teens that I would do what one woman could do to show that women had as much brains as men and could do things as well if she gave them her undivided attention."[78] The next year this militant for women's rights climbed Mount Coropuna, another Peruvian peak, and planted a flag with the slogan "Votes for Women." The national cause was not the only one that motivated the ascent of the peaks.

SEVEN

Exporting and Acclimatizing Regional Planning Models to the Tropics

Colonial expansion went forward with several simultaneous objectives and in several congruent modes. In the last chapter we presented two such modes that made abundant use of the notion of the mountain. In the first place, the colonial power optimized its occupation and control of the territories it had claimed by identifying and mapping all the landforms that served as natural ramparts and obstacles to free movement. Secondly, it circumscribed populations judged singular from the start, using the mountain environment as a social indicator and as a vehicle for naturalizing the peoples encountered there. But colonization also, and perhaps especially, proceeded through the adoption of modes for managing places and developing resources. Here again references to the mountain were common and usually conceived in terms of Western models.

Colonization, whether conceived solely as occupation or also as the exploitation of resources, was thus based on the deployment of forms of geographical knowledge that had originated in the West. That deployment preceded or went hand in hand with the deployment of men, capital, and techniques for development. The colonial powers approached the worlds they had discovered and made their own by means of categories that standardized descriptions and conditioned practices. The "tropical mountains," along with other objects, emerged in the wake of that colonial expansion and territorial appropriation. They were the product of a set of images, facts, resources, and projects that articulated the European model in terms not only of similarities but also of well-understood differences. The nation-states that emerged from decolonization usually retained that highly standardized model to configure the mountains. In this chapter we will make that argument through a journey to ten or so colonies and young independent states.

Rethinking Peruvian Territory from the Vantage Point of the Andes

The Peruvian Andes provide an early example of colonial and postcolonial configurations of a mountain chain, in this case, the Andean cordillera. They illustrate the three modalities we have just identified: territorial control, differentiation of peoples, and specification of modes for exploiting resources. Pre-Columbian societies had made the most of their environment by privileging complex forms of spatial organization. These often depended on the complementary exploitation of different regions, located at different elevations and with different orientations and climates.[1] That practice required seasonal migrations and political control of the entire area.[2] The term "vertical archipelago" has sometimes been used to describe this original system.[3] Researchers seeking to understand the location and repetition of these arrangements later portrayed them as specifically Andean, but the indigenous people did not need to conceptualize them in that way. Since all the regions were inside the cordillera, it was superfluous to identify the Andes in characterizing how the regions were related to one another.

In Peru as in the rest of Latin America, the Spanish had adopted a different conception of their environment. They had demarcated the cordillera from the coast as a totality divided into zones on the basis of average temperature (*tierra caliente, fría, helada,* and *templada*). But in colonial geographical descriptions prior to the eighteenth century, the term *cordillera* designated a singular entity that few authors had sought to link to a general knowledge of the "mountain." Furthermore, the Spanish vocabulary then in use did not really lend itself to that type of exercise. *Montaña* could designate a modest isolated hill, a forest, or, like the word *sierra,* a chain of peaks within or outside the cordillera. Autochthonous geographies and colonial geographies had thus adopted different conceptions of the landforms and environments, and neither sought to situate the knowledge acquired within a system of universal knowledge.

That dual territorial conception changed when the country became independent in 1821, at which time a so-called republican geography reconfigured Peruvian territory. Primarily the work of Peruvians, that geography expanded the scientific gains made by Westerners' analyses in the late eighteenth and early nineteenth centuries, especially the work of Humboldt (see chapter 1).[4] It structured Peruvian territory as three large topographical entities: the coast (*costa*), the highlands (*sierra*), and the Amazonian forest (*selva*). Textbooks adopted and popularized that ternary differentiation but simplified its representations.

Peruvian geographers, following the dominant practices in Europe and North America, established an elevation threshold to mark out the *sierra*. They fixed a modest cutoff (2,000 or 2,500 meters, depending on the case) so as to circumscribe a single topographical unit. If a higher elevation had been chosen (3,000 meters, for example), the unity of the *sierra* would have been undermined, since the northern third of the chain has some sectors that are lower in elevation.[5]

Republican geography linked these compact entities to the Peruvian populations in an effort to understand them. On the basis of the determinist arguments prevailing in the academic literature of that time, the indigenous peoples tended to be assimilated to the Andean landscapes and environments: "The Indians became the people of the highlands, the highlands the place of the Indians."[6] The physical and cultural characteristics of these populations were presented as the consequence of their natural environment. The indigenous peoples of the coastal region and of the Amazonian forest sometimes either disappeared from geographical descriptions and public documents or were reclassified as "primitive tribes" or "aborigines" to differentiate them clearly from the Andean populations.

That regional repartition of Peru and the assimilation of the indigenous peoples to the cordillera also made it possible to spatialize a discourse on modernity and progress. In the late nineteenth century, the Andes began to be described as an obstacle to the spread of modernity and to the rational development of the territory.[7] In that context the Peruvian regime often portrayed the indigenous people as a traditionalist pole of resistance.[8] This conception was common in academic studies throughout the twentieth century:[9] "Peru, two countries that bear the same name. . . . On one side, the slender coastal ribbon, a desert zone with its industrial oases and its cities, with populations of six million; on the other side, in the highlands of the cordillera, as many Indians continue to live nearly as they did a thousand years ago. Two separate worlds, two countries that know nothing of each other."[10] The term *indios*, having first designated all the autochthonous peoples, then primarily the Andean populations, later came to refer specifically to the highland populations, considered more traditional and closer to the state of nature.[11]

Throughout the nineteenth century, therefore, the Peruvian *sierra* was simultaneously objectified by a set of scientific practices and problematized in accordance with a modernist and utilitarian vision that the public authorities adopted for their national territory as a whole. The *sierra*—an "Indian" realm, a problem region, and an obstacle to development—was early on demarcated as a sphere of intervention that required specific policies and measures. This was particularly true during the agrarian reforms that began in 1950. The reform of 1969 had the objective of integrating the Andean populations into national society as a whole, linking them to the country's development, and reducing social inequalities.[12] It consisted of dissolving the large colonial estates, established in the early days of colonization, and restructuring small properties, judged inadequately productive, into cooperatives. The reform, however, did not intend to revive modes of farming that combined various ecological zones, a "millennial complementarity" that was considered "backward or inefficient."[13]

In 1981 the Peruvian government expanded its development effort with a program conceived on the scale of the *sierra*, which strengthened the identity of that region: the Programa Nacional de Manejo de Cuencas Hidrográficas y Conservación de Suelos, or PRONAMACHCS (National Program for Managing Water Resources and Conserving the Soil).

Acclimatization: Science and Practices

The Europeans transposed to the Andes conceptions of the mountain and of modernity borrowed from Western culture, but they also discovered plants and animals that they then took back to Europe, based on the same assumption of an equivalence among the world's mountains. In particular, several generations of travelers and colonial administrators, struck by the ruggedness of the llama and alpaca and the quality of their wool, became convinced that it would be helpful to export these animals to the mountains of Europe. Several attempts were made in Spain from the eighteenth century on. Buffon and Bernardin de Saint-Pierre wanted them for the Alps and the Pyrenees. The concept of acclimatization, which Daubenton invented in the mid-eighteenth century, gave a scientific form to that intuition.

It was in the nineteenth century, however, that interest in acclimatization took on greater scope, as a result of debates, scientific initiatives, and the effort to rationalize the colonial enterprise. Contradictory theories about the evolution of species motivated scientists to conduct a multitude of experiments. The proponents of Lamarck's transformism, numerous in France, thought that the acclimatization of exotic species would serve to prove the anatomical and physiological adaptation of species and the hereditary transmission of characteristics acquired in contact with the natural environment. Botanical gardens multiplied, becoming laboratories dedicated to that pursuit (fig. 15). In "Alpine gardens" especially, efforts were made to reconstitute biotopes and plant communities from different

Le Rocher des Mouflons.

15. The Mouflon Rock at the Jardin d'Acclimatation in Paris in 1905. In 1860 the Société Zoologique d'Acclimatation opened a "Jardin d'Acclimatation" at the entrance to the Bois de Boulogne in Paris. Penned up around an artificial mountain were wild sheep and Angora goats imported from Algeria. After llamas, species of sheep and goats—particularly what were called "rustic mountain" species—were those that most interested specialists in acclimatization in the early nineteenth century. In the early twentieth century the Jardin d'Acclimatation had a "Mouflon Rock," drawn here for a children's book. Bibliothèque Nationale de France.

mountain regions. In the introduction to a book he published in 1844 to promote acclimatization practices, Sabin Berthelot, former director of the botanical garden of La Orotava in the Canary Islands, used the mountain to illustrate his point: "Many plants that grow in the mountainous region of the torrid zone can be acclimatized to the temperate zone, it being understood that a gradual lowering of the temperature . . . subjects them to a cooler climate, as they are moved higher up the mountain."[14] This practice was institutionalized shortly thereafter. In 1854 Isidore Geoffroy Saint-Hilaire, professor at the Muséum National d'Histoire Naturelle in Paris, founded the Société Zoologique d'Acclimatation, the first of several dozen organizations created in the major European capitals and in the French and British colonies.[15]

The success of these organizations in the colonies can be explained less by scientific curiosity than by an interest in developing the export crops the metropolis needed. In Algeria, for example, theories of acclimatization raised hopes for a time that sugarcane could be grown on the plains and coffee in the hills.[16] These associations were also enlisted to rationalize stock-breeding techniques in the mountains of the Maghreb at a time when such practices were also being radically modified in the metropolis (see chapter 5). Auguste Hardy, director of the Jardin d'Acclimatation d'Alger (known as the Jardin d'Essai [testing garden]) in the mid-nineteenth century, wrote that "the whole of colonization is a vast deed of acclimatization."[17]

The Indictment of Slash-and-Burn Agriculture

Colonization, especially in its dealings with the mountains, was thus guided by two complementary attitudes: the stigmatization of traditional agriculture, as in the Peruvian Andes; and the promotion of an enlightened scientific agriculture, especially through theories of acclimatization. That dual attitude was at work in colonial policies on forest management, particularly in the mountains.

In the highlands of Africa and Southeast Asia, European administrations were confronted with a technique unknown to them: slash-and-burn agriculture. Unable to understand its logic or its effectiveness in terms of yield and the maintenance of the ecological balance, they went on the attack, conducting policies whose aim was to circumscribe or indeed eradicate it.

The colonial administration's repeated assaults on slash-and-burn production can be explained in several ways. The administration saw it as a squandering of timber resources and consequently as a factor in erosion. It was also wary of an agricultural practice whose methods and revenues were difficult to control. The plots of land were sometimes called "runaways,"[18] like those who cultivated them, thus emphasizing that both eluded colonial control. Finally, the colonizers, accustomed to perfectly laid-out fields, perceived the lands subject to slash-and-burn techniques as a landscape disaster.[19] In short, slash-and-burn agriculture was condemned because it did not give rise to any social or economic form, or any landscape, with which the Europeans were familiar. It was not practiced by a sedentary class of farmers, it produced an indistinct and constantly shifting landscape, and it was not consistent with the notion of a productive or protected forest.

That lack of comprehension led the administration to fight the practice of slash-and-burn farming and to adopt more or less sophisticated substitute policies. On the peaks of Madagascar, for example, the French waged war early in the twentieth century against the agricultural practice of occasionally clearing areas of the forest and grazing livestock there. In a style attesting to the administration's contempt, they denounced in passing these populations "who wander through the forest, ignoring and fleeing any organization into villages . . . drawing from the forest, where they live like wild animals, only the resources it produces."[20] At the same time, France declared its desire to create large forests. It replaced the chaos of traditional practices and the indetermination of modes of appropriation with rational production and land ownership, but also with a topographical knowledge constructed to make the environment and the resources it contained "legible."[21]

The Netherlands proceeded similarly in its East Indian colonies throughout the twentieth century.[22] In the Dutch Celebes, for example, the colonial administration, which had been established for decades in the coastal areas, was at first primarily interested in guaranteeing the supply of rice to its garrisons. Therefore, it initially confined itself to dealing with farmers from the highlands. The administration then undertook to rationalize development of the interior of the island and in particular to modernize the techniques of agricultural production on the slopes. Next, the colonizers sought to take over existing production and develop specific export crops, taking advantage of the different environments whose exploitation they sought to optimize. In the 1830s the metropolis began to promote wet rice cultivation (*sawah*) on the valley floor, sugarcane on the plains, and coffee plants (introduced to the island in 1822) in the highlands, where the climatic conditions appeared optimal. Here as well theories of acclimatization suggested the idea that coffee plants could prosper in all the wet mountain regions of the tropics.

At about the same time, the traditional systems of production were called backward or obsolete, especially slash-and-burn farming and the practice of letting fields lie fallow. In a book written in French and intended to make the work of colonizers in the region better known in Europe, a regular visitor to the Celebes wrote: "Celebes Island is very fertile, particularly the regencies of Maros and Sopeng. But in general, the farming industry there is still very backward, since the population is very attached to its ancient customs and superstitions. . . . The inclinations of a naturally bellicose and very energetic people lead them to prefer sailing, fishing, and trade to the quieter labors of agriculture."[23] The agrarian law of 1870 gave colonizers a free hand in the matter. Citing the threat of soil erosion from field burning, the law authorized the colonial appropriation of the mountain forests as a way to preserve them. At the same time, the colonial authorities, in order both to have the needed labor force at hand and to better control it, set in place a policy of outmigration. The populations from the highlands were forcibly moved to the valleys, which up until then they had carefully avoided to escape endemic illnesses.

As in most colonies at the time, especially British India and French Indochina (see chapter 6), Dutch colonization in the Celebes (now Sulawesi) began on the coast and moved toward the highlands. By taking over the mountains, the Dutch were able not only to acquire total control of the territory but also, and above all, to develop the complementary resources that these high elevations could provide to European businesses. A naturalistic reading of the highlands by explorers and administrators preceded the colonization of these areas. High elevations were often apprehended as a morphological totality, in terms of both the ecological zones that conditioned modes of agricultural development and the obstacles to travel that the slopes represented. Colonization, and the early globalization that accompanied it, usually proceeded by assessing local conditions in accordance with European codes and terminology and by transferring management methods. Some of these methods had been tested in Europe, while others were adopted in the early days of European expansion. In both cases, the notion of the mountain played a key role.

Production Forests and Forest Policies

The condemnation of slash-and-burn agriculture led to the promotion of alternative modes of mountain management in light of scientific knowledge. The first of these, which we have just seen at work in Sulawesi, followed the model of rational export agriculture. The second, to which we now turn, stemmed from the model of the production forest.

With colonization came a greatly increased demand for timber. To meet the needs of the metropolis, operations focused initially on forest species to supply the market with timber via the major specialized ports of Europe, particularly Hamburg. In addition, goods were needed to provide infrastructure in the colonized countries. In India, for example, the phenomenal demand for railroad ties as the rail system was being built put pressure on the forests. There as well, commercial exploitation went hand in hand with a desire for tax revenues. Precolonial institutions had usually taxed agricultural products but were much less likely to levy taxes on forest products.[24] The colonial administrations had greater expectations and increased their control of production.

In the first decades of colonization, the capacity of the tropical forests to satisfy the need for timber was judged to be considerable, in the mountains as elsewhere. In colonial and economic circles, the idea spread of an abundant natural world and of "the myth of the inexhaustibility of resources."[25] Those who took on the task of popularizing the colonial adventure promoted that image. As a result, many concessions were granted to private companies, which adopted modes of exploitation that showed little concern for the renewal of resources.

But this time scientists did not generally share that image and that myth. Very early on they became aware of the fragility of the tropical forest and its soil and of the risk of depleting them. Their writings often condemn traditional practices and forest concessions simultaneously. The engineer Louis Lavauden, for example, in a 1935 book devoted to the forests of the French empire, denounces slash-and-burn agriculture and the grazing techniques practiced by the autochthonous peoples, who are called "permanent enemies of the forest." A little further on he also rails against the indiscriminate logging techniques adopted in the large forest lands.[26]

The colonial administrators sought to ensure production over the long term. To that end, in 1864 the British set up a forest administration designed to optimize management of the forests in the Western Ghats and the central part of the Western Himalayas.[27] The region had previously had a great variety of management modes that were sensitive to the different contexts. The forest administration made a clean sweep, adopting a conception that claimed to be scientific and transferable from one environment to another. It borrowed from the natural sciences a knowledge of universal scope that could be used to establish what the administration called "scientific forestry" or "rational forestry" to replace practices called "unscientific, unconcerned with production imperatives, or irrational."[28] In the Indian subcontinent as a whole and in Southeast Asia, the scientific and operational category of the forest came into being, attesting to an interest in standardizing knowledge and management on a vast scale.[29]

In many cases that scientific knowledge and technology had been forged in contact with European forests and by virtue of forest policies set in place in the second half of the nineteenth century, in response to flooding (see chapter 4). That was not the case, however, for the British colonies, since the British Isles had not experienced a comparable need. Some even expressed surprise that England, at times called "neglectful" of its own forests, could develop such sophisticated forest policies in its colonies.[30] By contrast, when a colonial power undertook ambitious policies in its own territory, it often transferred to its colonies the scientific models and duplicated modes of intervention. Such was the case for France and Germany and also, later on, for the United States. U.S. forest policies, greatly influenced by the German and French model, were in turn exported, especially after World War II.[31] It has been said that the Western training of foresters "largely predetermin[ed] the diagnosis as well as the solutions implemented."[32]

Exporting the French Forestry Model to the Colonies

Early on France exported its farming techniques and forest policies to its colonies. In Algeria a forest service was created on the metropolitan model in 1877.[33] It duplicated a technical management model that, according to its promoters, was beginning to prove itself in the metropolis. This was also a way to spread the European model of civilization. M. Trottier, an agronomist posted in Algiers, wrote in 1878: "The gradual destruction of a forest can change forever the character of a region and of its inhabitants. . . . The connections established between the forest, nations, and civilization are obvious here. . . . It is through reforestation that our race will conserve its European faculties."[34] The forest service's work was supported by the Ligue du Reboisement de l'Algérie (League of Algerian Reforestation) and by an ambitious reforestation program adopted in the 1880s.

The forest service favored restricting access to the mountain forests (see chapter 4), which the administration judged gravely threatened by the intensive grazing practices of livestock breeders. In fact, the administration primarily wished to guarantee good conditions for developing the interior plains, which it handed over to *colons* (French settlers). It was therefore imperative to reduce the risk of flooding, regulate the water supply, and make available for colonial operations a labor force, considered to be supernumerary, brought in from the mountain villages. France conducted a similar policy in Morocco.[35] Independent Algeria and Morocco would later adopt this policy, which promoted agriculture on the plains and the regulation of water systems. Both countries nationalized forest policies, usually dedicated to reforestation and control of grazing practices.[36]

The French forest model was adopted somewhat later in sub-Saharan Africa and Madagascar. The forest administration was set in place there only after imperial borders were established (at the Berlin Congress in 1884) and territorial sovereignty achieved (in the early twentieth century). The forest service would continue to fight traditional practices of clearing and grazing as well as the farming techniques of the large colonial concessions.

With respect to the high elevations colonized by the French (the interior of Madagascar, Cameroon, and Fouta Djallon), a catastrophist discourse surfaced early on, a variant of the discourse the forest administration had forged in the Alps and Pyrenees. Once again traditional practices in the colonies were universally condemned. But authors usually argued as well that the topography and the climate exacerbated problems and that the local situations had to be understood in terms of their mountainous character.

In Fouta Djallon, for example, authors in the first half of the twentieth century insisted on the role of the massif as a "water tower" for western Africa as a whole. Influenced by the problematization adopted in the Alps, they accused Fulani shepherds of having burned down the forests to clear high-altitude grazing lands, thereby triggering erosion processes harmful to the entire region.[37] Later research would invalidate that analysis, which scientists had abandoned by the 1980s. For decades, however, the theory would continue to inspire international organizations, development aid, and the Guinean administration, because it was such an intellectually satisfying and convenient justification for reforestation policies and the forced settlement of the nomadic populations: "Whenever a tree burns in Fouta Djallon, the rate of carbon dioxide increases in the atmosphere and Timbuktu is bound to have run out of water by the end of the dry season."[38] In Fouta Djallon and many other regions, the analysis of reality on the ground was entirely subordinated to a geographical rhetoric—the discourse on the mountain—that had been thoroughly tested in the metropolis: "The trait that dominates and governs all the others is that the mountain is king here."[39]

German Forest Science and the Management
of the Tanzanian Mountains

A similar conclusion may be drawn for Germany, the other colonial power in possession of a great tradition of forest policies, this time in East Africa. An aggregation of highlands cooled by airstreams from the Indian Ocean traverses the territory of what is now Tanzania. These highlands enjoy a climate favorable for the development of mature forests, somewhat rare in that region, where savannas predominate.[40]

Many Germanic travelers who ventured into the region compared it to the mountains of Central Europe. The missionary Johann Ludwig Krapf, in 1852 probably the first European to cross the Usambara massif, compared the region to Switzerland and to Germany's Black Forest.[41] His successors readily adopted the comparison for their own uses, referring to the region as the "African Switzerland." In 1888 the Austrian cartographer Oscar Baumman, in describing his own exploration of the Usambaras, divided the region into three natural groups. The primal forest (*Urwald*), he emphasized, had a wealth of flora; the high pasturelands had a harmonious landscape comparable to that of the Austrian Tyrol, which he greatly appreciated; as for the zone inhabited and cultivated by the Waro, he presented it as degraded and overpopulated.[42] Baumann's description and the images associated with it had a determining influence on the colonial imaginary the Germans adopted when they seized the region in the wake of the Berlin Congress. Initially the Germans entrusted the exploitation of the Usambara mountain region to a private company, the Gesellschaft für Deutsche Kolonisation (Society for German Colonization). The colonists found a relatively cool climate and conditions favorable for agriculture, deemed similar to those they had known in Europe. The highlands of British Kenya had a comparable appeal; part of that region even came to be known as the "white highlands." The Germans initially promoted coffee growing in the Usambaras, but the results were disappointing. The apparent luxuriance of the hill forests concealed relatively poor soil, which the coffee plants quickly depleted. The Germans then turned to harvesting forest products.

Having learned from the ancient and widely recognized experience of German forest science, they began by circumscribing traditional practices, which consisted of tree cutting, livestock grazing, and slash-and-burn.[43] Several successive decrees, along with the forestry order of 1904, banned these practices and established the means for repressing them. Between 1906 and 1914 Germany established 231 forest reserves.[44] Behind the declared objectives to protect the environment, they essentially guaranteed the colonial state's supply of wood and water. To that end the colonial administration arranged to turn the existing forest into an organized and profitable cultivated forest. Drawing on the many acclimatization models and techniques that botanists had developed throughout the nineteenth century, the administration introduced plants and animals both from Europe and from other tropical regions. The restriction of access to these cultivated forests, where the local populations had previously grazed their herds, made livestock breeding impossible there.[45]

That mountainous tropical forest turned out to be more complex than anticipated and even resistant to German forestry models. The administration then adapted its instruments and promoted large reserves such as the one on the plateau of the West Usambaras. The lower boundary for the reserve was set at 1,600 meters—that is, between the highest settlements of local communities and the forests located on the peaks, which had strong commercial potential.[46]

This approach, motivated by a naturalistic and productivist conception of the mountain, was at odds with that of the local populations. The mountains, in fact, were the core of their identity. The region of Mount Meru, for example, was populated by Waro who had come there to escape conflicts with the shepherds on the plains. The name they gave themselves attests to the conditions that led to their settlement there: it means "the people who climb" and was supposed to designate all who sought refuge there. Foreign observers, who frequently called these populations "the Meru" after the region's principal peak, were struck by the connection between that people and their environment: "One becomes *Meru* not by sharing a common ancient history, but through arriving to the mountain and being accepted into the existing society. It is the mountain and the historic ties that the Meru have to it that help define them as people."[47] Essentially, that connection was denied by the establishment of the reserve in the western Usambaras, designed to ease the pressure that the local populations were putting on the forests the foresters considered of most interest. The colonial administration denied the local populations' customary rights, declared the lands vacant, and legitimated its own initiatives.

The transfer of colonial authority to England in 1919, followed by Tanzania's independence in 1961, would change little in that mountain configuration or in the management models promoted there. Successive regimes clung to an environmental vision, favoring restricted access to the forests, exclusion of the local populations, and subordination of the mountains to objectives of development and conservation, the rationale for which was determined in the capital cities and in the administrations.[48]

Thailand's Adoption of a Forest Policy
Inspired by the European Model

The circulation of normative models of mountain forest management also reached countries that had never been colonized, such as Siam. In 1896 King Chulalongkorn, having returned from a trip to Europe where he was able to study existing forest policies, decided to create a Royal Forest Department.[49] That initiative sought to combine the objectives of productivity and conservation of the forest cover, primarily for the sake of the hydrological system. Furthermore, it was meant to contribute to the construction of a modern state anxious to increase its control of the entire territory. The creation of the Royal Forest Department made it possible for the Siamese government to do business with foreign forest companies that exported teak and to receive the royalties resulting from their presence.[50] This allowed the government to intervene between local leaders and forest product harvesters.

Later the state also organized the fight against slash-and-burn agriculture on the slopes. Once again this measure was intended to increase state control over the territory as a whole. In 1959 Thailand (as Siam was now called) established a Central Hill Tribe Committee responsible for converting slash-and-burn farming into sedentary agriculture. But behind the ecological arguments advanced, the government had other political objectives: fight opium production, which was prospering in the northern mountains thanks precisely to slash-and-burn techniques; crack down on the Communist insurrection that had taken root in the northern part of the country and was the chief beneficiary of existing modes of production;[51] and promote intermarriage among different ethnic groups so as to build a unified population across the entire country, primarily to the benefit of the Thai ethnic group. In fact, the mountain populations were at first grouped together under wide-ranging and territorializing designations: *chao-khao* in Thai, the equivalent of the English "hill tribe"; or *pa*, which means "forest" and is more explicitly pejorative because it refers to the idea of the wilds and of primitive societies.[52] In Thai these highland peoples, who were systematically accused of poor environmental management and of deforestation, were contrasted with the *muang*, the populations settled on the plains, where the "Thai" ethnic group, engaged primarily in agriculture, dominated. The farm and forest policies conducted in recent decades have encouraged the *muang* to settle on the slopes and have tended to displace the *chao-khao* to the valleys in order to transform the ethnic mosaic into a truly national society.

In their water regulation policies the Thai government adopted the conception prevailing in Europe. According to this view, a mountain forest acts like a sponge, absorbing the surplus during the rainy season and postponing the flow of water until the dry season. In 1964 the Royal Forest Department established forest reserves. In 1989, after catastrophic floods throughout the 1980s, the Thai state decreed a ban on cutting down trees and listed a quarter of the mountain lands as forest conservation zones. A system for classifying the catchment basins, based on the criteria of slope and elevation, determined the land uses possible. The law mandated the maintenance and protection of the forest cover on the highest slopes, even reforestation in some cases. That measure came into conflict with traditional practices, sometimes in densely populated regions. In the 1990s, for example, a third of the villages in the northern provinces were located inside forest reserves. Given its multiple methods and implications, the forest policy in Thailand, under cover of environmental management, can be interpreted as a decisive element in the system of territorial control of the fringes of national territory, intended to "make the uplands more legible and governable."[53]

The Thai example has equivalents in all the rice-growing societies of Southeast Asia. They adopted similar views and attitudes toward their respective highlands, the populations who lived there, and the area's resources. Because of the common geopolitical and territorial fate of these mountain regions—marginalized, marked by the seal of otherness, and often hotbeds of resistance—researchers have sometimes lumped them together under the single term "Zomia."[54]

The Theory of Environmental Degradation in the Himalayas

The best-documented example of a transfer to the colonies of analytical models and forest policies conceived in Europe is also the example that has given rise to the sharpest polemics within the scientific community: the Nepalese Himalayas.

For centuries the kingdom of Nepal encouraged agricultural development in the foothills of the mountain range, controlling uses and taxing activities there. At the same time, it took measures to protect the forests on the plains, known as the Terai, located at the base of the mountains.[55] In the mid-1920s Nepal sought out the expertise of the British India Forest Service in its effort to exploit economically the forest resources of the Terai. It then established its own forest administration, the Kathmahal, which copied the model adopted in India.[56]

In the 1950s Nepal began to accept international aid. The forest question was reformulated at the time and was now focused on the mountains and on the objectives of environmental conservation.[57] The United Nations Food and Agricultural Organization (FAO) appointed a group of experts, headed by Ernest Robbe, who observed serious erosion on the slopes and many landslides. The FAO, using for its own ends the causal model that had been adopted in Europe a century earlier, viewed these phenomena as consequences of deforestation. It therefore blamed the traditional cultural and grazing practices and recommended soil conservation, forest protection measures, and even the reforestation of the hills. Far from condemning the local populations, Robbe's assessment pointed out their powerlessness in the face of the constraints on them and their imprisonment in a vicious cycle of overexploitation and resource degradation. In that respect, the analysis recalled the view of French foresters such as Surell and Briot in the nineteenth century, who had taken pains to understand the role of traditional practices in the erosion of the slopes and to assist the local populations in becoming an integral part of forest restoration projects (see chapter 5).

The government of Nepal opted for radical measures. In 1957 it nationalized the forests, and in 1959 it created a Ministry of Forestry. Nationalization, which was supposed to support the local populations in their forest operations, instead led to a destabilization in local modes of production and management.[58] In addition, the FAO's desire, expressed in its recommendations, to conserve the forest canopy often competed with strategies of commercial exploitation.

In the 1970s a succession of serious floods on the Ganges plains and delta revived the question. Once again these catastrophes were attributed to the deforestation of the highlands, and that deforestation was itself linked to population growth, poverty, and modes of production in the Himalayas.[59] The conceptual model prevailing at the time had been introduced at a conference organized by the Deutsche Gesellschaft für Technische Zusammenarbeit (German Agency for Technical Cooperation). The conference, titled "Development of Mountain Environments," was held in Munich in 1974. Development and cooperation organizations had met with scientists, especially members of UNESCO's Man and Biosphere program devoted to mountains (see chapter 3), and the organizers of the conference persuaded Erik Eckholm, a science reporter for the *New York Times*, to disseminate the information to the public. Eckholm's report, published after the first United Nations Conference on the Human Environment (Stockholm, 1972), which supported the idea that population growth was the cause of environmental degradation, made a considerable impact. In a work with the revealing title *Losing Ground*,[60] he predicted an imminent catastrophe between the Ganges and the Himalayas. In a comprehensive proposal, he clearly examined the model that the experts were inclined to validate at the time. Later critics baptized it "the theory of Himalayan environmental degradation."[61]

That theory is based on a series of causal relations and consists of eight propositions:

1. As a result of an improvement in the health system, the Nepalese population is increasing.
2. That population explosion, exacerbated by emigration from India, is increasing the demand for cultivable land and the pressure on forest resources in the mountains.
3. That pressure leads to deforestation, which sets in motion the process of environmental degradation.
4. Deforestation brings about soil erosion, loss of productivity of the soil, landslides, and the disturbance of the hydrological cycle.
5. These disturbances cause an increase in surface runoff during monsoons and flooding on the plains. Conversely, water resources are diminished during the dry season.
6. The increase in sediment transported by the rivers of the Himalayas affects the region of the Ganges delta and increases the number of alluvial islands in the Bay of Bengal.
7. The loss of agricultural lands to erosion leads to the appropriation of other regions for farming, setting in motion a second cycle of deforestation. Deforestation entails an increase in the work required to procure wood and greater use of cattle dung for fuel.
8. Dung, because it is now used as fuel, can no longer be used as fertilizer. The yield of farmlands is reduced, leading to the conversion of more land to agricultural uses.

The persuasive power of the theory of Himalayan environmental degradation lies primarily in the simplicity of the propositions and their causal relationship to one another. It also resides in the future catastrophe that is announced:[62] the theory predicts an "environmental collapse." In the early 1980s various institutions, including the World Bank, declared that the forests of the Himalayas would be gone by the 2030s.[63] Once again the mountain and the mountaineer served as handy categories for making the diagnosis, articulating the problem, and adopting public policies intended to correct the denounced state of affairs.

That analysis was contradicted by a group of complementary studies beginning in the mid-1970s.[64] Some explained that the greater part of that notorious environmental crisis could be explained by long-term biophysical processes resulting from the very pronounced tectonic activity in the region. Others observed that the major changes in practices that affected soil quality in the second half of the twentieth century had taken place in the foothills, where, compared to the mountain slopes, a higher proportion of trees had been cleared. In addition, these hills were usually cleared by commercial forest operations with no relation to traditional modes of production. Finally, though deforestation practices were documented for the mountain regions, an increase in the volume of the forest canopy had been observed in certain valleys. Researchers also discovered that, in the middle hills of Nepal, farmers themselves triggered landslides to increase the fertility of the soil (fig. 16). Such observations were an invitation to take into account the great diversity of situations in Nepal, particularly in the Himalayas.

In addition to disputing the facts, scientists disagreed about the theory itself. A scientific conference held in 1986 at Lake Mohonk, near New York City, was wholly devoted to the "Himalaya-Ganges Problem." It marked a turning point in the conception of the problem, the facts cited, and the underlying approach to the mountains. The prevailing view among scientists at the conference was that the theory of the degradation of the Himalayan environment was based more on postulates than on scientific evidence. That claim also appears in a book explicitly titled *Uncertainty on a Himalayan Scale*,[65] the report of a study the authors conducted on behalf of the United Nations Environmental Program (UNEP). On the strength of their analysis, they called into doubt "environmental orthodoxy," which according to them was behind the contested propositions. One of the theory's opponents presented it as a mélange of facts and moral considerations, "an amalgam of ostensibly scientific findings, casual observations, assumptions, opinions, and moral imperatives."[66] Others said that the theory's success could be explained by the many advantages it offered the governments and development assistance agencies concerned.[67] In fact, it offers a simple interpretive model for a very complex reality. It also designates a convenient scapegoat: the mountaineers. The political implications are therefore considerable for all the nation-states in the region. On the one hand, countries located downstream and subject to flooding found it easy to assign responsibility to the countries and populations located upstream. On the other, Nepal, in presenting itself as the victim of a situation

16. The authors Bruno Messerli and Jack Ives are among those who disputed the "theory of the degradation of the Himalayan environment" in the 1970s. In a book published in 1989 they explain, using photographs (here, the region of Kakani in Nepal) to support their argument, that landslides are among the characteristics of the natural environment with which farmers in the region have always had to contend. Photo: Jack D. Ives, from Ives and Messerli, *The Himalayan Dilemma: Reconciling Development and Conservation* (London: Routledge, 1989), 88.

it had played no role in producing, could hope to secure development aid from international organizations active in that field.

That theory, which itself borrowed from the explanatory framework forged in the mountains of Europe in the previous century, was exported to other tropical regions supposedly susceptible to comparable environmental crises. Traces of that theory can be found in the forest policies implemented in several regions of the world:[68] in Thailand in the 1980s, in Indonesia, in East Africa, and in the northern Andes. The global scope of the theory attests to the tendency to treat the mountains of the tropics as a relatively homogeneous whole, susceptible to comparable problems and requiring similar policies. It was not until the late twentieth century that forest policies would distance themselves from that type of universal scheme and take seriously the always unique relationship that local populations maintain with their environment. Nepal would not lag behind the others: it has promoted community forest management since the 1990s, a practice that grants particular attention to autochthonous expertise and traditional practices.

Nature and Heritage Protection in the Matopo Hills

The model of national parks and nature reserves in the mountains of colonized regions spread in a manner similar to that of forest policies. Following approaches already taken in the United States and Europe, the colonial powers—and after them, the independent nation-states—promoted the constitution of natural spaces, privileging mountain regions. According to current estimates, between a third and a half of the surface area of protected areas is located in mountain regions.[69] The appeal of mountain regions in the tropics owes a great deal to the work of naturalists and artists in the twentieth century, who touted the lush vegetation and extraordinary beauty of the peaks.

Protection measures often apply to the same spaces that were previously protected as hunting or forest reserves, where traditional practices had already been restricted or prohibited. These spaces were thus readily redefined as emblems of virgin nature. Mount Meru provides a good illustration.[70] The creation of such protected areas has often been justified by a critique of the community management modes the local populations had adopted, sometimes accompanied by forced settlement policies targeting the nomadic peoples.[71]

The national park in the Matopo Hills of Zimbabwe is a particularly good illustration of that process of reassessing a mountain region.[72] The first descriptions of the region attest to the interest of British colonists in the site's geological curiosities and in the forest potential of the highlands. They also indicate a lack of interest in the relationships the local populations maintained with their surroundings. In 1896 an insurrection of the autochthonous populations, called the Ndebele, forced the colonial administration to take seriously the presence of the indigenous peoples and to arbitrate between the various interests and conceptions in place.

Cecil Rhodes, founder of the British colony, chose to negotiate with the

Ndebele regarding the status of the highlands. He acknowledged the sacred value that they attached to the site and arranged for protection of the highest elevations. In so doing he also offered Europeans who loved the wilderness a "white playground" and pleased scientists who had pointed out the interest of the region.[73] Conversely, in placing himself on the side of the scientists and the Ndebele, Rhodes frustrated the desires of a portion of colonial society, which was seeking to exploit the site.

After Rhodes's death in 1902, however, the protection policy governing the site changed. The colonial administration decided to create a nature reserve. Geological formations and archaeological relics were recognized for their heritage value, and the uses the autochthonous peoples made of the region at the time were judged harmful to its forest and to its scenic landscapes. The rights of use of the local populations, though initially preserved, were abolished with the creation of a national park in 1926, and the populations were completely excluded from the park in 1952.

The case of Matopo Hills demonstrates the preeminence of a naturalistic and environmentalist conception in assessments of the qualities of a mountain site and in the adoption of a protection policy. Ultimately that policy turns out to be indifferent to the interests of the autochthonous populations.[74] That was the most common scenario in the colonies throughout the twentieth century. The actions of the French colonial administration in that regard are less well-known than those of the British. But the French administration too expressed a concern with protecting mountain sites to preserve their scenic quality and defend them against the threats that indigenous practices posed for wildlife. Hence the Water and Forests Office in Algeria arranged for the creation of a dozen national parks in the regions of Algiers and Constantine, including several in Kabylie and the Djurdjura.[75] A symposium held in the mid-1930s compared initiatives in the various French colonies and contributed to forging a doctrine that gave a real place in colonial forest policies not only to reforestation but also to species preservation.[76]

Reconciling Nature Protection and Traditional Modes of Life in the Cordillera Blanca

Nevertheless, a few experiments that run counter to the dominant practices in the French and British empires view conservation of the environment and the maintenance of traditional activities as complementary objectives. The Cordillera Blanca in Peru provides such an example.

The highest Andean chain in Peru has been the site of recurrent catastrophes caused by seismic activity and the draining of glacial lakes. A huge flood caused five thousand deaths in the city of Huaraz in 1941, an event that caught the attention of the Peruvian administration. In 1951 it established a commission charged with managing the environment and the development of the region: the Comisión de Control de las Lagunas de la Cordillera Blanca (CCLCB).[77] The tasks of protecting the valley populations—those most vulnerable to natural disasters—and of assisting the highland populations in their development were integrated into a policy to develop hydroelectric potential, water resources, and tourism in the region.

Beginning in 1942 scientists aware of the value of the sites advocated the development of tourism in the cordillera. In the Grupo Andinista Cordillera Blanca, an alpine club of sorts, the voices of scientists counted a great deal. That group promoted both the development of tourism in the region and the creation of a national park.[78] The idea, championed by a senator from the region, gave rise to a first series of restrictions (on felling trees and hunting native species). Then, in 1967, the Peruvian government commissioned a feasibility study, which was carried out by two members of the U.S. Peace Corps. They proposed listing 85,000 hectares around the peaks as protected areas. Protection measures focused primarily on the high mountain Andean landscapes; on a rare plant, the *Puya raimondii*; and on an animal species, the vicuña, which has had a strong emblematic value ever since Simón Bolívar ordered its protection in 1825. Hunting of the vicuña was banned in 1940 and labeled a "crime against the national heritage."[79] Huascarán National Park was established in 1975.

The park covers the entire Cordillera Blanca, with the exception of the northern tip of the range, and protects the most spectacular landscapes and the most famous peaks.[80] The objectives declared at the time combined the classic concern for environmental conservation (protection of wildlife and the establishment of a geological and scenic heritage site) with an interest in valorizing the cultural and archaeological inheritance. Due attention was also given to the aspirations of the local populations.[81] With that aim in mind, the Peruvian government took several actions. First it used the 1970 earthquake as a pretext to implement the agrarian reforms enacted in 1969. Then it acquired the large properties located in the park zone. On the recommendation of soil productivity studies, the government made sure that this zone covered only lands above 4,000 meters, the upper limit for farming in that region. It thereby sought to limit as much as possible encroachment on the areas farmed by the local populations. Since the cattle rangelands extended above the 4,000-meter cutoff, grazing practices were allowed within the park itself. The government also made efforts to provide financial assistance as compensation for the ban it had imposed on a few traditional practices.

Huascarán National Park is an unusual case in that it sought compatibility among several uses and management notions in the Andean high mountains. At first the project clearly adopted the model of national parks in the United States. Over time, however, national policies were designed to differentiate among elevations and to make standards of use more flexible, allowing traditional Andean activities to continue while at the same time introducing a tourism economy.

The Case of Sagarmatha

Huascarán is not an isolated example. A similar concern for combining environmental protection and the development of local societies can be found in the process leading up to the creation in 1976 of Nepal's Sagarmatha National Park. The interest in protecting the site can be partly explained by the prestige Mount Everest acquired once Rawling's measurements showed it was the highest peak on earth (see chapter 3), and after Hillary and Tenzing scaled it in 1953. In addition, the Himalayan region appears to correspond to the model of high-mountain scenic landscapes promoted in Europe and North America—by means of protection measures, for example. That perception contributed to the rise of tourism in the region and later on to its designation as a UNESCO World Heritage Site for its natural landscapes: "Sagarmatha is an exceptional area with dramatic mountains, glaciers and deep valleys, dominated by Mount Everest, the highest peak in the world (8,848 m). Several rare species, such as the snow leopard and the lesser panda, are found in the park."[82] The invocation of the wilderness paradigm allowed that characterization of the high mountain to circulate. The protection zone therefore lies primarily on the peaks: 69 percent of the territory of Sagarmatha National Park is composed of uncultivated lands located above 5,000 meters, another 28 percent of alpine meadows, and only 3 percent of forest lands.[83] Nevertheless, the park also includes part of Khumbu Valley, populated by the Sherpa. Although they earn money as porters and tourist guides, the Sherpa make their living primarily from agriculture, livestock breeding, and timber resources.

In the park's early days the administration, zealous and trained in Western ways, controlled and restricted the traditional uses of natural resources.[84] At the time, theories about the deterioration of the Himalayan environment and the mountain populations' responsibility for that deterioration prevailed. The brush and shrub cover that extended above the villages of the Khumbu Valley were thus considered a degraded form of primal forest. Later on, specific studies in the Khumbu Valley invalidated the hypothesis of a primal forest and of a recent degradation attributable to herding. That revision led to a reassessment of anthropic pressures on the environment and to a better integration, within protection policies, of development, the maintenance of local populations, and the preservation of local cultures. Over time, as ecological analyses were updated and protection models transferred and adapted, Sagarmatha National Park became a frame of reference for those who wanted to protect the mountains (often in the name of a naturalistic conception) while at the same time promoting their social and cultural development, formulated as a function of more anthropological notions.[85]

The Scientific and Institutional Figure
of the Tropical Mountains

The colonial powers and the independent nations that replaced them thus tended to view tropical mountains as a totality, in accordance with a small number of analysis and management models. These models adapted policies and practices forged in the metropolis to a new reality and to the specific needs of the colonial enterprise. The very use of the expression "tropical mountains," which became widespread in the late nineteenth century, attests to that new and coherent objectification of topographical features long apprehended in disparate ways.

This family of geological formations, though heterogeneous and clearly distinguished from mountains in the Western world, was primarily studied and developed as a function of the naturalistic paradigms put to use in the northern hemisphere. Environmental assessments, generally by colonial administrations and later by international organizations and the scientific community, usually emphasized such qualities as landscape, flora, fauna, and ecology. Transformations of the environment and natural catastrophes were understood in terms of the schemes used to analyze environmental crises. Such analyses had previously led to forest policies and the establishment of protected areas in Europe and North America.

The national administrations of newly independent states, anxious to conform their management methods to the dominant models, largely adopted that reinterpretation of the tropical mountains, seeking in that way to mark their modernity and sovereignty and also to qualify for institutional, technical, and financial aid from countries in the northern hemisphere. The dual process of nationalizing mountain resources and achieving international legitimacy tended to reproduce the situation already observed in the temperate countries: many tropical mountains were converted into national or global community property. In many cases, such as Huascarán, Mount Kenya, and Sagarmatha, certification as a UNESCO World Heritage Site served to ratify these efforts.

The autochthonous populations, whose identities and territorialities were grounded in local concerns, often paid the price. The dominant norms depreciated or condemned traditional practices, which were blamed for environmental degradation. The populations were often expelled, leaving the field open for modes of appropriation and management methods consistent with the new functions of the space.

Southeast Asia specialists have shown the political, national, and international character of natural resource and environmental management. In analyzing the construction of the tropical forest as a generic analytical category and as an object of colonial and postcolonial policy, some have proposed the term "political forests."[86] After the colonies achieved their independence, forest policies constituted a decisive modality for constructing the new state's legitimacy and authority. Scientific expertise and international aid have played a large role in that process.

We have just seen that the same is true for the mountain. Scientists and experts frequently adopted the notion as an analytical category, while countries in the northern hemisphere and international organizations used it as a model to champion policies on scenic areas and the environment. That notion was forcefully enlisted to conceptualize tropical mountains and their management. It is equally appropriate to use the term "political mountains" to refer to them, because of the change in scale and in the models by which they were apprehended after the mid-nineteenth century.

Like the notion of the mountain in Europe in the nineteenth century, the category of tropical mountains was constructed by means of scientific orthodoxies and managed as a function of new problems defined on a global scale. Therefore, the issues identified in the African, Asian, and Latin American mountains must be understood not as problems specific to particular regions but as changes in the global discourse on the environment and as illustrations of a globalization of that discourse.

A generation of authors is now challenging those scientific and political orthodoxies with contextual analyses of environmental situations. Such analyses are better able to take into account the ecological knowledge and expertise of the local populations and their ability to regulate local situations. In so doing these studies strive to provide perspective on two forms of condemnation: "that of practices that supposedly prey on the environment [and] that of the ethnic communities who are accused of implementing them."[87] Depending on the point of view adopted, therefore, the enlistment of the notion of the mountain to account for the facts is undergoing a radical change.

EIGHT

The Globalization of Mountain Issues

On December 15, 1998, a small group of scientists from Europe, the United States, and Latin America began the ascent of Chimborazo. Among them were several researchers specializing in mountain environments, in particular Bruno Messerli from the University of Bern, president of the International Geographical Union. The ascent was planned as an adjunct to a Quito research seminar devoted to sustainable development in the mountain regions. The seminar had assembled researchers, representatives of international organizations (the FAO, UNESCO, the International Union for Conservation of Nature, and others), and regional organizations such as the Andean Mountain Association.

At an altitude of about 5,000 meters, the small group met up with Ecuadoran political and administrative leaders. Together they sang the Ecuadoran national anthem, "despite the oxygen-deprived lungs of the participants."[1] Then someone affixed a plaque to a stone pyramid erected in earlier times to the memory of Simón Bolívar. On the plaque was an homage to Alexander von Humboldt: "The Andean Mountains, especially Chimborazo, stirred the imagination and scientific labor of this great man. In addition to his many other contributions, it was in this tropandean landscape, beneath the eternal snows of this majestic volcano, where he laid the foundations of 'mountain geoecology' or 'montology' that continues to mold world society. The Rio de Janeiro Earth Summit of 1992 ensured international recognition of the importance of our mountains (and created an awareness that is finally transcending into action). This advance culminated in November 1998, when the General Assembly of the United Nations declared AD 2002 as the International Year of the Mountains." Then, in capital letters: "FOR A BETTER BALANCE BETWEEN MOUNTAIN ENVIRONMENT, DEVELOPMENT OF RESOURCES, AND THE WELL-BEING

OF MOUNTAIN PEOPLE." After the plaque was set in place, the little group read, in Spanish and English, the visionary text that Simón Bolívar is said to have written at that very place: *Mi delirio sobre el Chimborazo*.

Two centuries apart, in a strange and auspicious compression of time and space, scientists paid homage to another scientist, Alexander von Humboldt, in one of the places that contributed the most to his notoriety (see chapter 1). A scientific community, keeper of the flame for a discipline they proposed to designate by a new term, "montology," thus celebrated a high point in the history of science. But that same place, which put Humboldt on the path toward his global "plant geography," was also being proposed as an emblem for another kind of globalization, this one political, that of the Earth Summit and the movement for sustainable development. Grafted onto that dual celebration of the universal values of science and of environmentalism was the independentist romanticism of Bolívar and the nationalist enthusiasm of prominent Ecuadorans. Between that place and the rest of the world, the shadow of the colonies and the ascendancy of nation-states persist.

This expedition on Chimborazo, however anecdotal it may be, demonstrates what we propose to call the globalization of mountain issues.

The Emergence of New Global Political Objects

For the last few decades some of the principal types of planetary ecosystems have acquired the status of "global political objects." A specialist in the globalization of tropical rainforest governance introduced that expression to designate "object[s] of international negotiation leading to decisions likely to affect worldwide political and economic practices."[2] In addition to the rainforests, among the first entities to be identified as natural objects whose conservation, exploitation, and management deserve to be conceptualized and implemented on a global scale were the oceans, the wetlands, and the continent of Antarctica. In some cases—that of Antarctica, for example—the nation-states concerned signed treaties pledging to observe specific rules.

Thanks to many research projects in recent years, it has been possible to compare the conditions for the emergence of these new global political objects. These analyses show the decisive role that certain actors have played, especially those who have been able to put on the international agenda questions usually ignored in public debates or reserved for debates at the local or national level. On environmental matters, the activism of organizations such as the World Wildlife Fund (WWF) and Greenpeace is now well-known. For other questions, the powerful effect of a convergence between scientific groups and intergovernmental organizations has attracted attention: for example, an analysis of the World Health Organization's ability to raise questions at the global level has forged the concept of "epistemic community."[3] That term designates a convergence of actors of varying status, from different fields of knowledge and with different methods of action, around a way of problematizing these questions, making assessments, and agreeing on measures to be taken to treat the problems identified.

In the wake of these examples, it is possible to analyze similarly the international community's political recognition of the importance of mountains, by identifying key actors, alliances, and convergences of concerns. As now inscribed on the Chimborazo plaque, that recognition was achieved during the United Nations Conference on Environment and Development (UNCED), also called the Rio de Janeiro Earth Summit, of 1992. It was on that occasion that mountains, as a generic category of public action, entered the political sphere at the global level.

A group of scientists, developers, and political actors initiated this process of global political recognition about twenty years ago. All were concerned with identifying the principal environmental issues on a global scale and with promoting coordinated policies of environmental protection and sustainable development. They found it necessary to elaborate a new figure of the mountain, that is, a new way of circumscribing and problematizing mountains. The reframing of objects, scales, and problematics grew out of a new epistemic community organized around new scales of action and new institutional levels.

Putting Mountains on the International Agenda

Mountains were absent from the first global conference devoted to environmental issues, the United Nations Conference on the Human Environment, held in Stockholm in 1972.[4] The first major U.N. conference to shed light on mountain issues was the Earth Summit of June 1992. That conference marked a turning point in world governance of environmental issues, broadening the debate well beyond circles of experts and political leaders. Mountain advocates, taking advantage of that opening, seized on the summit as an opportunity to promote mountains in the international sphere and to ensure they found a place on the international agenda.

They succeeded: chapter 13 of Agenda 21, the plan of action that emerged from the conference, is dedicated exclusively to mountain regions. Although barely noticed by anyone but a few specialists on the subject,[5] that chapter provides the foundations for an international mountain policy. Titled "Managing Fragile Ecosystems: Sustainable Mountain Development," it affirms that mountains must be treated as a major issue in sustainable planetary development and calls for an implementation of appropriate policies.

Putting mountains on the international agenda was a major political act. It allowed advocates of the mountain cause to base their demands on the specificity of the territories dear to them. Tage Michaelsen, former FAO task manager chair at the United Nations, pointed out that "these are of value as internationally negotiated documents, which can be quoted and used as evidence of international support for programs on sustainable mountain development."[6] Various U.N. resolutions have repeatedly promoted mountains under the aegis of the FAO, which serves as project manager in implementing the initiatives backed by Agenda 21. In the Plan of Implementation of the World Summit on Sustainable Development (WSSD), a conference held in Johannesburg in 2002, paragraph 42 is devoted to mountains. Mountains are also mentioned in the declaration "The Future We Want," which emerged from the Rio+20 Earth Summit. A decisive step in that internationalization of mountain issues was the proclamation of an International Year of Mountains (2002),[7] followed by the adoption of an International Mountain Day (December 11).

For that international year, several intergovernmental organizations, in particular the FAO, UNESCO, the United Nations Environmental Program (UNEP), and United Nations University (UNU), agreed to hold a series of events and to bring out publications with the aim of shedding light on the chief issues relating to mountains (fig. 17). National committees devoted to organizing events within the framework of that year were set up in seventy-eight countries. In addition, international conferences were convened at the initiative of several particularly activist countries, especially Italy, Switzerland, and Kyrgyzstan.

Between these first initiatives undertaken in 1991 and the present, the mountain has been patiently constructed by the most active U.N. agencies, countries, and civil society actors as a privileged object in terms of which international development and environmental conservation policies need to be conceived and implemented. The scientific community especially—their efforts in that direction date back to the 1970s—has constantly raised awareness and driven in the importance of mountains.

17. Official logo of the International
Year of Mountains, with the motto
"We are all mountain people." FAO.

Conferences on the Environment and on Development

In the 1990s intergovernmental organizations understood the necessity of working with partners other than representatives of nation-states. They were aware of the expertise that worldwide NGOs possessed for identifying problems and raising public awareness. In recent years, therefore, they have adopted a procedure open to such actors, especially within the context of global conferences. The Earth Summit was decisive in that regard. It was in great part open to NGOs, thanks to the creation of a liaison bureau, and to scientific communities. Such openness attested to a desire to increase effectiveness in implementing whatever political decisions would be made during the Earth Summit but also to attain new legitimacy.[8]

In the case of mountains, that process occurred in an unprecedented way, thanks to parallel initiatives by scientists and the Earth Summit's organizing committee, and to the ability of both parties to hold a common view of the issues and of the strategy to be adopted.

Peter Stone and Ruedi Hoegger initiated two distinct measures addressed to Maurice Strong, secretary general of the summit, to ensure that mountains would appear on the conference agenda.[9] Stone, a British national and member of the Alpine Club, had been director of information for the Brundtland Commission, after which he was public information officer for the Stockholm Conference of 1972. Hoegger had served as assistant director of the Swiss Agency for Development and Cooperation (SDC) and as president of a regional organization devoted to development in the Himalayas, the International Center for Integrated Mountain Development (ICIMOD).

From the late 1980s on Strong encouraged these activists because "he deeply believed in the possibility of treating mountains as an exemplary case for implementing sustainable policies at a global scale."[10] Strong was no neophyte on mountain questions. He had been present at the inauguration of the ICIMOD (1983); served as honorary president of the Himalaya-Ganges Problem conference (Lake Mohonk, New York, 1986); and written the preface to the book *The Himalayan Dilemma.*[11] The conference and the book introduced innovative analyses of environmental degradation in the Himalayas (see chapter 7). Once Stone, Hoegger, and Strong were united in their convictions and initiatives, they approached Bruno Messerli. A professor of geography at the University of Bern in Switzerland, Messerli had earned a reputation for his knowledge of environmental changes in the mountains of East Africa. He had played a major role in the 1970s and 1980s in organizing programs and scientific conferences devoted to mountains. His fame within the discipline would be officially recognized somewhat later, when he was elected president of the International Geographical Union in 1996, a position he held until 2000. His involvement in the Earth Summit therefore opened the door to the international network of scientists specializing in mountain issues, especially Jack Ives of Canada and Jayanta Bandyopadhyay of India.

In anticipation of the Earth Summit, that network established itself within a group known as the "Mountain Agenda." This organization, created in 1990, was dedicated to sustainable development in mountainous regions. Its objective "was to put mountains on the world's environmental agenda (with the Rio Conference of 1992 as the initial focus)."[12]

The name "Mountain Agenda" originated at a meeting held in Appenberg, Switzerland, in 1990.[13] Although that meeting was an important step toward the elaboration of a plea for mountains on a global scale, it was dependent on earlier initiatives, as attested by the list of attendees: the informal group was composed of a mix of scientists and experts in international development with common professional interests, all devoted to sustainable development in mountainous regions, and included a number of pioneers of the global mountain agenda. Most already knew one another from their earlier activities in, for example, the Highland-Lowland Interactive Systems program of United Nations University, the International Mountain Society, and the ICIMOD.

At the time, Mountain Agenda's activities focused on raising awareness, in order to put mountains on the global environmental agenda. This was seen as merely a preliminary measure: "Then, perhaps, the mountain problems alongside other problems of the next century can be wisely and effectively addressed. This will never come about, however, until the mountain issue is clearly identified and until mountain people, scholars, practitioners and decision makers are working together on a fully international scale. The Earth Summit (UNCED) in Rio de Janeiro should prove the best occasion to attract the attention of world leaders and provide our mountains with their proper place on the twenty-first century agenda. One major task is simply to increase awareness of the mountain issue."[14]

The Earth Summit leadership as a whole still had to be convinced of the soundness of that priority. Initially the proposal met with various forms of opposition, which "challenged the stand that mountains constituted a viable unitary system, or formulation, and argued that the mountain issues were already covered in many of the other chapters that were being developed to become Agenda 21."[15] If this resistance was to be overcome, the proposal, because it did not originate with a government or a U.N. agency, had to be considered within the preparatory committees for the conference. The Swiss ambassador granted Messerli and Hoegger the right to participate in the third preparatory committee (August–September 1991) as part of the Swiss delegation: "Their advocacy at the beginning and end of the PrepCom agenda item secured a place for mountains in the Rio Conference agenda—and so in Agenda 21 itself."[16] The formidable lobbying efforts during the summit by the Swiss delegation, especially Ambassador Jenö Staehelin, and also by Jean-François Giovannini and Olivier Chave of the SDC, played a decisive role as well.

The scientists' successful cooperation with the Earth Summit organizing committee, and later with a few intergovernmental organizations, was possible because the different parties shared the conviction that they could complement one another by providing expertise and promoting effective policies. Intergovernmental organizations, for mountains and for many other matters, had long identified scientific expertise as necessary for the formulation of problems and the construction of legitimacy for their actions. Among scientists a dual conviction prevailed: the importance of disseminating scientific knowledge when its aim was to improve environmen-

tal protection and the living conditions of the mountain populations, and
a belief that public policies had everything to gain by attracting scientists
from the development stage on. Messerli and Edwin Bernbaum in particular articulated that view, believing that the "historic model" whereby "the
scientists analyse, interpret, evaluate and report" before the "politicians decide" was no longer viable, given how unsatisfactory that relationship had
proved to be in "solving highly complex problems." As an alternative, they
proposed an "assessment model" that incorporated "a dialogue between
scientists and policy makers in order to bring about consensus on scientific understanding" while keeping in mind "the time-frame of interest to
politicians."[17] Beyond that strategic and operational concept of an optimal
pairing of knowledge and action, scientists and intergovernmental organizations also shared a profound, manifestly emotional conviction that the
mountains deserved their personal engagement at the global level.

The leadership role played by scientists in putting mountains on the
international environmental agenda is also evident within Agenda 21 in
a few recommendations directly related to their activity. For example, of
the two "realms of activities" defined in chapter 13, the first concerns "the
acquisition and increase of knowledge about ecology and the sustainable
development of mountain ecosystems."

Mountain Advocates: An Epistemic Community

The process leading up to the drafting of chapter 13 highlighted the determining role played by scientists. But that coalition was in place even
before the UNCED. Back in 1973 UNESCO, within the framework of the
Man and Biosphere (MAB) program, had launched "Project 6," known as
"Impact of Human Activities on Mountain and Tundra Ecosystems." In the
opinion of several participants, that program, through its ability to spur
comparative research and to structure the scientific community, constituted a decisive stage in the promotion of mountain ecology on a global
scale.[18] Shortly thereafter, in 1974, an international conference called "Development of Mountain Environments" was held in Munich at the initiative of the Deutsche Gesellschaft für Technische Zusammenarbeit, again
in collaboration with UNESCO. As two scientists who played a key role
at that stage remarked, it "ensure[d] creation of a much larger number of
concerned mountain aficionados by linking the UNESCO MAB-6 academics with a more applied group drawn from both the private and public
sectors."[19] The same comparatist and constructionist concern for global,
standardized knowledge of mountain ecosystems governed the organiza-

tion of that event. Ample space was therefore allotted to analyses related to environmental degradation in the Himalayas and in other mountain chains (see chapter 7). The United Nations University program "Highland-Lowland Interactive Systems," launched in 1977, and the subsequent Lake Mohonk Conference on the Himalaya-Ganges problem in 1986, were part of the same process.[20]

For the fifteen-odd years preceding the Earth Summit, a set of initiatives thus made it possible to build a community of scientists on the foundations of shared methods and diagnoses and, moreover, on the shared conviction that mountain environments and societies raised a specific set of major issues that research ought to take into account. That community had equipped itself with institutions such as the International Mountain Society, founded in 1980, and had set up its own publications, the journal *Mountain Research and Development*, created in 1981, and the *World Mountains Newsletter*, which first appeared in 1990. In the early 1990s the scientific community was thus ready to launch initiatives outside its specific area of competence and to serve as a reliable and legitimate partner for intergovernmental organizations.

The Adoption of a Common Objective

Once the objectives and bases for cooperation between that group of scientists and the Earth Summit's organizing committee had been determined, both parties had to adopt and promote a common vision of mountains.

The first step was to argue the importance of taking mountains into account in analyses of several global-level issues. It was frequently claimed that mountains occupy about 20 percent of the earth's surface and are home to about 10 percent of the world's population.[21] In the early 1990s these figures were cited without a rigorous statistical base. To avoid criticism and acquire an explicit frame of reference, specialists promoted a precise definition of "mountain" and a clear methodology for collecting data (based on the scientific criteria of land area, number of inhabitants, and so on). The scientists were by no means unaware of the difficulty of the task; they had known for a long time that no consensus had ever formed around an objective definition of "mountain."[22]

In 2000, after consulting extensively with scientists, administrations, and political representatives, the World Conservation Monitoring Center, an office of UNEP (UNEP-WCMC), worked out a simple definition and global classification system for mountains. That definition and system responded to a dual concern: first, to remain as close as possible to the everyday meaning of "mountain"; and secondly, to use existing databases that are easily available and relatively homogeneous.

In order to adhere to the most common representations of mountains, the WCMC retained the concepts of elevation and slope and chose to apprehend them partly as a function of the constraints they imposed on human beings. For example, any point in space located above 2,500 meters might be considered mountainous because of the idea, broadly shared within the scientific community, that that value represents the threshold beyond which altitude (primarily because of reduced oxygen) begins to have physiological effects on human beings. For elevations between 2,500 and 300 meters, it was agreed that the criterion of slope would be given increasing importance as one moved into the lower brackets. That method formalized an idea, much more ancient than the concept of elevation (see chapter 1), that at lower altitudes a mountain is defined by differences in slope on the mountain itself and between it and its surroundings.

In the interest of using available and relatively homogeneous databases, the WCMC selected a U.S. Geological Survey database known as GTOPO30, composed of satellite images. It was developed in 1996 and, beginning that same year, became the core of an international program called the Global Mapping Project, which had sixteen member countries. The aim of the project was to make available to the international community, in the wake of the Rio conference, a world map of continents, rivers, plant formations, and land uses that could be employed as a universal frame of reference.

GTOPO30 was paired with a digital terrain model that made it possible to simulate Earth's relief. That model allowed scientists to estimate average elevations for areas of one square kilometer and, by comparing adjacent areas, to estimate average slope as well, a valuable and relatively precise tool for measuring elevation (averages per kilometric unit) and slope (between adjoining kilometric units).

As a result, it was estimated that 35.8 million square kilometers (24 percent) of the earth's surface was mountainous in nature.[23] Since then, that topographical objectification has become the most common standard of reference in global and regional descriptions by international organizations (table 1).

Many actors judged the method used to arrive at that new objectification of the mountain pertinent enough to adopt it themselves. Nevertheless, it differed greatly from the definitions adopted by the many nation-states that had felt the need to circumscribe the mountains within their territories. For example, the estimate of mountainous areas in Albania is

Table 1   Classes of Mountains

Class (elevation in meters)	Additional Criterion	Area (km²)	% of the Earth's Surface
> 4,500		1,774,987	1.2
3,500–4,499		2,704,557	1.8
2,500–3,499		6,903,118	4.7
1,500–2,499	slope > 2°	5,277,525	3.6
1,000–1,499	slope > 5° or LER > 300 m	6,160,158	4.2
300– 999	LER > 300 m	12,993, 092	8.8
Total		35,813, 437	24.3

Note: LER (Local Elevation Range) measures the maximum and minimum elevation in an area. The criterion of LER was selected to include low-elevation mountains and to take into account their relative elevation compared to the surrounding area. For each point of measurement, the LER is assessed over a seven-mile radius. The zone under analysis is considered "mountainous" when the LER is 300 meters or above. Table of classes of mountains according to UNEP-WCMC. Source: Valerie Kapos et al., "Developing a Map of the World's Mountain Forests," in *Forests in Sustainable Mountain Development: A Report for 2000*, ed. M. F. Price and M. Butts (Wallingford: CAB International, 2000), 4–9.

25 percent higher using the UNEP model than in the nation's official definition. In Croatia it is 25 percent lower. The self-evidence of the mountain, which seems so clearly rooted in shared representations, becomes much less so when it comes time to circumscribe it.

For estimating the number of inhabitants, the use of national population statistics turned out to be unavoidable, even though a number of them are known to be unreliable. In many countries, population censuses have numerous methodological weaknesses: imprecise counts, uncertainties about considering nomad and migrant populations, and so forth. Furthermore, administrative authorities sometimes deliberately underestimate the number of inhabitants in the mountainous regions when that undercount can serve their political objectives.

Despite these difficulties, the proposed estimates adopted the delimitations developed by UNEP-WCMC. The FAO chose data from a study it had commissioned that estimated the number of mountain inhabitants at 720 million (that is, 12 percent of the global population)—333 million of them in the Asia-Pacific region and 112 million in Latin America (see table 2).[24]

Shortly before that, another estimate, reached by comparable methods of geolocation, placed the proportion of the world population in areas the authors considered mountainous at 26 percent (1.481 billion).[25]

"We Are All Mountain People!"

A complementary means of highlighting the importance of mountains is to recall their influence over neighboring regions and populations. The official

Table 2 Population Estimates by Elevation Bracket

Class (elevation in meters)	Developing and Transitional Countries	Developed Countries	Total Population
> 4,500	4,137,000	—	4,137,000
3,500–4,499	13,320,000	5,000	13,325,000
2,500–3,499	45,602,000	263,000	45,865,000
1,500–2,499 + > 2° slope	138,388,000	3,080,000	141,468,000
1,000–1,499 + > 5° slope or LER > 300 m	136,461,000	8,722,000	145,183,000
300–999m + LER > 300 m	324,884,000	43,929,000	368,812,000
Total	662,792,000	55,998,000	718,790,000

Source: Barbara Huddleston et al., *Towards a GIS-Based Analysis of Mountain Environments and Populations* (Rome: Food and Agriculture Organization of the United Nations, 2003).

slogan of the International Year of Mountains was unambiguous: "We are all mountain people!" It suggests that all of Earth's inhabitants are influenced in one way or another by natural phenomena and resources located in the mountains and by the modes of production and events that take place there. The first lines of chapter 13 in Agenda 21 are more analytical: "Mountains are an important source of water, energy and biological diversity. Furthermore, they are a source of such key resources as minerals, forest products and agricultural products and of recreation. As a major ecosystem representing the complex and interrelated ecology of our planet, mountain environments are essential to the survival of the global ecosystem."[26]

Such a range of topics made it possible to develop mountain themes for a large number of planetary issues. Scientists and international organizations latched onto that way of globalizing their object. Several successive scientific symposia and public conferences emphasized the importance of mountains for the water supply, for global biodiversity, for the reduction of poverty, for gender equality, and so on. The vulnerability of mountain ecosystems to global change in general and to climate change in particular was regularly advanced to justify viewing these systems as a laboratory for the observation of the effects of such changes.[27] For each topic the same rhetoric was at work, and both scientists and international organizations made use of it: "The message has become clear: the mountains of the world, with their natural and human resources, are no longer only of local and national concern; they are a matter of global concern in the twenty-first century."[28]

That rhetoric was disseminated in large part through documents targeting specific readerships. Mountain Agenda published a series of brochures elaborating the issues related to mountains and their regional environment: on the global importance of mountains (1992 and 1997), water resources (1998), tourist resources (1999), forest resources (2000), energy and transportation issues (2001), public policies (2002), and so on. Most of these brochures were prepared for distribution during the main international conferences (Earth Summit, 1992; Johannesburg Conference, 2002), and annual meetings of the United Nations Commission on Sustainable Development.

Mountain Agenda also commissioned the first book for the Earth Summit, titled *The State of the World's Mountains*.[29] A summary in the form of a brochure called *An Appeal for the Mountains* was distributed in extraordinarily large numbers.[30] The book's objective was "to make an authoritative statement on the environmental status and development potential of the world's mountains" so as to ensure that mountains would appear on the international agenda.[31] Political concerns were therefore given the spotlight prior to a more thorough synthesis of the science, as the introduction of the book indicates: "For this first edition of a world mountain status report, the political goals outweigh the scientific."[32] These two publications were addressed less to the scientific community or to a readership interested in mountain questions than to the delegates to the Earth Summit: "The aim of both publications was to appeal to governments to put mountains on the world's environmental agenda, both at Rio and in the years to come."[33]

The book adopted a regional plan in describing the planet's mountain issues. It combined physical description, an economic assessment, a statement of social conditions, and political analysis for each of the principal mountain chains or regions (Africa, the Alps, the Himalayas, the Andes, and the Appalachians), but it sometimes also grouped mountains together as part of a geopolitical entity, such as the "former USSR." In so doing the book proposed a fragmented and diversified view of the world's mountains, for which it would be criticized during the Earth Summit: it did not construct an overall image or promote concerted policies at the global level. For example, some observers noticed that, when the book came out, national delegates to the United Nations eagerly looked up information related to their own countries.[34]

In any case, the authors of *The State of the World's Mountains* were keenly aware of the need to produce a thematic synthesis later on in the interest of advancing knowledge. "A more exacting second edition," which was to appear after the Earth Summit, was already announced in the preface.[35]

That second book appeared in 1997 and had a radically different structure. *Mountains of the World: A Global Priority* presents one by one a series of themes—forests, conflicts, water, poverty, and so on—all related to objectives of sustainable development in the mountain regions.[36] The rhetoric of the book consists of emphasizing, in theme after theme, the singularity of the world's mountains and the issues characteristic of them, even if that sometimes means using phrasing not often found in scientific publishing, such as "the mountains, paradise of biodiversity." The initiative for that book lay with the SDC, which, to coordinate it, had appealed to the Institute of Geography at the University of Bern, where Messerli was a professor.[37]

To sustain the plea for the mountain cause at the international level, the book was scheduled to come out in time for the meeting of the U.N. Commission on Sustainable Development, which would do a first assessment of the results of the Rio conference five years after it had taken place.[38]

The critical reception of *An Appeal for the Mountains* was comparable to that of *The State of the World's Mountains*. The cover page shows the Huascarán in the Peruvian Andes, with its western slope as it appeared following the earthquake of May 31, 1970. A giant flow of ice, rock, and mud had obliterated the city of Yungay, killing eighteen thousand residents. According to the authors, that cover allowed them to show the highly sensitive relationship between man and nature, or between human activities and natural processes in the world's mountains.[39] That image, then, had a primarily emblematic value. But it was received differently: many delegates saw the earthquake as an event that, though traumatic, was localized; its consequences, they said, were national at most and did not reflect any issue that was valid for all regions of the world. To avoid that stumbling block and promote a more generic vision of their object of choice, an effort was made in later publications to compose covers through a juxtaposition of images better able to show the diversity of regional situations. The cover of *Mountains of the World*, for example, is composed of photographs taken in Africa, Asia, and Europe. These various communication problems are a good illustration of the difficulties involved in proposing a generic definition of the mountain and of the importance of such a definition in political arguments.

Organizing the Scientific Community and Framing Applied Research

The scientific community, while working with intergovernmental organizations to constitute a joint mountain agenda, also strove to construct the foundations of its own organization and to achieve visibility on a global scale. To promote the specificity of its field of research within the scientific world as a whole, it relied on the scientific institutions and journals cre-

18. Mountain Pavilion website banner, Rio + 20 Summit. Lobbying
for a cause is a never-ending process. The tour de force of putting
the mountains on the world agenda during the 1992 UNCED would
be further advanced by the presence of mountain actors at world
conferences. At the Rio + 20 Conference, held in Rio de Janeiro, Brazil,
on June 20–22, 2012, Peru, with the support of other mountain
countries, had a Mountain Pavilion. The pavilion provided a place
where mountain stakeholders could promote sustainable mountain
development and showcase their contribution to sustainable
development. Its aim was to raise awareness and secure political
support for concrete sustainable mountain development activities
and programs. The lobbying carried out during the conference was
successful, since mountains were mentioned in the final document,
"The Future We Want" (paragraphs 210, 211, and 213).

ated in the 1980s (the International Mountain Society, *Mountain Research
and Development*) and published thematic analyses and notes on topical
subjects. That same concern led some scientists to promote the neologism
"montology" to designate the interdisciplinary production of mountain-
related knowledge.[40] But many had epistemological and strategic reserva-
tions about adopting that term.[41]

At the same time, the scientific community sought recognition from the
major research agencies and international scientific programs. It drew inspiration from the example of the Man and Biosphere program "Impact of
Human Activities on Mountain and Tundra Ecosystems." On the basis of
that model, it promoted the construction of specific modules on mountain
problematics within the major environmental science programs, as well as
global databases developed with the aid of intergovernmental initiatives—
for example, Diversitas,[42] the Global Terrestrial Observing System (GTOS),
and the Terrestrial Ecosystem Monitoring Sites. The Mountain Research Initiative (MRI),[43] whose aim was to promote global research on mountain issues, institutionalized the strategy with the support of several partners: the
International Geosphere and Biosphere Program (IGBP), the International
Human Dimensions Program on Global Environmental Change (IHDP),
GTOS, and UNESCO's MAB program.[44]

In addition, regional associations of scientists were created on a continental or mountain-chain scale—Africa (the African Mountain Association, 1986), the Andes (the Andean Mountain Association, 1991), the Alps (International Scientific Committee on Research in the Alps, 1996), the Carpathians (S4C, Science for the Carpathians, 2008), and others—so as to organize thematic research at those levels. These regional organizations all declared their desire to structure knowledge production while delimiting professional territories that combined field research sites, areas of expertise, meeting places, academic institutions in the participating countries, and so on.[45]

Finally, scientists played a central role in the creation of organizations and networks that spurred cooperation and the exchange of knowledge with NGOs and local actors. Such a system was inaugurated on a Himalayan scale with the creation of the International Center for Integrated Mountain Development in 1983. Based in Katmandu, the center has the objective of conducting applied research in the Hindu Kush–Himalayan region in coordination with regional development partners. The success of the initiative inspired the creation of a series of similar regional institutions, such as the Consortium for Sustainable Development in the Andean Ecoregion (CONDESAN),[46] as well as, at the global level, the Mountain Forum. This is an electronic platform, the idea for which emerged in the wake of the Earth Summit, that uses the Internet to make resources available to all mountain actors in the world and to establish communication among them through forums and listservs. The aim was to make it possible to consult with a wider circle of actors on chapter 13, which the rapid drafting of the text had not allowed.[47] In February 1995, just after the Rio conference, the FAO and the Mountain Institute, a nonprofit organization based in the United States, with the assistance of scientists and the cooperation of Switzerland held an international meeting of NGOs in Lima, Peru.[48] The event brought together representatives of 110 organizations and institutions, primarily NGOs active in mountain zones, but also governments and agencies from nearly forty countries, as well as most of the members of Mountain Agenda. One of the explicit goals of the talks was to "create an ongoing forum of mountain NGOs for information-sharing and mutual learning."[49]

Such initiatives, promoted or supported by scientists, clearly indicate their desire to encourage applied research and the sharing of knowledge. But they also attest to their concern to structure forces and jurisdictions targeting the defense of the mountain cause on a regional and global scale. As one of the scientists most active in the field said at the time: "The Mountain lands, unlike the tropical rainforests, the arid lands, the oceans, Antarctica, the ozone layer, the greenhouse effect, or acid rain, have not yet been able to develop a strong, well-informed constituency. This is urgently needed if there is to be the necessary fuller understanding of mountain dynamics and its effective linkage with the political processes of mountain resource management. The mountains, like the oceans, need their own Jacques Cousteau!"[50] The creation of the Mountain Forum in 1996 represented a clear response to that concern: "The Mountain Forum seeks an open constituency of groups and individuals committed to the mountain agenda and to mutual support and information exchange. . . . The Mountain Forum intends to create an innovative coalition of organizations bound together through mutual interest, support, and commitments."[51]

A Coalition in Defense of Mountain Interests

Between 1991 and 2002, then, a broad coalition of interests worked relentlessly to construct and promote the mountain as a global object, a specific but exemplary object for understanding a set of major world issues (protection of the environment, reduction of poverty, sustainable development, and so on).

During the International Year of Mountains in 2002, the Johannesburg conference gave rise to what is known in U.N. terminology as "Type 2" partnerships, which involve governments, local and regional authorities, intergovernmental organizations and NGOs, and the private sector. An International Partnership for Sustainable Development in Mountain Regions, commonly called the Mountain Partnership, was launched during the conference and established its headquarters in the offices of the FAO in Rome shortly afterward. It sought to take advantage of the enthusiasm created during the International Year of Mountains. Within a few years, the Mountain Partnership comprised 53 countries, 13 intergovernmental organizations, and 162 NGOs from either civil society or the private sector.[52]

Although clearly conceived as a continuation of the process begun in Rio, the Mountain Partnership's declared priorities differed from the initial objectives of the global mountain agenda. The two chief priorities of chapter 13 in Agenda 21 (protection of the environment and development) do make their appearance, but the partnership "addresses the second, and perhaps more challenging, priority of actually improving livelihoods, conservation and stewardship throughout the world's mountain landscapes."[53] In addition, the Mountain Partnership adopted the method of combining initiatives on thematic issues (agriculture and rural development in the mountains, stewardship of catchment basins, biodiversity, education, gender issues, sustainable modes of existence, policy and legislation, research) and a regional approach that sought to ensure efficiency (focusing on East Africa, Central America and the Caribbean, Central Asia, the Andes, Europe, and the Hindu Kush–Himalayan region). This second approach grew in scope over the years, despite what were considered disappointing results for the global initiatives.[54] Nevertheless, twelve years after its creation, the Mountain Partnership has not met all of its financiers' expectations. It found itself under fire from critics during the Rio+20 conference, during which the mountain question enjoyed much less publicity than it had at the two previous major conferences.

Repositioning by the Nation-States

In becoming the object of attention by international organizations and conferences, mountains did not necessarily lose their importance at the national level. In fact, several scenarios were played out. Nations that had adopted policies in conflict with the objectives declared at international conferences—especially policies promoting the intensive exploitation of resources or stripping away the rights of local populations—fought against the international groups or displayed a studied indifference toward them. States that had a long history of indifference toward their mountain regions and populations were sometimes receptive to the emerging credo. Those that had policies that were compatible, even exemplary from that standpoint, sometimes sought to take advantage of the new international context. But in most cases, nation-states had to reposition themselves in relation to a new arrangement of actors and proposals, which acquired their pertinence and legitimacy by virtue of the global scale on which they had all been configured.

During the 1990s several nations proved hostile to any internationalization of mountain questions, especially at the decisive moment of preparations for the Earth Summit. They readily argued that the management, stewardship, and protection of mountains fell solely within national jurisdiction.[55] The United States, for example, through its representatives reminded everyone that nothing about the situation of its mountains justified impinging on U.S. sovereignty. In other cases, such as China, Indonesia, and Myanmar, nations sought to shield the mountain regions from international attention. These regions were home to ethnic minorities whose desire for autonomy and recognition continued to raise problems. But the resistance of nation-states did not take the form of determined opposition to mountains being included in Agenda 21, no doubt because nothing in that text could have raised fears about a real encroachment on sovereignty.

In subsequent years international initiatives had a true ripple effect on public policies in a large number of nations. The International Year of Mountains spread to seventy-eight countries through national committees responsible for organizing political or cultural events or celebrations. France, for example, opened a debate on updating public mountain policies. Other countries, such as Poland and Romania, seized the opportunity to lay the foundations for their own policies. Mexico took advantage of the event to relaunch a sustainable development plan for the mountain regions and to prepare a long-term strategic forestry program and a national hydrological program targeting in particular the mountainous part of the catchment basins. Spain, Turkey, and Madagascar adopted national mountain strategies.[56] In recent years, about fifty countries have been involved in the Mountain Partnership. They attest to the aspiration to relaunch and institutionalize reflections on mountains in keeping with a model of governance linking very diverse partners.

Of the few dozen countries that have decided to become engaged in the globalization of mountain governance, three countries stand out: Switzerland, Italy, and Kyrgyzstan.

Switzerland has sought to assert itself as a leader in putting mountains on the global environmental agenda. We have already seen that Swiss representatives at the United Nations and the SDC proposed many initiatives in an effort to influence the content of Agenda 21. The Swiss delegation also promoted the mountain question behind the scenes of the Earth Summit, notably rallying Bolivia, Peru, Nepal, Bhutan, Lesotho, and Ethiopia to its cause. These nations then either received financial aid from the Swiss Confederation or became members of the International Monetary Fund group run by Switzerland. In many cases Switzerland later proved to be a decisive financial partner: it provided a large share of the budget for several events, such as the Bishkek summit (2002), and for a few scientific conferences. It is the chief investor in the Mountain Partnership and finances a scientific review, *Mountain Research and Development*, and organizations such as the Mountain Research Initiative.

That wide-ranging engagement can be explained in several ways. It is among the international initiatives Switzerland has taken that give precedence to a few themes, either out of conviction or in the interest of transparency and efficiency. On the question of human rights, for example, the mountain has for years been a theme privileged by Swiss diplomacy and the SDC, which offered its aid to mountain regions in Eastern European countries after the fall of Communism. Finally, the leading role played by the Swiss Confederation in building a global forum dedicated to mountains from Rio to Johannesburg can be linked to Switzerland's desire to participate on the international scene as an equal partner, in keeping with priorities set decades ago: Switzerland did not become a member of the United Nations until 2002, after decades of a radical neutrality policy.

Another reason for Switzerland's international engagement on mountain issues is its concern to advance, export, and further develop the assessments of its own agencies and researchers in that area (fig. 19). In fact, the considerable importance that the mountain has played in Switzerland's national imaginary and public policies since the nineteenth century (see chapter 2) have influenced the administrative and scientific culture to such a point that these policies are now on the leading edge internationally. The Rio and Johannesburg conferences were useful forums for consolidating these images and the initiatives aimed at reinforcing them. The predominant role of Swiss researchers and institutions in organizing the scientific community from the 1990s on constitutes one of the achievements of that policy.

Italy also invested a great deal of energy in institutionalizing that globalization of mountain issues. Italian investment primarily took the form of ambitious conferences within the framework of the International Year of Mountains and a large financial contribution to that event and to the Mountain Partnership. The reasons for that commitment are not as clear as they are for Switzerland, though it does seem to express a concern with extending national strategies beyond the Italian border. Among the European nations, Italy has been in the forefront since the 1970s in designing and implementing mountain-related public policies. The place and autonomy of mountain communities, as well as the measures taken to compensate for their frequent demographic and economic weakness, constitute recurrent themes in Italian political life. The Italian state and its regions have made a concerted effort to create research centers dedicated to mountains in Lombardy and Alto Adige in particular. Finally, Italy has expended considerable energy since the late 1990s to bring its full weight to bear on the Alpine Convention, which binds all the Alpine countries in an effort at joint supranational regulation (see chapter 10).

A third state, Kyrgyzstan, is also intensely involved in international mountain-related initiatives. Having emerged in the wake of the Soviet Union's dissolution in 1991, Kyrgyzstan established a special relationship with Switzerland. It gained international recognition and has belonged to the group Switzerland runs within the International Monetary Fund. It is also a top priority among the nations targeted by Swiss aid. It was in fact Switzerland that prompted Kyrgyzstan to submit an official proposal to the

Mountains and People

An Account of Mountain Development Programmes
supported by the Swiss Agency for Development and
Cooperation (SDC)

19. An SDC brochure presents the agency's activities for the International Year of Mountains
in different mountain regions of the world. In the preface, the director of the SDC writes:
"Switzerland is known as the mountain country '*par excellence.*' More than 75% of its land
area is mountainous. . . . As the source of major rivers and streams that wind their way
through Europe and empty in four different seas, Switzerland gladly lays claim to its status
as the 'water tower of Europe.' Switzerland's early development as a nation took place
around the Gotthard Pass, one of the most important north-south European trade routes.
Consequently, the country has a long history of experience and tradition related to sustainable
development in mountains. . . . Switzerland's strong commitment to sustainable development
in mountain regions at the global level . . . is therefore understandable, particularly with
respect to exchange of experience, communication, advocacy, and network and capacity
building." Brochure of the Swiss Agency for Development and Cooperation, Bern, 2001.

United Nations General Assembly to mark an International Year of Mountains, shortly before Switzerland asked to join the United Nations. This led to the decision to have Kyrgyzstan's capital city play host to a conference held at the end of 2002, the Bishkek Global Mountain Summit. Immediately afterward, the Kyrgyz parliament passed a national law regarding mountain territories. The government also negotiated with the neighboring states of Kazakhstan and Tajikistan in drafting a Central Asian Mountain Charter. That charter, signed on the first day of the Bishkek summit, set common objectives for the countries of the region. Federal assistance agencies in Germany and Switzerland supported and partly financed that process.

The Political Globalization of Mountains

The introduction of mountain issues into organizations and international conferences has been a major innovation of the last twenty years, the result of a reconfiguration of stakeholders: the initiative fell to a heterogeneous group of individuals, organizations, and institutions that included scientists, national administrations, and various international organizations. It proceeded to recompose the levels of analysis and recommendation and called for a redefinition of the category of the mountain and of the objects that compose it.

The conditions that allowed that reconfiguration are themselves very heterogeneous: the desire on the part of various actors to promote new initiatives in the name of a certain vision of the mountain and the interest it represents for world governance of the environment and sustainable development; the concern among some, especially scientists and NGOs, to position themselves as key players through a reshuffling of the cards; and the emotional attitude that several have shown toward the mountains, an attitude that has allowed actors from diverse backgrounds and of varying status to arrive at shared positions. That odd assemblage of individuals and institutions has led to significant results (networks, specific institutions, emerging research centers), which on several occasions have achieved notable recognition from the United Nations.

Nonetheless, despite these undeniable successes, the initiatives have constantly run up against a limit, which is posed as an alternative: can and should mountains as such be considered a global issue? Or should they be conceptualized and managed at a regional level, so as to remain as close as possible to local realities and issues and to optimize public action? We have seen that for certain actors the inclusion of mountains among the global issues of environmental conservation and development was an objective in and of itself. That required adopting a new discourse, iconography, and rhetoric and even entailed an evolution in the structure of books intended to promote mountains at that level (from *State of the World's Mountains* to *Mountains of the World*). But any global initiative or organization in that realm will have difficulty taking root if it is not adopted at the level of the mountain regions themselves and if the diversity of environmental, socioeconomic, and political situations is not taken into account. The Mountain Partnership has suffered from an inadequate implementation of such regionalized activities. The International Year of Mountains sidestepped that danger by enlisting nation-states that have the ability to capitalize on the efforts made by intergovernmental organizations. By contrast, the articulation of the global and regional levels was decisive from the start in organizing the scientific community and establishing the very structure of certain networks, such as the Mountain Forum. Such joint efforts seem particularly fruitful for regional organizations that are part of continental or global cooperation networks (for example, the ICIMOD and the Alpine Convention).

It is therefore undeniable that the identification of mountain issues on a global scale has become an increasingly powerful force, as have the global organizations tasked with apprehending these issues. That globalization is part of a contemporary trend toward adopting a global scale as a means of emancipation from a primarily national representation of issues. And yet, the still decisive weight of the actors at intermediate levels and the extreme diversity of mountain situations throughout the world require a multiscalar approach to each of the problems and a multilevel approach to the modalities for solving them.

NINE

Mountain Men and Women
of Globalization

Globalization has now obliged individuals and social groups in the mountains, as everywhere else, to reposition themselves economically, culturally, and politically. At the economic level, the resources of their environment (water, timber, scenery, metals, and so on) are increasingly sought after. The internationalization of business and of markets, as well as the growing interest by a number of countries in developing territorial resources, has tended to whet appetites and intensify power relations. Global competition has identified market niches. At the cultural level, all forms of circulation (of information and especially of people) have intensified, thanks to telecommunications networks and access infrastructures, which have reached regions that until now have been more or less isolated.[1] Finally, at the political level, as regional or global international treaties increase in number, they sometimes incite the populations directly affected to take a position vis-à-vis the issues on the table.

Needless to say, what is at stake in economic, cultural, and political globalization is comparable for every population on Earth. Writings in the social sciences over the last two decades have focused on identifying a few common effects and reactions to these issues: in one place, an assimilation of the new rules of the economic game and intentional participation in the global market; in another, protest, which may take the form of anti-globalism or of "globalization from below"; elsewhere, a recomposition of identities that reinvents the local or the global.[2] A large number of other adaptations and reactions also exist. But always, or almost always, the re-shuffling of levels and scales may lead the individuals and collectivities directly concerned to take a stance.

These globalization issues often have a particular intensity, as well as a few unique characteristics, for populations living in the mountains. Their ways of life and modes of production are sometimes very different from the dominant models now in use, making the challenge of adaptation all the greater. In addition, their identities are strongly conditioned by the propensity of outsiders to associate them with the mountains. In fact, the proponents of globalization, true to the Age of Enlightenment and the Industrial Revolution, tend to link them in one way or another to their natural environment and to cultivate the image of the "mountaineer." As a result, these mountain populations have an incentive to reconcile themselves with these images, that social identity, and to take that factor into account in presenting themselves to the outside world, especially on the political stage and in public debate. We have already seen that process at work in the political organization of mountain populations within a nation's society as a whole (see chapter 2). It is also at work at the global and regional levels. Economic and cultural globalization, as well as international initiatives, has made mountain populations a key player in the areas of environmental protection and sustainable development. Mountain identities, understood here as a collective sense of belonging (a collective identity) and as a mode of engaging in public debate (a political identity), have also taken on new regional and global forms.

World Conferences on the Mountain and Local Populations

Since 1992 the major international conferences designed to identify environmental, global, or regional issues (desertification in arid regions, deforestation in tropical regions, and so on) have understood the role of the local populations in two opposing ways. Some have pointed the finger at traditional practices, judging them to be inconsistent with the objectives of environmental protection. We have seen that representation at work in the promotion of nature conservation and forest management over several decades (see chapter 7). Others, by contrast, have proved more attentive to the needs and modes of behavior of these populations and have sought to bring them in as partners in the promotion of "good practices."

The Earth Summit (United Nations Conference on Environment and Development, 1992) clearly opted for the second attitude, which is certainly more favorable to the local populations. Agenda 21 attests to this stance. The general objectives of sustainable development are combined with others: recognition of the rights of the autochthonous communities (chapter 26), efforts to combat poverty (chapter 3), and the promotion of gender equality (chapter 24). These same concerns governed the drafting of chapter 13, which is devoted to the mountains. Chapter 13 states that the populations who live in such environments ought to be the primary beneficiaries of public policies, including those dealing with the environment. From that standpoint Agenda 21 constitutes a turning point for mountain regions. Their populations, held responsible for environmental problems (deforestation, floods, degradation of protected spaces, and so on) from the late nineteenth century to the 1980s (see chapters 4 and 7), are now being recognized, at least in official texts, for the specificity of their cultures and modes of life. They are even being supported in their demands for autonomous decision making.

That new view of the mountain populations owes a great deal to a general evolution in the attitude of international organizations. This evolution is not unrelated to the leadership role played by agencies primarily responsible for economic and cultural questions (the FAO, UNESCO, United Nations University, and others) in promoting chapter 13 of Agenda 21. But it is also related to the involvement of NGOs that work to combat poverty, endemic disease, and the cultural challenges mountain populations face. Finally, the shift can be attributed in part to the empathy of the many scientists engaged in preparing that chapter of the agenda. In fact, though many of them were trained as natural scientists, they are often won over to the cause of the mountain people. They develop an almost sentimental attachment to these communities and a desire to carry out applied or even participatory research.[3]

The Limits of Chapter 13 and Its Later Elaborations

It is impossible to deny, however, that the Rio conference and Agenda 21 were primarily guided by environmental concerns, particularly among representatives of the countries in the northern hemisphere. Social, cultural, and political questions were often problematized in light of naturalistic knowledge and the objectives of environmental protection. That is especially true for chapter 13, which strongly emphasizes an approach to mountain issues that centers on catchment basins and promotes modes of management integrated with them. From the vantage point of the populations who live in the mountains, catchment basins are not always perceived as relevant. Any desire to partner with populations to manage properly an entity that does not have great significance in their eyes obviously expresses a "nature-centric" point of view.[4]

Follow-ups to the Earth Summit made an effort to correct that state of affairs. The extensive talks with nongovernment organizations in the mid-1990s and the creation of the Mountain Forum that followed (see chapter 8) placed greater emphasis not only on the populations, their needs, and their projects but also on their capacity to grasp the issues being debated at the global or regional level. The declared objectives of the Mountain Forum in particular attest to that shift. The organization presented itself as "a global network for mountain communities, environment and sustainable development."[5] In the talks and international conferences that have taken place since Rio, mountain populations are increasingly mentioned both as the beneficiaries of the policies promoted in mountain regions and as key actors in defining and implementing these policies.[6] Sometimes considered the best guarantees of local, national, and international interests, they are readily granted the role of stewards of the resources located in mountainous territories.

This is one of the factors that spurred the Mountain Forum to specialize in the production of information and its dissemination to its vast network of members and, beyond them, to Internet users, a priority based on the conviction that information is "perhaps the most powerful force shaping the world today," "a critical tool for empowerment and a catalyst for social change."[7] The managers of the network say they are persuaded "that one of the most pressing needs in advancing the mountain agenda is sharing information."[8] Among the top priorities announced by the network are cultural diversity; sustainable development; systems of production and modes of life; energy supply and demand; tourism; water resources; biodiversity; natural risks; and the spiritual, symbolic, and sacred meanings of the mountains for their populations. Many of these priorities move toward refocusing the agenda to respond to the local populations' concerns.

The World Summit on Sustainable Development (Johannesburg, 2002) furthered that evolution. Although several of the objectives of Agenda 21 can be found in the final declaration, the paragraphs devoted to the local populations are more substantial. In particular, they emphasize the objectives of skill acquisition and institutional reinforcement. The same intention is evident in the documents devoted to mountains. Paragraph 42, which concerns these populations, defines six fields of action: eradicating poverty and promoting sustainable development, combating environmental degradation, battling social inequalities, promoting sustainable modes of life, encouraging participation by the populations, and promoting research and the development of skills.[9]

Motivations for Involving the Mountain Populations

At conferences and in international declarations, several justifications are cited for that growing emphasis on the local populations.

First, these populations are portrayed as the primary stakeholders. It is primarily they who suffer from poverty, health problems, and the lack of public services, especially in the countries of the southern hemisphere. They are also on the front lines of environmental degradation, which is now attributed to growing exogenous pressure: "All over the world, expanding economic pressures are degrading mountain ecosystems while confronting mountain peoples with increasing poverty, cultural assimilation, and political disempowerment."[10]

Secondly, they are understood to be the prime agents for the implementation of sustainable development policies. The desire to involve the mountain peoples also drives the interest in understanding local cultures, practical knowledge, beliefs, and social structures as a means whereby development programs may better adapt to local characteristics: "Programmes of sustainable mountain development need to take cultural values, traditions and preferences into account: if they do not, they will fail to engage local communities and other stakeholders whose support they need to be truly sustainable over the long term."[11]

That is how we ought to understand the attention of scientists and international development experts to traditional knowledge, and in particular, to traditional ecological knowledge. The invocation of such grass-roots knowledge has constituted a motif that unites autochthonous populations around the world. That is true at a global level for a number of environmental issues. It has even been said that "the *holder of indigenous and traditional knowledge* has emerged as a new actor in international discourse, with legally recognized claims to sit at international bargaining tables."[12] It is also true for the mountain regions in particular.[13] More precisely, the importance of the spiritual connections between the mountain populations and their natural environment has drawn the attention of researchers and NGOs.[14] They have demonstrated that some religious prohibitions (regarding certain plant species, for example) have contributed to maintaining the local ecological balance.[15] To understand ecological dynamics and make recommendations related to environmental management, one would therefore do well to take traditional ecological knowledge into account, particularly the role of beliefs and religion.[16]

For all these reasons, the prevailing representation is that the "mountain people, rather than being part of the problem, are part of the solution."[17]

They are clearly thought of in two different ways: as a category in and of themselves and as a collection of singularities. In fact, a large part of the literature produced during these years has invoked the similarities among these populations to justify a totalizing discourse about them. At the same time, however, the diversity of these populations is often noted as a characteristic worthy of interest. Their contribution to the cultural diversity of humanity is regularly cited and linked to the diversity of the biocenosis.

Alternative Interpretations

This interest in taking local populations and their cultural specificities into account no doubt stems from good intentions and humanist concerns. But some critics of that approach have pointed out its limits and dangers. They argue that, when such curiosity is subordinated to objectives of environmental or patrimonial conservation and to an interpretation of human ecology, there is a risk of instrumentalizing the populations, placing them in the service of a cause that is not necessarily their own.

That search for efficacy appears clearly in many documents. Chapter 13 of Agenda 21 presents the participation of the autochthonous communities as "a key to preventing further ecological imbalance." The integration of local populations is sometimes justified as an "effective way of ensuring that policies are implemented."[18] Cultural diversity has explicitly been invoked in service of the same ends: "[mountain] biodiversity can only be conserved when equal attention is given to cultural diversity";[19] "support is needed to recover and foster the cultural expression of mountain populations because mountain cultural diversity is a strong and valid basis for sustainable use and conservation of mountain resources."[20] Such remarks may give the impression that communities are conceived "as objects and/or implements, rather than formulators, of policy and a resource to be employed in defense of the environment."[21]

To support their thesis, those who criticize the instrumentalization of the autochthonous populations readily recall that these populations were largely absent from the various world conferences and events, and were left by the wayside when the general objectives of global mobilization were defined: "The voices of mountain people are largely silent in the present mountain movement."[22] Those who make that observation quite logically recommend that "innovative venues be created to make mountain people central to the planning for the future of mountains."[23]

Other social science researchers, especially those who have adopted constructionist paradigms in place of those of human ecology, have formulated an even more critical view. These researchers claim to be very close to the local populations whose systems of values, representations, and motivations they study. They are therefore wary of any attempt at social engineering that would place these populations in the service of objectives not their own.

> We have tried to identify the mechanisms that might reinforce the capacity of the mountain peoples to resist the massive transformation of their regions, their products, and their landscapes into raw materials in the service of the logic of capitalism. That work, however, can be carried out only on the condition that we give up claiming to speak in the name of the mountain peoples, that we give up confining their representations within rigid and stereotypical images, on the condition, finally, that we remain vigilant about the risks of potential constraints generated by permanent interventions in their traditional activities. In fact, we must remain aware of the dangers of the idealized view, largely imposed by outsiders, that the mountains constitute a natural environment of great purity, that they are inhabited by populations of great cultural originality, since that view leaves aside the complex role played by mountain peoples and the transformations they undergo. . . . In fact, as scientists, we must also analyze the situated nature of our own knowledge, which is necessarily partial and multiple, and must make explicit our motivations and policies as researchers inscribed within disciplinary traditions and institutional contexts.[24]

Such scientists have generally preferred to work with the local populations and their representatives, and with NGOs rather than intergovernmental organizations. To date, few efforts have been made to incorporate such researchers and their ideas into specialized scientific organizations.

Conflicting analyses of the motivations and modalities for taking mountain populations into account have thus constituted a bone of contention within the scientific community. A similar division exists in the different attitudes that the populations most affected have themselves adopted.

The Variety of Forms of Engagement by the Affected Populations

The various local populations who have sought to position themselves in relation to the global mountain agenda have adopted very different attitudes.

Some have played by the rules of the regional and global networks, whose aim has been to combine the skills of scientists with those of the principal stakeholders. The International Center for Integrated Mountain Development (ICIMOD) for Himalaya and the Consortium for the Sustainable Development of the Andes (CONDESAN) are regional examples of that option.[25] Both have sought to establish a network of complementary relationships among the actors promoting sustainable development in their respective regions. At the global level, the Mountain Forum, intent on supporting or publicizing local initiatives that can be transferred elsewhere, has given their promoters a certain visibility and new resources.

Other mountain populations have become associated with exchanges initiated by scientists, who have set up comparative programs leading to tangible results for the locals. The U.S. Agency for International Development (USAID) financed a program that has allowed scientists and local actors from Ecuador, the Philippines, and Nepal to work together. The HimalAndes program, founded in 1991 by the Peruvian anthropologist Alejandro Camino, has facilitated contact between people and organizations based in the Himalayas and the Andes. Researchers in rural development undertook that collaboration, then opened it up to local actors in the interest of promoting concrete projects. According to the founders, direct cooperation between two mountain regions in the southern hemisphere offers potential beneficiaries "a more realistic and pragmatic option" than international conferences, "generating income and employment opportunities to the local mountain communities."[26] Another anthropologist, Robert Rhoades, a partner in the scientific research program Sustainable Agriculture and Natural Resource Management (SANREM), complemented his activities with an initiative, Mountain-to-Mountain Exchange, whose aim was to promote exchanges between local communities.

These operations have often run into difficulty in trying to enlist local actors in transnational or even transcontinental exchanges, since their framework is usually quite remote from these actors' practices. The Glochamore program, set up by UNESCO and financed by the European Union to facilitate the exchange of knowledge and experience between biosphere reserves in mountain regions, has regularly had to contend with the disinclination of the managers of these reserves to participate.

In other cases local organizations have taken the initiative for exchanges and collaboration. Foxfire, an organization based in the Appalachians that is active in environmental education and the promotion of cultural traditions (see chapter 5), has organized exchange visits in collaboration with a similar Ecuadoran organization, UNORCAC. A large number of mountain communities throughout the world have sought out twin cities, mak-

20. Opening ceremony for the Bhutanese suspension bridge in the Bois de Finges,
May 15, 2005. Within the context of the International Year of Mountains, the canton
of Valais in the Swiss Alps established a special partnership with Bhutan. Among
other projects, that exchange led to the construction of a suspension bridge, built
in conformance with Himalayan techniques, within the zone of a regional nature
park under way in Finges (Valais, Switzerland). Local workmen benefited from the
advice of Bhutanese participants invited for the occasion. The opening ceremonies
happily combined symbols of both cultures. Photo: Pascal Vuagniaux.

ing possible exchanges to seek common solutions to common problems
(fig. 20).[27] Such initiatives have sometimes benefited from the interest of
intergovernmental organizations and of the Mountain Partnership in promoting decentralized cooperation. It is in this context that the Val d'Aosta
in Italy provided aid to the Moroccan mountain territories. National committees for the International Year of Mountains also encouraged such initiatives, sometimes providing technical support and financial assistance.

The contributions of local communities to the global mountain agenda,
sometimes with decisive support from scientists and organizations involved in the design of that agenda, have been many and varied. Even so,
can these communities be considered major participants in that agenda?
Have they influenced its orientation? Such questions raise other, more political questions: How much weight do these communities really have? Are
they truly representative? What is their level of institutional involvement?
Beyond the local community, at the regional and global levels, who can
claim legitimacy to speak "in the name of the mountain people"?

Scientists for the mountain global agenda have not dodged such questions. In the conclusion to the principal scientific work promoting that

agenda (*Mountains of the World: A Global Priority*), the editors, Jack Ives and Bruno Messerli, in conjunction with the anthropologist Robert Rhoades, propose an "Agenda for Sustainable Mountain Development." They formulate "seven prerequisites for a twenty-first century agenda." The first prerequisite is called the "Mountain Perspective." They justify using that expression to designate how mountain people see the world: "Most of what has been written about mountains, or performed upon them in the guise of aid and development, has been undertaken from the outside by flatlanders and mainstream institutions, whether government or private, business or academic, conservationist or political. Very often a mountain perspective has been absent."[28] NGOs have formulated similar views: "Agendas for action in mountains need to fully represent the people who live there, because agendas made by outsiders usually fail to incorporate the local knowledge and priorities that are essential to sustainable development."[29] But the recurrent nature of such recommendations is sufficient to demonstrate that most experts in the field have not yet gotten the message. In a critical analysis of the movement that led to the political recognition of the mountains, an anthropologist congratulates the epistemic community behind the approach, a community to which he himself belongs. But he speaks of "unfinished business," arguing that "the voices of mountain people are largely silent in the present mountain movement."[30]

Strategies of Resistance or Alternatives: Who Can Speak in the Name of the Mountain People?

The expectation that mountain populations will express themselves directly has been met with several types of responses. One NGO, the Panos Institute, conducted a specific program, known as Mountain Voices, whose aim was to collect, preserve, and disseminate artifacts of collective memory and individual narratives in many mountain regions. The initiative came in response to an observation: "[During the UNCED,] much of the focus has been on environmental issues. Mountain peoples have had insufficient opportunities to speak out for themselves. And yet they are a vital key to understanding mountains and to their conservation."[31] In a sense, the Panos Institute set itself up as a mouthpiece that literally amplified the voices of the local populations.

In a large number of other cases, however, the inhabitants themselves have decided to raise their own voices. Vista 360, an organization based in Jackson Hole, Wyoming, announced the objective of promoting mountain identities and cultures. It portrays itself as "the only global mountain organization with a primary focus on cultural development."[32] To that end it developed exchanges between the region's inhabitants and other mountain populations throughout the world. The initiative is not without political motivations, however. It explicitly militates for the self-determination of these communities and for a growing initiative on their part "to take advantage of globalization rather than fall victim to it."[33] In the same vein, local community associations have emerged in various mountain regions of the world: the Alliance in the Alps, the Alliance of Central Asian Mountain Communities, and the Mountain Alliance of the Caucasus. These three networks attest to a globalization of mountain issues; they also maintain relationships with one another and hold joint meetings.

Finally, a worldwide organization has formed to give mountain people a voice, following the political principles and rules of representative democracies. The World Mountain People Association (WMPA) had its beginnings at a conference called the World Mountain Forum, held in Chambéry, France, in 2000. That event, which welcomed nine hundred participants from seventy countries, at the time presented itself as the "first meeting of the world's mountain people." It was planned at the initiative of the National Association of Elected Representatives from Mountain Areas (ANEM), an important French political lobby (see chapter 5), and played a decisive role in forming a comparable lobby across Europe as a whole (see chapter 10).

From the start the association reported its fears about the initiatives of scientists and U.N. organizations, and especially about their modalities. The interest groups representing mountain populations having been absent from the Earth Summit and related conferences, the Chambéry forum sought to give visibility to the populations themselves and to publicize their demands. It proposed to "bring recognition to the social, economic, and cultural issues that the mountain represents and to enlist the inhabitants of the highlands in writing a charter representative of all these aspects."[34] Early on, the WMPA, which emerged from that forum, declared its wish to become the official representative, the institutional spokesman, for the mountain populations in discussions about the future of the mountain regions: "Elected officials, representatives of the population, and economic as well as cultural and associative actors in the mountains could not leave it to the experts, whatever their areas of competence, to establish the desirable orientations for our territories."[35]

The WMPA's second reason for being wary has to do with the declared or perceived objectives of international initiatives. Like many other observers, the WMPA believes that the Earth Summit took more heed of environmental concerns than of the economic and social development of the regions concerned. The controversies that arose in Europe in the 1970s and 1980s (see chapter 4) led a number of European lobbies to view initiatives relating to environmental protection as interference from exogenous groups, most of them made up of city dwellers, in the affairs of the mountain communities and as a threat to their pursuit of economic development. According to them, a comparable risk looms at the global level because of how ecological problems encountered in the mountains are treated.

The Chambéry forum thus initiated a dual process of publicizing and institutionalizing the representation of mountain populations. The WMPA, which existed in embryonic form at Chambéry in 2000, officially came into being at the second meeting of mountain populations held in Quito, Ecuador, in 2002. The organization, divided into regional and national branches, now has a presence in more than seventy countries on four continents. In accordance with its founding objectives, the WMPA brings together representatives of local and regional communities. But NGOs involved in development projects also belong to the group, as do scientists who explicitly reject the collaboration with intergovernmental organizations undertaken by some of their colleagues.

In the years following its creation, the WMPA has clarified its objectives and expanded the range of its initiatives. The initial priority—to organize the political representation of mountain people on a global scale and to promote their autonomy in decision making—has remained at the top of the list. For example, during the 2002 summit in Bishkek, Pierre Rémy, general secretary of the ANEM and one of the founders of the WMPA, recalled that "the first thing that unites all people living in the mountains is that they should govern their mountain regions. This is probably one of the foremost factors for establishing cooperation between all mountain people of the world."[36] The association thus reaffirmed the right of mountain communities to take charge of the development of their territory: "Our alliance should enable very different people to come together on a common project that is very dear to them: taking the destiny of their country in hand."[37] The organization's strategy of promoting local autonomy on a global scale has also been exercised at intermediate levels, particularly the national level. Calls for local autonomy at international venues are directed less at intergovernmental organizations than at national delegations, especially for countries where the mountain populations' singularity is denied and their democratic representation deficient or overlooked. In the guise of globalism, the WMPA's real aim is to promote mountain lobbies in the countries of the world that lack them. In this it resembles the ANEM in

France, which has supported the project from the beginning. In that sense the WMPA complements the coalition to promote the global mountain agenda, whose ambition is to support mountain policies in countries currently without them (see chapter 8).

To construct its legitimacy, however, the WMPA had to be able to argue that a collective entity asking to be represented actually exists, thus reviving the idea of a global community of "mountaineers" assembled around an emerging political identity. At best, however, that collective exists only as a scientific category (the mountain populations of the world) and in popular representations—certainly not as a community with a sense of itself that would guarantee its cohesion. It may have a social identity; it certainly does not have a collective identity. The WMPA showed that it was perfectly aware of this when it listed, as one prerequisite for its objectives, the need to "construct the [global] community of mountain men and women."[38] All other things being equal, that situation and its acknowledgment recall the case of Italy in the mid-nineteenth century when, in the wake of unification, a Piedmont politician declared that the task of "making the Italians" still remained (see chapter 3). That is one reason the WMPA has promoted a series of ideas and slogans: the "shared problems" of mountain people everywhere, "mountain pride" throughout the world, and so on. But the WMPA's invocation of a mountain identity must not be understood solely as the means to the political end of setting itself up as the sole legitimate representative of the targeted population. It is also an end in itself, at least for those within the association. Since the beginning they have earnestly supported that humanist vision of a mountain internationale moved by the same attachment to a type of environment and by the same concern for solidarity and mutual curiosity.

Finally, the WMPA has economic ambitions. Intent on remaining as close as possible to the everyday problems of the communities it aspires to represent, it has sought to encourage the development of certified labels for agricultural and artisanal products, since such labels increase the product's value. Among its other projects, the WMPA promotes economical energy technologies and supports farmers threatened by the battle against the cultivation of illicit plants (coca in Bolivia, kef in Morocco, poppies in the Middle East).

Despite its activity over more than a decade, an overall assessment of the WMPA is difficult to make, and its list of achievements calls for some qualifications. The WMPA eventually overcame the distrust or skepticism of some intergovernmental organizations, which had had trouble understanding its motivations and approach, and has been accepted as a member of the Mountain Partnership. In addition, several issues close to the association's heart, especially that of illicit-plant cultivation, have been brought before intergovernmental organizations. Yet the WMPA still does not have a base that can guarantee its legitimacy. The "mountaineer community" of its dreams has yet to come into existence. Collective identities and engagement remain securely rooted in the local, the regional, and the national. It is true that many historians, geographers, and political scientists have shown that the nation-state was constructed around "imagined communities,"[39] especially through symbols that guaranteed connections among individuals. But however "imagined" the nation may be (that is, not built on individuals who know and depend upon one another), it exists through the set of images, institutions, *lieux de mémoire*, maps, and museums that tell its story and give it its substance.

Nothing similar exists for the mountain community. And an organization with a modest budget obviously cannot be expected to set up an apparatus akin to what the state has patiently constructed. Lacking adequate means, the WMPA struggles to hold regular meetings. In addition, though the association has spread to a large number of countries throughout the world, its leadership remains very European and even quite French. To date, for example, it has failed to gain much of a foothold in North America or East Asia, and a large number of organizations with similar objectives have remained outside it, finding no real benefit in a partnership. And because it is not supported by a "mountain community," the WMPA cannot serve as an umbrella organization federating associations and NGOs with comparable aims.

The global mountain community is still only a prospective political subject. So far, such collective and political mountain identities have been constructed in only a few nation-states (see chapter 5) where it has been possible to combine such identities with national identities and place them in the service of political strategies directed at public debate and public policies. Mountain identities also seem to be taking shape in a few transnational contexts (see chapter 11). But despite the diverse forms that globalization currently takes, it has not given rise to comparable political recompositions and reformations of identity at the global level.

Do Mountain Women Have a Place on the Global Agenda?

Within that context of a globalization of mountain issues and a recomposition of collective and political identities, some of those involved have endeavored to grant a place to a particular category of mountain dwellers: the mountain woman.

For some thirty years international conferences have given increasing space to questions relating to the condition of women and their role in society, thus taking up on a global scale debates and policies that had arisen in a large number of regional and national contexts. That systematic consideration of gender questions, known as "gender mainstreaming," has particularly marked major conferences on the environment and development.[40] Principle 20 of the Rio Declaration and chapter 24 of Agenda 21 are both devoted to women. That concern also appears in the implementation plan for the Johannesburg Summit.

Such a shift is also perceptible in initiatives relating to the mountains. Chapter 13 of Agenda 21 deals with that general question, adapting it to the context of the regions concerned. The motivations for invoking women and women's condition as part of the global mountain agenda are related to the reasons that led to inclusion of the local populations in general. In some cases they stem from social and economic concerns, especially the fight against poverty and social inequalities. In others they result from an interest in optimizing the environmental benefits of sustainable development policies. Researchers and development experts have noted for many years that the division of roles by gender makes women strategic interlocutors. In particular, women's contribution to agriculture and to the exploitation of water and timber resources appears decisive for understanding local modes of production and the pressure of traditional activities on the environment. Global and regional organizations have concluded that, to solve environmental problems, a preferential place must be given to women. Some believe, therefore, that a "natural convergence" between the environmental agenda and the women's agenda is taking shape.[41] That analysis has played a decisive role in promoting a gendered approach to development projects in the mountain regions: "Since mountain women play a central role in the sustainable use and management of resources, the specific needs and constraints of mountain women must be addressed."[42] That discourse has even found its way into the U.N. General Assembly, which emphasizes "the need for improved access to resources for women in mountain regions as well as the need to strengthen the role of women in mountain regions in decision-making processes that affect their communities, cultures and environments."[43]

These concerns gave rise to a series of initiatives whose aim was to organize and globalize the debate on the category of the mountain woman. In 1988 the Mountain Forum created a discussion group called the "Mountain Forum Discussion List—Women." Its eight hundred members corresponded about "issues related to women living in mountain environments."[44] Within the Mountain Partnership, a working group on questions of gender comprised thirteen countries, five intergovernmental organizations, and twenty-six groups[45]—forty-four members in all. The initiative announced it was seeking "to ensure that gender equity is mainstreamed in mountain development policy and action."[46]

In addition, several conferences on special topics have addressed the question of women: in Katmandu in 1988; in Orem, Utah, in 2007 and 2011; and, most important of all, in Paru and Thimphu, Bhutan, in 2002, where the Celebrating Mountain Women conference was held (fig. 21).[47] Two hundred and fifty participants from thirty-five countries gathered at that conference, the proceedings of which were published in 2012. It sought above all to offer a forum for mountain women around the world, a regrettable absence at earlier events.[48] The announcement for the conference called it a "unique opportunity to highlight the realities of life in the mountains and put women on the mountain agenda."[49] The final document, known as the "Thimphu Declaration," was presented as "a message from the mountain women" to the global mountain community.[50] That message was meant to be passed on to the Bishkek summit, which concluded the International Year of Mountains. Celebrating Mountain Women also resulted in the constitution of an organization, the Global Mountain Women's Partnership, whose declared purpose is "to promote the interests and perspectives of mountain women and contribute to an improvement of their livelihoods."[51] The organization thus aspires to lobby for the cause of mountain women at the global level.

But many of these initiatives went nowhere. The Global Mountain Women's Partnership was never a functioning entity. Several factors may explain that state of affairs, in particular the lack of involvement by global feminist organizations and the suspicion that the cause was being instrumentalized by people generally motivated by environmentalist considerations. Another factor sheds light on the limits imposed on the globalization of mountain issues. Like "mountain populations" considered as a totality, the category "mountain women of the world" never managed to lead to collective action, since the people so designated were unable to constitute themselves as a political subject. The category has remained an intellectual construct.

It is not that no efforts were made to delimit the category and to test its relevance. Such was in fact the objective of the first discussion devoted to women at the Mountain Forum. Participants were instructed to discuss the similarities of mountain women's condition throughout the world. They proposed a number of conditions that such women share: a heavy

21. Cover illustration of a publication produced by the ICIMOD
within the context of the Celebrating Mountain Women conference. Phuntshok Tshering and Ojaswi Josse, *Advancing the Mountain Women's Agenda: A Report on a Global Gathering "Celebrating
Mountain Women" in Bhutan, October 2002* (Katmandu: International Center for Integrated Mountain Development, 2003).

workload, the seasonal outmigration of the men, poverty, a lack of infrastructure, and a specific knowledge of natural environments. The U.N.
General Assembly recently took a similar stand: "Mountain women face
many of the same challenges that are faced by women throughout the developing world, but their work is intensified by altitude, steep terrain and
isolation."[52]

In addition to the difficult living conditions, another similarity became
a rallying point: mountain women belong to marginal societies. As Phunt-

shok Tshering and Rosemary Thapa point out, even as "women all over the world continue to struggle to be accepted as equals, to have their values recognized as relevant, and to overcome the multiple burdens of home and employment, . . . mountain women are further challenged as they belong to societies that are already marginalized and often cut off from mainstream society."[53] That double marginalization—as a subaltern group within already marginalized societies—has been a recurrent theme of transnational discourses that assert the specificity of mountain women's condition and constitutes one of the key points of the Thimphu declaration: "Without effective policies, networks, partnerships and alliances at the local, national, regional and international levels, mountain women's economic, social and political marginalization will continue to hamper their development and the development of their communities."

Another recurrent concern has been the upheavals in mountain societies as a whole. Particularly worrying is the deterioration in women's status in many communities throughout the world. In many regions, it is said, women long enjoyed a relatively favorable status: "Because of the predominance of less rigid religious beliefs within the indigenous systems, and the dominant role of women in the livelihood systems of the mountains, mountain women have traditionally been afforded more freedom of movement, greater independence in decision-making, and higher status than women of the lowlands."[54] But the integration of mountain societies into national and international economies may have threatened these advantages:[55] "Indications are that women's value in their households, communities, and societies is declining as traditional mountain societies are being transformed by the prevailing values belonging to lowland religious, nationalistic and cultural paradigms."[56]

But are all these assessments enough to define the category and make it operational? Certainly not: women's conditions in the mountain regions are as diverse as the mountain societies themselves.[57] "Does the mountain woman exist?" a Swiss researcher wondered.[58] It must be noted that, beyond official declarations, diversity remains the rule. Even more important, no shared condition, problem, or interest has arisen that could provide a common representation for the interested parties themselves. Most of the local organizations dedicated to women's issues have not perceived any validity in coordinating their actions on a global scale, or any potential for doing so. As a result, the collective identity of mountain women is weak; other forms of collective identity appear much more relevant.

Such situations have been the object of lengthy debates in the social sciences. The discipline of gender studies in particular has often discussed the conditions under which women could conceive of themselves and mobilize as a group on the basis of several social markers.[59] Many studies have concluded that "sisterhood," like the mountain populations, is heterogeneous.[60] It is now agreed that the term "women" covers a great diversity of experiences, needs, concerns, and interests that cannot be unified into a single female identity. In particular, studies have pointed out the difficulty of defining the categories of "Third World" and "rural" women,[61] a difficulty that undermines the validity of invoking a collective, monolithic, homogeneous, and universal subject. The notion of being a "mountain woman" may seem to have meaning within the context of global mobilization and among activists with a remote and synthetic view of regional situations. But for local actors it is not self-evident. A network of women active in Asia, when questioned by researchers seeking to measure what identities informed the women's work, declared: "We do not emphasize mountains in our agenda since most of us are in the mountains anyway but we do not recognize the particularities of the mountain environment by which our cultures have actually developed."[62]

It must therefore be said that declarations of (good) intentions to improve the status of women in the mountain regions have not met with the desired response. Such declarations have been unable to turn the imperative to feminize international initiatives into a collective force for action. The voices of these women have trouble making themselves heard at a global level for the same reasons that the voices of the mountain populations as a whole falter. The globalization under way has not (yet) led to solid new forms of identity that could move the debate about the future of the world's mountains in the direction of collective action.

Transnational Recompositions of Identity and Identity-Based Strategies in the Mountains

We see, then, that various attempts have been made to construct a community of mountain men and women at the global level in the wake of the institutional globalization of mountain issues. Such efforts have relied on discourses whose aim is to homogenize the category or, more realistically, to gather local and regional singularities around a common frame of reference—namely, a type of environment. They have sometimes invoked shared problems and interests that might transform that group into a collective political subject. Thus far, however, this strategy of transformation remains at a formative stage. If it hopes to succeed in positioning that collective subject within a globalized mountain policy, the populations concerned must clearly conceive of themselves in global terms. Thanks to popular science, the image of a mountain system structured at the global level has become widespread; it has even made its way into ordinary geographical representations. So far, however, it has played only a marginal role as a common frame of reference for constituting a shared global identity.

The process under way does seem to have had more noticeable effects in certain regional contexts, however. In fact, the proliferation of transnational and international initiatives has provided political alternatives to a representation of the mountain and of mountain populations circumscribed within modern nation-states. It has facilitated the emergence and spread of reconfigurations of identity that take the mountains as their frame of reference on the scale of transnational massifs and transborder regions. But that movement does not affect all mountain regions to the same degree. It is weak in the Caucasus, strong in the mountains of North Africa, and uneven in the Andes, where each geopolitical entity assumes a different role.

The Caucasus: Identities Thwarted in the Mountains

Russian expansionism and the colonial policies adopted by the czarist empire and the Soviet Union (see chapter 3) constructed a unified representation of the Caucasus. Scientific development, driven by the conquest, imposed the structured representation of the Caucasus as a single mountain range. Authoritarian policies targeting the Caucasian populations allowed an image to take hold of a mosaic of populations quick to resist Russian ambitions. In that new geographical and geopolitical context, the local populations themselves adjusted their self-representations and interpreted the figure of a mosaic in their own ways.

Beginning in the mid-nineteenth century several attempts were made to federate the populations of the Caucasus in order to counterbalance their dispersal by the colonial power. The war waged by czarist Russia in the 1830s gave rise to an independent state in the eastern Caucasus that lasted for more than two decades (1834–59).[63] In the western Caucasus the Circassian tribes, as they were known, also attempted to unify and received support from Westerners. David Urquhart, a young Englishman, sought to organize their resistance, even writing for them a "declaration of independence" that he sent to all the courts of Europe. These first germs of the idea of a confederation of Caucasian peoples would resurface well after the resistance was crushed.[64]

The 1917 Revolution revived the idea. In May 1917 the Bolsheviks held the first North Caucasus Congress. Although they saw the North Caucasus as an entity to be integrated into revolutionary Russia, they promoted a special political status for the region. That same year in Vladikavkaz, nationalist leaders in the region formed a Union of Mountain Peoples, a unified force bent on opposing the Bolshevik policy of incorporation. The group mounted its resistance in 1928, demanding the independence of North Caucasus within the Mountain Republic.[65] The Soviets regained the upper hand, but they also granted concessions. Federalism was officially promoted from 1924 onward. That principle, along with the constitution of Soviet republics in the southern and eastern Caucasus (Armenia, Georgia, and Azerbaijan) and of autonomous republics in the north (Chechnya, Dagestan, and others), maintained appearances. Nevertheless, the Soviet Union opted to partition the region along ethnic lines—though in practice, its notion of ethnicity was approximate to say the least—to the detriment of unification among the mountain peoples.

With the fall of the USSR, the question once more arose of what would become of these federated and autonomous republics. The breakup of the Soviet Union occurred along its internal frontiers. Georgia, Armenia, and Azerbaijan achieved independence, turning a good part of the ridge line of the Caucasus into an international border. The autonomous republics retained their status within the new Russian Federation. But age-old tensions resurfaced: demands for independence in Chechnya, separatist demands in Ossetia and in Georgian Abkhazia, tensions in Dagestan. They gave rise to a new wave of military operations intent on repressing these movements or on modifying international borders—during the war between Russia and Georgia in the summer of 2008, for example.

What figures of regional unity were enlisted in that context? The figure of a mosaic of peoples regained currency as a result of worsening interethnic tensions, sometimes encouraged by the countries of the region. The figure of Islam acquired new force. Although Islam had been used as a rallying point for some who resisted the Russian invasion, the Soviet regime had put a damper on the expression of religious identities. These resurfaced with the dismantling of the USSR and the rising influence of Islamic fundamentalism. Although not all the populations of the Caucasus are Muslims, most are, and some advocate unifying different populations on that basis.

The figure of the mountain also enjoyed a resurgence in interest. For several parties active in the region, it had the advantage of not being associated with memories of intercommunity tensions based on ethnic or religious identities. Like other organizations in the region, the Mountain Alliance of the Caucasus, created in 2006 on the model of the Alliance in the Alps (see chapter 11), made use of that figure in its search for frames of reference common to all the populations of the region. The Caucasian alliance favored the exchange of information and expertise among local actors; its objective was to implement sustainable development at the local level and throughout the Caucasus Mountains. Members include communities from countries on either side of the chain: Armenia, Azerbaijan, Georgia, and the Russian Federation. The Deutsche Gesellschaft für Technische Zusammenarbeit (GTZ) widely supported its establishment and operations in 2002, the year Germany chaired the Alpine Convention.

Berber Identities and the Rediscovery of the Mountains

Colonial France contributed greatly to naturalizing the distinction between Arabs and Berbers in North Africa. The identity and singularity of the Berbers were frequently associated with the mountains, where many Berber speakers lived (see chapter 6). The political elites who took control of the young independent nations of Morocco (in 1954) and Algeria (in 1962) sought to deny both the distinction and its geographical basis, in the name of a unitarist concept of the nation.[66] The status of Berber language and culture, however, remained a sensitive issue throughout the second half of the twentieth century. In the early 2000s Berber speakers reportedly constituted 20–25 percent of the population of Algeria and about 50 percent of that of Morocco. In regions where Berbers are in the majority, the issue sparked social movements, strikes, and riots, such as those in Kabylie in the 1980s.

That question acquired newfound importance a few years ago within the context of increasing demands at the national level by the minority autochthonous peoples, the political transformation of the political regimes in place, and the globalization of mountain issues. In particular, these changes have led to the emergence of an international organization, the World Amazigh Congress. It claims to be the representative of the Berber populations (renamed "Amazigh" to avoid an exogenous designation) living in the Canary Islands, Egypt, the Mediterranean, and the southern Sahara. The mountain has also reemerged as a frame of reference for the formation of these populations' identity.

Militants of the Berber cause are well aware that the spatial distribution of Berber populations does not coincide with that of mountains, however these may be defined. Before the Arab invasion, Berbers settled in all regions of the Maghreb, and historians have demonstrated that the invasion did not cause a mass migration of the Berbers to the highlands. The mountains of the Maghreb are thus not refuges. By contrast, the Arabization of Maghrebi societies, through the adoption of Arabic especially, did occur to a lesser degree in the mountain regions. Berber singularity is therefore not a fact of nature but a linguistic fact, a result of cultural resistance within a context of strong acculturation. Furthermore, a number of Berberophone populations have lived for centuries in the oases of southern Algeria and in regions of the Sahara, sometimes very far from the mountains. It is on this basis that researchers generally refute the idea that the Berbers of the Maghreb are mountain dwellers and that the mountain people are Berbers. In addition, current specialists in the social sciences still remember the instrumentalization of ethnography and geography by the colonial powers and the political aims that led the occupier to associate Berbers with the mountain for strategic reasons.

Under these circumstances, why was the Amazigh movement so quick to make use of the figure of the mountain to justify its cause? It seems that the principal reason lies in the symbolic and expressive resources of that figure. Indeed, though for at least two centuries the mountain regions have occupied an increasingly marginal position economically and culturally in both the West and the Arab Muslim world, their favorable symbolic image has grown on both sides of the Mediterranean. The positive value associated with mountains in the monotheistic religions has been complemented by their aesthetic appeal for European tourists. Aestheticizing representations of the Atlas Mountains and the Djurdjura Range, for example, proliferated in the first half of the twentieth century, often at the initiative of the Maghrebis themselves. The visibility and the symbolic charge associated with the mountain make it all the easier to enlist as a form expressing the identity of the place and of its communities.

In addition, a few particularly well-identified massifs in the Maghrebi mountains have often served as emblems for the region as a whole: Kabylie, the Rif, and the Atlas. These massifs include some of the highest mountains in the region. They are densely populated and have very structured farming societies, whose members often practice a moderate form of Islam. For these reasons they were not greatly affected by Islamic fundamentalism in the 1980s–90s. The historic prestige conferred on them by the great dynasties that originated there (such as the Abbasids in Morocco) and by their fierce resistance to the occupying forces and normalization has allowed their champions to present them as a regional variant on well-established myths, in particular those of the free mountaineer and peasant democracy.[67]

But not all the mountains of the Maghreb lend themselves to that simplified image symbolized by a few particular massifs. Other regions have very different characteristics: the Middle Atlas of Morocco and the Ouarsenis of Algeria, for example, are nomadic herding regions with looser social structures. The mountains of the Tell and of southern Morocco have low population densities, greatly reduced by recent emigration and impoverishment. In these regions Islamic fundamentalism has thrived, an indication of a certain degree of social disintegration and despair that stands in contrast to what can be observed elsewhere. The mountains of the Maghreb are thus home to various societies with different trajectories. They do not easily lend themselves to the construction of a regional version of the positive myth of the mountaineer.

In the WMPA the mountain rhetoric of the Amazigh movement found a privileged vehicle for institutional expression and visibility at a global level. Early on, that organization sought a foothold in the Maghreb, taking advantage of the many connections between France, where the WMPA originated and where it has its headquarters, and its former colonies. Soon members of the Amazigh movement became the principal supporters of the national delegations of the WMPA in Morocco and Algeria. The two associations use the same political rhetoric: they argue for the singularity of the mountains and for the identity of mountain people in order to achieve increased decision-making autonomy for these populations. As a result, the publications and declarations of the two groups converge and reinforce each another.

A brochure published by the WMPA, titled "The Mountain People Are Organizing and Innovating in Service of Their Mountains," describes several experiments conducted for the purpose of social or economic development. One case study deals with the resurgence of beekeeping in Kabylie at the initiative of a local organization. The first of the ten paragraphs devoted to the question does nothing more than attest to the mountainous character of Kabylie through the use of descriptions of the landscape, toponymical references, and historical narratives.

In November 2006 the WMPAP, the World Amazigh Congress, and the Coordination des Berbères de France gathered in Nador, Morocco, for what was billed as "the first joint meeting of the mountain populations of North Africa." The declaration that emerged from that conference repeatedly instrumentalized the equation between the mountains and Berber identity. According to the preamble, "the mountain populations of North Africa wish to report their institutional and economic marginalization and the permanent degradation of their living conditions and standards. . . . In addition, their identity has been subject to exclusion, primarily targeting

Amazigh culture, language, and history. The ability of these populations to control and manage responsibly the natural resources of their territories is not recognized. And yet, over the course of centuries, they have been able to develop skills, behaviors, and practices that have allowed them to live in these difficult environments. They are proud to have created a civilization specific to the mountains of North Africa." The equivalence between mountain environment, civilization, and marginalization having been established, the declaration continues: "We, actors and representatives of the populations of the mountains of North Africa, want an acknowledgment of our identities and cultures; official recognition of the Amazigh language; preferential treatment and specific means earmarked for mountain territories, which will allow them to make up for the delay in access to basic services . . . ; a recognition of the right to manage the territories . . . ; and so on." Along the way the declaration appropriates some of the terms of the Rio and Johannesburg conferences: "recognition of the fundamental place of women in mountain society . . . ; implementation of convention 169 of the ILO and of the United Nations declaration on autochthonous populations; full exercise of democracy; and so on."

The Andean Populations: Indigenous and Mountain Identities

Comparable political recompositions of identity have occurred in the Andes, especially the central Andes, between Ecuador and Bolivia. That part of the range is densely populated, and a large proportion of the population is considered indigenous. In the nineteenth century the national imaginary of some newly independent countries, such as Ecuador and Colombia, promoted the idea of a unitary nation, while the imaginary of Peru (see chapter 8), for example, led to a tripartite division of the nation based on a naturalistic reading of its territory (*silva*, *sierra*, and *costa*). Nonetheless, these imaginaries always inscribed identities within a national context rather than an Andean or transborder one.

In the 1970s the indigenous peoples of North and South America began to play a real leadership role in constructing and publicizing in international arenas the political demands of the autochthonous peoples.[68] In some of their battles and forms of action, they were also on the leading edge of the antiglobalist campaign. They mobilized against many of the legacies of colonization and the nineteenth century, especially latifundia, and against the indifference of other nations and multinational corporations toward their rights and territoriality. Indigenous people protested the mining concessions being granted to global businesses in the sector, the constitution of new large landed estates, and state management of water resources, which disadvantaged the local populations. Taking advantage of the recent affirmation of the rights of autochthonous peoples, they were able to build alliances with international organizations and NGOs that supported their actions.[69] That movement has also benefited from the increase in the number of tourists to the Andes, who are curious about the specifics of the indigenous peoples.

To defend their cause, these populations prefer a symbolic frame of reference that is ethnic or cultural in nature. While accepting the social partition adopted by most national administrations (indigenous, mestizo, and Spanish), they demand equal rights for all, including economic and cultural equality, and call for the defense of individual ethnic identities. In several countries the indigenist movement has demanded that the constitution recognize the "plurinational" character of the populations. That term is used in the new Bolivian constitution, adopted by referendum in 2009, and the new Ecuadoran constitution, approved by the electorate in 2008. The people in these countries can draw on an extensive range of identities. Depending on the issue, they may claim to be *indios, indígenos, aborigenes, campesinos,* or, more recently—and the most relevant here—*andinos.*[70]

To what extent did these populations need the mountains in order to promote their cause and defend their interests? In a large number of cases the power of the indigenist movement and the ethnic foundations on which it rested were able to dispense with making reference to the mountain. The most famous case is that of Chiapas.[71] This very mountainous region in southern Mexico was the site of the most emblematic indigenist protest movement, which targeted both the place of the indigenous populations in Mexican society and the incursion of multinational corporations, made possible by the neoliberal policies of the government in place. The mountainous character of Chiapas has played only a marginal role in the struggle of the indigenous peoples of the region. Their rhetoric was constructed exclusively on ethnic, antiglobalist, and (specific to Mexico) Zapatista foundations.

Other regional movements have referred explicitly to the Andes in asserting their identities and expressing their demands. That is especially true of the Aymara in the border regions of Chile, Bolivia, and Argentina. In that transborder zone, where the indigenous ethnic group is in the majority, an organization known as Aymaras sin Fronteras (Aymara without Borders) has managed to combine autonomist demands and cultural affirmation with the contestation of colonial borders. The image of the Andes has allowed them to refer both to the historical prestige of pre-Columbian civilizations and to the topographical connection that the mountain range seems to establish among the indigenous peoples.[72]

So far the new configuration of the mountains that emerged on a global scale in the late 1980s has not really led the populations most concerned to constitute themselves as a political subject. Global spokespersons are having difficulty finding their voice. For the most part, regional identities and demands have their roots in the rich soil of cultures and ethnic groups, enlisting the image of the mountain only on the margins, for its rhetorical virtues. The mountain populations do not readily identify themselves as such; when they do, they are not very involved in identifying shared points of reference, problems, and battles. They do not (yet) constitute an "imagined community."

TEN

The EU Mountain: Nowhere to Be Found?

In 2001 Valéry Giscard d'Estaing, former president of the French Republic, assumed the presidency of the newly established Convention on the Future of Europe (European Convention), which was given the task of drafting the treaty that would establish a constitution for Europe. On February 27, 2003, he received three officials at their request: Luciano Caveri, member of the European Parliament representing Italy; Michel Bouvard, French deputy; and Nicolas Evrard, secretary general for the European Association of Elected Representatives from Mountain Regions (AEM). All had been, were at the time, or would later become elected representatives of mountain municipalities or regions: Giscard for Auvergne, Bouvard for Savoie, Caveri for Val d'Aosta, and Evrard for a municipality in the region of Mont Blanc. Bouvard, Caveri, and Evrard were very involved in associations of elected representatives from the mountain regions. Caveri had just transferred the presidency of the AEM, the association for which Evrard served as secretary general, to Bouvard, former president of the French association (ANEM).

The three men requested the meeting to ask that the constitutional treaty being drafted make reference to the mountain. The term "mountain" was absent from all previous treaties. For ten years the AEM had been doing everything possible in Strasbourg (seat of the European Parliament), Brussels (seat of the European Commission), and other European capitals to make the case for considering European mountains within EU policies. Giscard d'Estaing promised them his support provided they could prove that the mountains were an EU and a European issue, not simply a local, regional one.

In November 2003 an informal meeting was held in Taormina, Sicily, of ministers in charge of regional planning for the member states and for the Central and Eastern European countries that were to join the European Union a few weeks later. With them were officials from the European Commission. Italy, president of the European Union at the time, was determined to promote recognition of the mountains in all member states. The participants at that meeting, citing the International Year of Mountains and the Johannesburg conference (see chapter 8), agreed to that objective.

In Rome on October 29, 2004, the twenty-five member states of the European Union signed the treaty establishing a constitution for Europe. That treaty stipulated that "particular attention shall be paid to rural areas, areas affected by industrial transition, and regions that suffer from severe and permanent natural or demographic disadvantages, such as the northernmost regions with very low population density, island, cross-border, and mountain regions." The lobbyists proclaimed victory: the mountains, largely absent from previous EU policy, within two years had shed their marginal status. Despite the rejection of the draft of the constitutional treaty by referenda in France and the Netherlands, a turning point had been reached. The wording was adopted in the Treaty of Lisbon, ratified in 2009.

The issue may seem trivial in view of the energy expended. Does the presence of one apparently insignificant word in a document—even a "constitutional" document—merit so many precautions and negotiations? It certainly does. Once adopted and ratified, such a document, similar in this respect to international declarations, can be cited by any and all to argue their point of view and the legitimacy of their proposals. To mention the mountains—as well as the cities and coasts—in a European treaty is quite clearly to facilitate the work of those who want the European Union to adopt these categories of thought and, if need be, give them a political content. That was truly the objective of the lobbyists who in the early 2000s increased pressure on the European Commission and member states. For half a century the mountain and, more broadly, a large number of territorial questions had been left aside in the plan for a unified Europe or had in fact been deliberately effaced.

The European Project: Jurisdiction over Public Policies and Their Objects

The history of the European project is characterized by three complementary movements designed to deal with the geographical entities within its purview.

The first, which is also the best-known and most obvious, is the redistribution of authority to different institutions. The adoption of an agricultural policy on the scale of the European Union belongs to this first movement. It is not simply that jurisdictional authority has been transferred, since the member states and sometimes the regions, Länder, and provinces have held on to their administrations. They still have a political margin for maneuvering in such matters. But the Europeanization of agricultural policy set out to rescale decision making and administration, with the European Union acquiring greater authority over time.

The second movement has to do with the geographical scale on which one reaches a diagnosis and conceives of the terms of a policy. The geographical rescaling of decision making and administration in agricultural policy has gone hand in hand with a reformulation of descriptions, problematics, and recommendations (geographical and economic in particular) in order better to take the institutional changes into account. That is also the case for EU environmental policy: the categories that member states were using to conceptualize the diversity of ecosystems and biotopes had to be adapted to the change in scale of environmental analyses, which resulted from the new EU jurisdiction in the matter.[1] Water policies, when treated at the EU level, moved in the direction of integrated management policies of catchment basins, which often straddle borders.[2] The transportation policies of different countries were also brought into line, at least in part, so that transnational projects could be designed and implemented throughout the Alps, for example.[3] Models of urban governance also evolved a great deal as a result of the increased power of the European Union and the adoption of a European scale of analysis and comparison.[4]

The third movement characteristic of EU policies has been the progressive shift from sectoral policies (agriculture, energy, transportation, the environment, and so on) toward coordinated or cross-sectoral policies (on landscape, rural development, sustainable development, and territorial development). This is a recent and as yet very incomplete shift. The still dominant influence of sectoral policies has dictated that geographical entities and the objects associated with them (communication routes, coal fields, ports, and forests) have been more clearly apprehended than more complex entities, such as the cities or the coasts, where many disparate factors must be taken into account.

The history of the European project is thus that of a phenomenal reshuffling of the type of objects judged pertinent, the level of analysis and reference, and jurisdictional authority. The specifically institutional issues, which concern jurisdiction, are not independent from the question of how the objects of these policies are identified. It is therefore possible to speak of a "politics of scale" and of a political construction of objects of knowledge (in this case, on the scale of the European Union).

For a long time the European mountain remained apart from that range of objects and that reconfiguration of scales by which the EU project came into being. The mountain has gained visibility in the last decade, however, because of several shifts in EU policies and institutional power relations. It thus lends itself particularly well to an analysis of the modalities for constructing an object of EU policies and, at the same time, to an analysis of ways of redefining that object, circumscribing it, even while ensuring that it is "Europeanized."

Promoting a European Space Free of Obstructions

There are several very different explanations for why European policies have not taken the notion of the mountain into account to any great extent. In the first place, the European project was originally motivated by a vision that was not territorial. True to the conception of its originators, especially Jean Monnet, the founding states agreed to set in place a free-trade zone, the Common Market, which they expected would reduce the differences among the nation-states and promote peace, democracy, and human rights. That free-trade zone was destined to expand to neighboring countries that shared the same political ambition and the same strategy. The European Community, established in Rome in 1957, was thus not a political project comparable to that which had created the founding states. Its purpose was not to circumscribe a people or a coalition of peoples within a common territory. In fact, it lacked a territorial imaginary as such. Free trade is based on the idea that, within a certain zone, commodities first and money and people second circulate freely. Any interference with that free circulation is perceived as an obstruction.

In the representations inherited from previous centuries, mountain ranges constitute such obstructions. The many international borders drawn along the chains break up the continent. Border checkpoints are hindrances to free circulation. The free flow of commodities transported by train is complicated by topography and obstructed by national standards. Some are technical in nature: electric voltage, for example, or rail track gauge, which is different in France than it is in Spain. Others are fiscal and regulatory. And the absence of a territorial imaginary comparable to that possessed by nation-states in the previous century deprived the mountains of their emblematic role as a political image. That spatial conception of Europe allows us to understand the important role that the European

Community played in financing major transportation infrastructure from the 1970s on. The promotion of continent-wide traffic corridors made the crossing of mountain massifs a particularly important issue. The geopolitical configuration inherited from the nineteenth century also made the mountain regions the privileged beneficiaries of a transborder cooperation program (Interreg), which, beginning in the 1970s, sought to reduce the differences in border regions across the European Community. Granted, the mountain is never described as such in the European Community establishment treaties. But in its first article, the Treaty of Rome affirms the desire of the founding states "to ensure the economic and social progress of their countries by common action to eliminate the barriers which divide Europe." The mountain could easily serve as a privileged illustration, since it had so often taken on that function in previous centuries.

A Common Agricultural Policy and the Mountain as Handicap

The mountain was also absent from common agricultural policy in its early days. The priority placed on agricultural policy in the early 1960s can be explained by two factors: an obsessive concern with food security in a Europe that had only recently emerged from war restrictions, and an interest in modernizing an economic sector that had lagged behind the major transformations of the interwar period. The first years of common agricultural policy thus emphasized productivity, profitability, and farming competition—the guarantee, according to its promoters, of a virtuous modernization. Contending with overproduction for the first time, Sicco Mansholt, the first commissioner for agriculture, proposed to cut aid to producers and to reduce the acreage cultivated by five million hectares. Mountain farmers felt particularly threatened: their farms had low profit margins and low productivity, for reasons both natural (a short growing season, topographical obstacles to mechanization) and socioeconomic (the size of traditional farms).

A European Economic Community directive adopted in 1975 took into account geographical differences in production conditions, particularly targeting "mountain and hill farming and farming in certain less-favoured areas."[5] These were deemed eligible for financial assistance. But rather than allow the European Commission to decide which regions would benefit based on its own definition of the mountain, the directive states that it is up to national governments to determine their own zoning of mountain regions. Different governments proceeded in very different ways: countries that had already demarcated a mountain zone within the context of their own agricultural policy (Italy in 1952 and France in 1961, for example) generally proposed the existing parameters. The definition could therefore be broad, encompassing all the farms that had the slope and elevation judged to be a handicap. It could also be narrow, as it is in Austria, which has been a member of the European Union since 1995. That country's official definition of the mountain, inherited from the Austro-Hungarian Empire, initially included only high-elevation grazing lands. Other countries, such as the United Kingdom, did not even make use of that opportunity, despite the fact that most Britons consider the Highlands of Scotland, a large part of Wales, and a few regions of northern England to be clearly mountainous. As a result, the European definition of the mountain appears strangely heterogeneous and at odds with dominant social representations. When jurisdiction over agricultural matters was transferred to the European Community, the scale on which agricultural problems and specificities were defined had trouble keeping pace.

Thanks to subsequent regulations, national definitions of the mountain have tended to converge in their principles and modalities. To circumscribe their mountainous zones, most member states have adopted elevation thresholds, higher on average in the southern part of Europe (1,000 meters in Spain, 800 meters in Greece and Mediterranean France) than in the northern part (200 meters in Ireland, 240 meters in the United Kingdom, 300 meters in Belgium), as well as criteria relating to slope. The countries that joined the European Union in 2004 and 2007 have followed suit.

Despite the scales and standards for defining the mountain, the criteria used in that definition were subordinated to sectoral policy. The features selected were motivated solely by the concern to grant compensation for agricultural handicaps. Article 18 of the most recent EU regulations in that area (Council Regulation EC 1257/1999 on support for rural development) is very explicit:

> Mountain areas shall be those characterised by a considerable limitation of the possibilities for using the land and an appreciable increase in the cost of working it due to the existence, because of altitude, of very difficult climatic conditions, the effect of which is substantially to shorten the growing season; at a lower altitude, to the presence over the greater part of the area in question of slopes too steep for the use of machinery or requiring the use of very expensive special equipment, or to a combination of these two factors, where the handicap resulting from each taken separately is less acute but the combination of the two gives rise to an equivalent handicap.

The EU mountain, a juxtaposition of national mountainous zones, is thus mainly considered an agricultural handicap zone whose defining criteria stem from the concept of handicap itself. Note, for example, that regions without noticeable differences in altitude have been included within that definition because of their short growing season. According to common agricultural policy, the zones at high latitudes in Scandinavia and Finland obtained the same classification as mountainous regions: "areas north of the 62nd Parallel and adjacent areas shall be treated in the same way as mountain areas."

EU Environmental Policy

The same holds true for another European Community policy that was expanded in the 1970s: environmental policy. In that case, the European Community did not seek to take the place of member states in creating protected areas. But it was often through "nature reserves," "natural parks," and "national or regional parks" that European governments demonstrated their interest in mountain landscapes and, later, ecosystems and biotopes (see chapter 4). The European Union has been primarily interested in defining uniform standards for protected areas; in obtaining a better representation of ecosystems that had heretofore drawn few benefits from these preservation measures; and, even more, in promoting environmental policies of general import (on air and water quality, waste treatment, cleanup of polluted sites, and so on). At best, the mountain appears here and there as a type of ecosystem, but more often as a set of different ecosystems that deserve the attention of the European Commission, especially in its role of protecting biodiversity.

To that end, the commission divided the territory into eleven biogeographical regions, "each with its own characteristic blend of vegetation, climate and geology."[6] Within the context of the Habitats Directive, Natura 2000 zones were selected in concert with member states to create a group of protected sites that attest to the variety and quality of the European environment. One of these, called the Alpine region, initially included the Alps, Pyrenees, central Apennines, and Scandinavian mountains; the Carpathian Mountains and a few Balkan ranges were added when the European Union was expanded. The other European massifs belong to more compact zones, characterized primarily by climate: the Scottish Highlands and the Cantabrian Mountains of Spain were attached to the Atlantic region; the southern Apennines and the Greek and Iberian mountains, apart from the Pyrenees and Cantabrians, to the Mediterranean region; and the Vosges, Hartz, and Ore Mountains to the Continental region.

Thus did the European Commission demonstrate its interest in mountainous environments. In fact, the surface area covered by the Natura 2000 sites is proportionally greater in the mountains than elsewhere. A study commissioned by the European Environment Agency defines 36 percent of European territory as mountainous but estimates that 43 percent of Natura 2000 sites are located in these zones. In all, the Natura 2000 sites are said to cover 14 percent of mountainous zones as defined by the study.[7] The mountain is therefore a particular target of European nature-protection instruments. But EU policy uses two different models for treating these regions: the highest mountains and those that display the most marked vegetation belts enjoy special attention; the others are assimilated to the other landforms in their respective climate zones.

Regional Policy, or Reading the Mountain between the Lines

In 1975 the European Community also adopted a regional policy, whose aim was to reduce the differences in wealth and development within European space. Initially a policy of marginal importance, in 1986 it became one of the European Commission's principal instruments, known as the cohesion policy.

The policy adopted several tools to respond to different objectives. Objective 1 was to arrange financial transfers to "regions whose development is lagging behind." To be so designated, the regions, corresponding to an intermediate level of partition (called "nomenclature of territorial units for statistics," or NUTS 3), had to have a per capita gross domestic product (GDP) that was less than 75 percent of the European Community average. The beneficiaries were primarily the regions located on the periphery: initially Ireland, the Scottish Highlands, and southern Italy, and subsequently, with the integration of new Mediterranean countries, a large part of Spain, Portugal, and Greece. In the following decades, that rule would allow large financial transfers to the Länder of eastern Germany, northern Scandinavia, and all the regions incorporated during the expansions of the European Union in 2004 (the Czech Republic, Slovakia, Poland, Slovenia, and the Baltic states) and 2007 (Romania and Bulgaria).

Many beneficiaries of the European Union's regional policy are mountainous, at least in part of their territory. But they do not receive European funds by virtue of that characteristic. Most of the Alps, a large part of the Pyrenees, and most of the average-sized massifs between Great Britain and Germany receive no funding, since the corresponding NUTS 3 regions have per capita GDPs higher than the threshold. The scale on which the European Commission identified the eligible regions played a determining role in that calculation. Mountain ranges with little in the way of resources were sometimes left out because they were part of NUTS 3 regions whose GDP, which was largely determined by urban areas outside the mountain regions, was higher than the threshold. For example, the east slope of the Massif Central and the southern Alps in France were ineligible because of the economic weight of the urban areas of Lyon, Marseille, and Nice; so too were the Piedmont Alps in Italy, thanks to the economic activity of Turin.

Regional policy also granted funding, but at a much lower level, to address two other objectives. Objective 2 targeted "areas experiencing structural difficulties"—for example, regions affected by sectoral economic crises (such as in the coal and steel industries). Objective 3 was designed to deal with a few special situations, in particular marginalized urban sectors and inequalities resulting from the existence of international borders. Within the framework of that last objective, the Interreg program allowed many mountain regions, often border areas, to receive special funding. It has been estimated that in the late 1990s 95 percent of mountain zones in Europe were beneficiaries of European funding under objectives 1 and 2. But in none of these cases were the regions made beneficiaries because of their "mountainous condition" as such.[8]

A Regional Planning Policy in Europe?

Although the aim of the regional policy was to make up for inequalities of wealth throughout the European Community (which became the European Union in 1992), it was not a planning policy. Its principal modality—financial assistance—did not pursue objectives of territorial balance, complementarity, or organization. Although European treaties gradually introduced the concept of cohesion, it was conceived exclusively in economic and social, not territorial, terms.

Nevertheless, the need for a concerted reflection on regional planning quickly became clear. In the 1960s and 1970s most of the member states conducted ambitious regional planning policies that contributed to the construction of new cities, the planning of transportation and irrigation infrastructure, the adoption of a comprehensive safety net of public services, and the economic development of outlying regions. But nations were anxious to keep that jurisdictional authority in their own hands and did not wish to transfer it, even partially, to the European Community level. At the same time the European Parliament, composed for the most part of elected representatives wishing to increase the authority of the European Community (and then the European Union) at the expense of its member states, on several occasions pronounced itself in favor of a European regional planning concept (1983) or of a "coherent perspective on the use of Community territory" (1990). The reluctance of many member states, however, blocked the initiatives.

It was only in the 1990s that the aggressive actions of the European Commission and a few member states broke the deadlock. In 1990 the commission launched a series of studies—called "Europe 2000" (1991), "Europe 2000+" (1994), and so on—that provided an ongoing diagnosis of European regional planning, documenting the spatial effects of sectoral policies (agriculture, transportation, and energy) and the growing concentration of wealth in the largest cities and in the economic and political heart of the European Union. This zone, often called the "Pentagon," is delimited by the urban areas of London, Hamburg, Munich, Milan, and Paris. The studies familiarized the general public and sectoral administrations at the national level with regional planning analyses and with transnational issues. In 1991, with the encouragement of Jacques Delors, president of the European Commission, France also called informal meetings with ministers in charge of regional planning. These meetings allowed participants to become familiar with the diversity of regional planning conceptions prevailing in Europe and gradually led to a common vision.

About ten years after the dual process initiated by the European Commission and the regional planning ministers of the member states, the European Union was able to adopt a universally shared position, which took the form of a European Spatial Development Perspective (1999). That document, purely descriptive and not binding, identified three major regional planning priorities: the promotion of "polycentrism" within the urban European network, a balance between the cities and the rural areas, and consideration of natural and cultural heritage in EU policies. The commission and the member states, however, did not establish any EU jurisdiction in the matter, and no instrument for implementing that vision was set in place. But the cautious construction of that consensus had an impact on national regional planning policies, which adopted several of the objectives and concepts.[9]

Nevertheless, the shift that began with consideration of regional planning questions in European administrations did not lead to any clarification of EU representations and strategies vis-à-vis the mountains. Although certain generic concepts (especially heritage and polycentrism) and categories (such as urban and rural areas) were chosen to describe the issues and to outline the orientations for European territory, the category of the mountain was still apprehended heterogeneously.

When reference to the mountains did become central in regional planning analyses or initiatives, it was within the context of a particular region and not across the European Union as a whole. That was especially true for the Alps: the Europe 2000+ report, published by the Commission in 1994 to reflect on regional planning in Europe, identified a few major supranational regions within which to conduct converging initiatives. These large regions were then used to organize one of the sections of transnational cooperation policy (known as Interreg IIC). Funds were earmarked for collaborations on that scale. Among the regions identified, only one area was clearly demarcated as mountainous by its parameters and name: Alpine Space. These parameters extended far beyond the usual definition of the mountain range, since they included a number of large cities and agricultural plains in Central Europe. Nevertheless, most of the supported programs brought together mountain collectivities for projects that were clearly motivated by a common vision of the Alps and by problematics defined as being specific to the mountain. Even now, no other mountain chain occupies an equivalent place in EU regional policy, though some have advocated the creation of a Carpathian region on the model of the Alpine Space.

The Turn of the Twenty-first Century:
The Mountain in the Name of Territorial Diversity?

This overview of the principal policies of the European Union in its first decades has shown that, though no one in European agencies denies that mountains constitute important aspects of the natural context and spatial construction of Europe, the mountain as such has never become a category of public action. Common agricultural policy is only the exception that proves the rule: in refusing to define "mountain" on the basis of supranational criteria, the European Union left the effective use of the category to other jurisdictions. In treating the difficulties of mountain agriculture as merely a consequence of diverse natural handicaps, the European Union privileged categories other than that of mountain in designing its public policies. That approach is even more obvious for other policies, including regional policy and the coordinated planning and development of European space.

The 2000s marked a turning point in this respect. One reason is the appearance and increasing influence of mountains as an object of worldwide conferences on the environment and on development (see chapter 3).

The Earth Summit (1992), the Johannesburg conference (2002), and the International Year of Mountains (2002) contributed to raising awareness and facilitating effective mobilization across Europe, once again demonstrating how adjustments in focus come about through an articulation of different scales.

During these same years the question of Europe's territorial cohesion and diversity assumed a central place in public and institutional debate within the context of the European Union.[10] The concept of territorial cohesion originated in the Assembly of European Regions in the mid-1990s.[11] There, regional actors seeking to counterbalance the neoliberal philosophy of the EU growth model promoted it as a complementary aspect of the social and economic cohesion policy. In contrast to the model for spreading growth across space solely by market mechanisms, which had prevailed since the origin of the Common Market, the concept of territorial cohesion is part of a vision to promote an aggressive policy to reduce regional imbalances caused by the European Union's market philosophy and sectoral policies. Several commissioners in succession, Michel Barnier in particular, supported that vision on the European Commission. In 1997, when he was French minister of European affairs, Barnier had sought to introduce the concept into the Treaty of Amsterdam.[12] When he became commissioner for regional policy (2000–2004), he made it a major analytical concept in the second report on cohesion ("Unity, Solidarity and Diversity for Europe, Its People and Its Territory," 2001) and again in the third report ("Towards a New Partnership for Growth, Jobs and Cohesion," 2004). And as a representative of the European Commission at the Convention on the Future of Europe, he made it an element of the Constitutional Treaty Plan (2004). Despite the rejection of that plan, the concept would reappear in similar terms in the Treaty of Lisbon (2007) and in the Community Strategic Guidelines on Cohesion, adopted in 2006: "Under cohesion policy, geography matters. Accordingly, when developing their programmes and concentrating resources on key priorities, Member States and regions should pay particular attention to these specific geographical circumstances."[13]

These initiatives clearly reflect France's interest in orienting regional policy toward a more aggressive conception of the structure of European territory, consistent with its strong tradition of regional planning. The French strategy, though shared by countries such as the Netherlands and Luxembourg, has not been accepted by all the member states. Some, such as the United Kingdom, continue to back an essentially liberal EU strategy, one less concerned with ensuring the continued existence of costly financial-assistance policies (agricultural and regional) than with promoting strategic investment policies (research, innovation, competitiveness, and so on).[14] Others, such as Germany and Poland, wish to maintain a clear differentiation between a regional policy of financial redistribution under EU jurisdiction and national or subnational planning policies. The variety of viewpoints has been aired in discussions and in the adoption of various, sometimes very different documents that promote either strategies of competitiveness (Lisbon Strategy, 2000), sustainable development policies (Göteborg Strategy, 2001), or objectives of territorial cohesion (2008).[15] These diverse views have also found expression in the debates of the European Parliament, the European Economic and Social Committee, and the Committee of the Regions, which, however, all lean clearly toward territorial cohesion.

In the mid-2000s, then, the European debate on EU strategy was eminently political and ideological. But to support their arguments, the participants needed to promote a geographical vision of European space favorable to their point of view and to enlist ad hoc categories. For the promoters of territorial cohesion, it was imperative to identify clearly the types of regions that justified such a policy.

That work has often fallen to pressure groups, which constitute or reorient their strategy in that context. Such groups already existed, especially for the coastal regions (Conference of Peripheral Maritime Regions), the border regions (Association of European Border Regions), and subnational entities located at great distances from the European continent (Conference of Presidents of Ultra-peripheral Regions).

For the mountains, several organizations existed in the mid-1990s.[16] The European Association of Elected Representatives from Mountain Regions (AEM) was created in 1991, primarily at the initiative of the French and Italian national associations. It encouraged the formation of national associations in countries that did not yet have them (such as Bulgaria, Slovakia, and Poland) in order to represent the interests of the mountain populations and defend their desire to decide the future of their regions. Luciano Caveri, one of the group's previous presidents, clearly stated its ambition to become an influential pressure group: "We must succeed in constituting a true Europe-wide network. . . . Hope lies in that lobbying effort, which will allow us to make representatives of the mountain regions key partners for dealing with a large variety of issues."[17] That association played a leadership role in campaigns to raise awareness among members of the European Commission, elected representatives of the European Parliament, and representatives in the various committees meeting in Brussels.

Euromontana, an association of socioeconomic interest groups from the agricultural sector, had been formed well before that time. In 1996, however, it became a true umbrella group, a place for reflection and proposals on the future of the rural mountain regions (fig. 22). By 2014 it comprised more than eighty organizations from twenty European countries. Since 1996 it has promoted an integrated conception of rural development in the mountain regions. For Frank Gaskell, its president at the time, "Europe's mountain areas are among the last reservoirs of diversity, not only biodiversity but diversity of culture, craft and local products."[18] A particular focus of Euromontana's activities has been on quality products from mountain areas; twelve years of work led to a clear success in policy terms in 2012, when the European Parliament approved a new regulation protecting the use of an optional quality term, "mountain product."

Finally, the Mountain Forum (see chapter 8) structured its European network of experts and local development professionals into a European Mountain Forum. These three organizations, possessing only modest means, moved closer together to some extent, proposing, among other

22. One of the most recent initiatives of Euromontana has been to conduct research on the added value, as perceived by distributors and consumers, of mountain agricultural products. This effort to promote quality products from mountain areas has led to the elaboration of a Charter of Mountain Products. The photograph shows the introduction of the charter at the European Parliament in 2005. Years of work led to a clear success in policy terms in 2012, when the European Parliament approved a new regulation protecting the use of an optional quality term, "mountain product." Courtesy of Euromontana.

things, that the European Mountain Forum should become a communication and information tool for sustainable development in the European massifs. Its operations have come to a standstill, however, because of a lack of resources and the restructuring of the Mountain Forum on a global scale.

While these lobbies were being organized, and even before, other initiatives in the same direction came into being. Back in 1988 a European Charter of Mountain Regions was adopted in Trento, Italy. Its aim was to point out the shared problems faced by such regions. The same year, the European Economic and Social Committee formulated a first opinion urging that European Community policies take the mountain regions into account. In the 2000s the lobbies launched similar initiatives. In 2002 the European Economic and Social Committee sent out a second report that called for instilling a "European vision of upland regions" and promoting official recognition of them in EU treaties: "Such recognition is justified by the disadvantages and challenges facing these areas, which could be given the right to solidarity, difference and experimentation."[19] In addition, the report appropriated the principal argument of promoters of a global mountain agenda (see chapter 8): it pledged to make a European policy for mountain regions a "model of fair and sustainable development."[20] In an opinion formulated in 2008 (2008/C 120/11), the committee pointed to "the need for a specific EU policy for upland areas [and] a harmonised EU definition. . . . Upland areas require a cross-sectoral and territorial approach to their sustainable development."

During the same period the European Parliament, the European Economic and Social Committee, and the Committee of the Regions also formulated opinions and resolutions in support of mountain regions, demanding specific consideration of their singularities.[21] The European Economic and Social Committee even went so far as to propose a conception of the mountain, adopted by the Committee of the Regions,[22] that claimed to serve the objectives of territorial cohesion: "An upland area is a physical, environmental, socio-economic and cultural region in which the disadvantages deriving from altitude and other natural factors must be considered in conjunction with socio-economic constraints, spatial imbalance and environmental decay."[23] The Council of Europe Conference of Ministers Responsible for Spatial/Regional Planning (CEMAT) was not to be outdone. In the guidelines for sustainable territorial development of the European continent, which it published in 2002, the council noted that "mountain regions represent an exceptional potential for Europe and fulfil numerous ecological, economic, social, cultural and agricultural functions."[24]

Furthermore, documents emerging from the European Commission, though they have not yet been translated into public policy, take these concerns into account. Three successive reports on economic and social cohesion mention mountains alongside islands, the coast, and other regions.[25] In them the mountains are portrayed as "geographical barriers" but also as problem regions, though the texts sometimes add nuances to the generality of that observation: "While some mountainous areas are economically viable and integrated into the rest of the EU economy, most have problems."[26]

The Construction of a Common
Vision of the European Mountain

In the context of that political turning point within European debates and the stepped-up action of pro-mountain lobbies, a wide-ranging discussion of the mountains has begun in European administrations. In reality, the discussion has less to do with the mountains themselves than with the advantages and disadvantages of adopting "mountain" as a geographical category in EU actions. Recall the terms in which the 2003 discussion between Caveri, Bouvard, Evrard, and Giscard d'Estaing raised the problem. At the time, an initial threshold was about to be crossed: the emergence of a consensus that mountains be mentioned in European treaties. It was reached through a broad, systematic, and vivid diagnosis of the mountain regions, which defined and characterized the category for European decision makers.

The Prodi Commission (2000–2004) took the decisive step. Romano Prodi, a former prime minister of Italy, seems to have been won over to the cause, encouraged by the commissioners for agriculture (Fischler) and regional policy (Barnier). Both these commissioners came from the Alps, the first from Tyrol in Austria, the second from Savoy in France. And both had already had occasion to demonstrate their interest in European mountains. On October 17 and 18, 2002, Prodi, Fischler, and Barnier all took the stage for a major event called "Community Policies and Mountain Areas," held in Brussels by the European Commission itself. That event, organized within the framework of the International Year of Mountains, brought together researchers, elected representatives, and supporters of mountain-region projects. The conference gave the commissioners the opportunity to declare their conviction that the European mountains must be taken into account as a totality in EU debates. Prodi's speech praised mountains as a "natural place of diversity"—biodiversity, to be sure, but also cultural diversity, both of them a focus of the European Spatial Development Plan's efforts to manage natural and cultural heritage sites. In that same speech the president of the Commission defended an agricultural policy adapted to mountain contexts, one that could combine farm aid to compensate for "natural handicaps" with rural development and environmental conservation. Above all, Prodi's speech insisted several times on the European Union's own interest in concerning itself with the mountain regions. In so doing he sought to convert a regional question into an EU issue. Commissioners Fischler and Barnier used similar arguments, as did elected representatives from mountain regions: Caveri vaunted the "cultural biodiversity" of the European mountains and launched a defense of mountain communities, arguing that they were guarantors of the high quality of mountain environments and mountain products.

This use of the argument of diversity seemed a clever strategy. In fact, certain EU authorities had long been reluctant to treat mountains as a totality because of the great diversity of their natural characteristics and demographic and economic situations. The policy of allocating structural funds had been unable to take into account the category as a whole, primarily because of disparities in wealth. Several documents, even in the 2000s, mention this diversity, some proposing a typology to convey a better understanding of it. In 2003, for example, the Committee of the Regions proposed that four types of mountains in Europe be distinguished: "1. mountain and similar areas (arctic areas) of northern Europe; 2. mountain areas in temperate regions; 3. Mediterranean mountain areas (incl. major islands); 4. outlying and outermost island regions."[27] On that day in Brussels, the invocation of diversity—not the first such invocation for promoters of the cause—thus led to a shift in values. The favorable connotations that had come to be associated with biodiversity and cultural diversity in international declarations of the 1990s made it possible to convert what detractors saw as the principal weakness of the category into an asset.

At the time of that Brussels conference, the Directorate-General for Regional and Urban Policy (DG Regio) had already launched a study, the first of its kind, to come up with a diagnosis for the European mountains. They had engaged Nordregio, an international consortium sponsored by an agency for spatial studies and analyses located in Stockholm, to carry out the study.[28] Its aim was to enlighten member states and the commission on the reasons for adopting the category of the mountain for the next phase of regional policy (2007–13) at a time when that policy had to address the consequences of expansion and a steep rise in poverty in the east.

The first phase of the study constructed a definition of the European mountain, to be validated by members of the Commission, elected representatives, and representatives of national administrations. The only existing EU definitions at the time were those that had been adopted within the context of agricultural policy. We have seen that these definitions, whose aim was to demarcate handicaps to competitiveness, had selected criteria associated with limits on farming and stock-breeding productivity. The study commissioned for DG Regio aspired to be broader in scope and less dependent on the implementation of a single policy, and to have a greater capacity to shape strategic choices. Experts and officials agreed to take as their starting point the method UNEP had adopted to define the mountains of the world (see chapter 8). The study, which became the frame of reference for intergovernmental organizations after the International Year of Mountains, estimated that the mountainous areas in Europe (between the Atlantic and the western borders of the former USSR and Turkey) cover 1.7 million square kilometers (fig. 23).[29] It was agreed, however, that the study should be adapted to the specificities of the European context: "In several parts of Europe, including the Iberian Peninsula, the British Isles, Greece, and Fennoscandia, there are mountains along the coasts, extending down to sea level."[30] The definition of "mountain" would come to include slopes below 300 meters in elevation, which had served as the threshold in the global study. Furthermore, at lower elevations, a difference in altitude greater than fifty meters between two points a kilometer apart was deemed a condition indicating that "the landscape is sufficiently rough to be considered as 'mountain' despite the low altitude."[31]

The authors of that study stated that several statistical trials had been conducted to adjust the criteria and the threshold values to mountains as national administrations and specialized organizations conceived them. Once again the statistical definition of the mountain emerged as the result of a negotiated and conventional assessment of the object in question. It was also agreed that, consistent with earlier decisions, the topographical identification of mountains would be complemented by a climate-based identification for the Nordic regions of Scandinavia and Finland, on the grounds that they too suffer from marginality and climatic constraints.

A last adaptation of the method of the global study linked the zone identified as mountainous (including the Nordic zone) to municipalities so that the statistical data available at that level could be used. This adjustment led to the incorporation of "'non-mountain' land falling within the boundaries of mountainous municipalities," those whose territory was located primarily in the mountains as statistically defined. It also left out "'mountain land' within nonmountainous municipalities." As a result of this tinkering with categories for the needs of the method, 1.32 million square kilometers in EU-15 (that is, 41 percent of the territory as a whole)

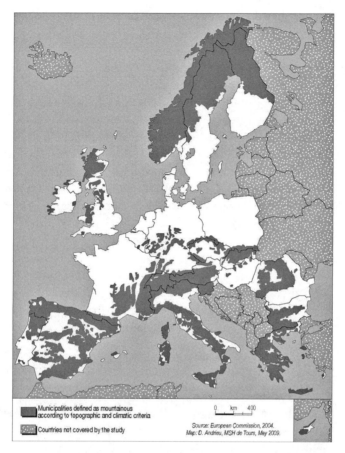

23. Delineation of mountain municipalities in Europe. Courtesy
of Dominique Andrieu. This delineation is not official, however. It
was proposed in a report for the European Commission, whose aim
was to provide statistical and geographical information necessary for
describing and analzying the situation in the European mountain
areas. The delineation integrated three criteria: altitude, slope, and
local elevation range. European Commission, *Mountain Areas in
Europe: Analysis of Mountain Areas in EU Member States, Acceding
and Other European Countries* (Stockholm, Nordregio, 2004).

were identified as mountain areas, and 0.24 million square kilometers
were so designated in the twelve expansion countries (22 percent of their
area). Twenty-seven percent of European municipalities lay within the parameters defined by the study.

In addition, the use of statistics at the municipal level made it possible
to do a much more reliable census of the inhabitants than had been done
at the global level: 75.9 million inhabitants of EU-15 were listed as living

in districts designated as mountainous (20 percent of the total), 19.4 million inhabitants in the twelve countries that later joined the union (18 percent of the total).

Thanks to that work and the preexisting databases, the European Union now possessed the most precise and most systematic study ever done on the mountain regions of the world. The authors were well aware of its limits, however: the size of the municipalities, which varied a great deal by country, led to heterogeneous demarcations of mountain areas. In addition, large urban municipalities, such as Zurich, Switzerland,[32] were classified as mountainous because of topographical features in their territory, though nothing about their social composition or activities coincided with the usual notion of the mountain. According to the Belgian participants in the European Commission study, "In a certain sense, it is Europe that has placed Wallonia up against 'its' mountain zone." Belgium, though aware it had a few hills (about twenty, some of them less than 100 meters in elevation) as well as the "high plateaus" of the Ardennes, only then discovered that it was a mountainous zone.[33] Thus, despite the precautions taken in selecting the method to ensure that the final demarcation would be in line with conventional representations of the European mountain, notable exceptions attest to the limits of the exercise.

From the Mountain to the Massifs, by Necessity

In 2004 defenders of mountains prevailed on two points: the mountain appears in the most recent European treaties, and many of the actors agreed on a definition. But the place to be granted to that entity in EU policies still had to be agreed upon. The results in that area provide fewer reasons for satisfaction among the principal stakeholders.

In fact, the strategy of Michel Barnier and of the specialized lobbies did not succeed in all areas. The cohesion policy program for the period following expansion (2007–13) did not adopt the mountain as a spatial category for allocating structural funds. For a time Manuel Barosso, acting president of the Commission, supported the idea of opening a serious public debate on the mountain in Europe in the form of a green paper. But when that idea was recycled in the green paper on territorial cohesion (2008), equal weight was given to all regions presented as peripheral in one capacity or another (maritime, border, outermost, mountainous, and so on). Furthermore, in a context of growing world competition, certain actors privileged the objective of competiveness, in keeping with the Lisbon strategy. As a result, initiatives with cohesion objectives lost ground.

The mountain plays a key role in the document the European Commission drew up to promote its initiatives on territorial cohesion. A photo in the December 2008 issue of *Panorama inforegio*, the official quarterly for regional policy, bears the caption "Geographical elements such as mountain ranges challenge our previously held notions of borders."[34] The priority given to cohesion and competitiveness has led advocates of the European mountain to reproblematize the notion of the mountain to fit the circumstances. In November 2008 a European conference titled "Innovation, Growth, and Jobs: What Sort of Competitiveness for the Mountain Regions?" demonstrated how the most pragmatic of these actors are attempting to reconcile a certain vision of the European mountain with the dominant objective. Complementary studies are also considering how Europe's territorial diversity, and especially the singularity of the mountain regions, can contribute to the objective of competitiveness. But that is not really the chosen field of those working on behalf of European mountains.

The alternative is to abandon to some extent the generic category of the mountain and to rally behind the "massifs," entities that can be incorporated into the regional and transborder initiatives of the European Commission. The European Association of Elected Representatives from Mountain Areas in particular is committed to that strategy. In the mid-2000s it piloted a program called "Interact Pro-monte" that compared transborder initiatives in mountain regions and drew a few lessons that could be put to use elsewhere. Others took inspiration from regional initiatives that rely on a broad partnership of actors with various skills to develop an exemplary system of governance.[35] The Baltic Basin serves as a model: nation-states, regions, and associations collaborate to develop territorial projects that everyone can support.[36] In the future, the same model may be adapted to the Alps.

Can all the institutions that mobilize around the theme of the mountain feel at ease within the confines of massifs—transborder and transnational massifs in particular? Has that entity taken on greater importance because of its geographical characteristics or because of the jurisdictional zones and strategies of the various players? No one can say. International treaties have already imposed a structure on the Alps and the Carpathians. The Alpine Convention and the Carpathian Convention (see chapter 11) give the greatest role to national governments, followed by a few quite militant associations that promote that field of collective action.[37] But subnational entities, such as (in the case of the Alps) French and Italian regions, German and Austrian Länder, and Swiss cantons, play no role in that plan. These collectivities have preferred to become involved in Euroregions

(Alps-Mediterranean, Alpe-Adria, and ARGE-Alp) that for the moment share characteristics not confined to the mountains.

The French regions, the Spanish autonomous communities, and the principality of Andorra did found a Working Community of the Pyrenees, but other institutions and associations are not really involved. In addition, the regional council of Franche-Comté and several Swiss cantons (Bern, Vaud, Neuchâtel, and Jura) established a Jura Working Community in 1985. Because French partners have far fewer avenues for action than their Swiss counterparts (the French state holds most of the authority in such cases), that institution of transborder dialogue was reorganized in 2002 into a Transjurassian Conference, in which the French state and the Swiss Confederation are major players. It is thus clear that there is no consensus in the various European countries about how to deal with issues on the scale of the massif. In France the notion has clearly been enlisted by pro-mountain lobbies and national administration, but neighboring Spain has not undertaken anything similar.

In the end, it does seem that representations of the mountain in Europe are too heterogeneous to permit the imposition of a unique vision and a common model as a political strategy. The object itself has several faces, and political actors borrow one or another of these to advance their own strategy. But they are unable to coordinate with one another in a manner sufficient to impose their point of view on all political actors in the European Union, despite an apparent consensus on the theme of the diversity and cohesion of European territory.

ELEVEN

The Unifying Mountain

From September 16 to September 18, 1997, the city of Somerset West, near
Cape Town, South Africa, hosted the first global conference devoted to
transborder protected natural spaces. Called "Parks for Peace," this conference illustrates the leadership role of southern Africa in such initiatives.
For several years already, heads of state in the region, encouraged by Anton
Rupert, regional representative of the WWF, had been involved in several
operations of that kind. At the opening ceremony for "Parks for Peace,"
Pallo Jordan, the South African minister of environmental affairs and tourism, welcomed his guests: "The rivers of Southern Africa are shared by
more than one country. Our mountain ranges do not end abruptly because
some 19th century politician drew a line on a map. The winds, the oceans,
the rain and atmospheric currents do not recognize political frontiers. The
earth's environment is the common property of all humanity and creation,
and what takes place in one country affects not only its neighbours, but
many others well beyond its borders."[1]

Eighteenth-century naturalists could have made a similar observation,
except for the reference to the environment as "common property," which
was not an Enlightenment idea. In the following two centuries, however,
governments and the military preferred to use rivers, mountains, and
oceans to mark political borders (see chapter 2). To be sure, at Somerset
West the countries of southern Africa did not propose to give up their
borders, not even those lying along natural boundaries. They did suggest,
however, that the regions, ecosystems, and biotopes through which these
borders passed could be the object of coordinated management, protection, and development measures.

In the years following the Somerset West conference, the idea of "peace
parks" elicited enthusiasm and a large number of concrete initiatives,[2] ben-

efiting from a growing globalist and environmental awareness everywhere in the world, which gave precedence to the imperatives of the earth's environmental quality over geopolitical considerations and the defense of national spheres of influence. The idea found fertile soil in the mountain regions: international borders had consistently been located there, and natural scientists just as consistently promoted the mountains as a scientific laboratory. But with the far-reaching changes in the status of the borders after World War II and the proliferation of transnational forms of cooperation, the naturalistic view of the mountain sometimes made mountain massifs political objects of a new kind. Political entities and management zones took mountain ranges and massifs both as a context for their projects and as their frames of reference.

War and Peace in the Cordillera del Condor

The borders of Peru and Ecuador are the most emblematic example of that recent shift in the political meaning of the mountain. Ever since they achieved independence in the early nineteenth century, the two countries had had a continuing border dispute in the region of the Amazon that on several occasions degenerated into military conflict. In 1942 the two countries reached an accord, ratified in the Rio Protocol,[3] in which they recognized a line of demarcation that followed the ridge lines separating the catchment basins of two rivers in the region.[4] That solution conformed to prevailing practices—that is, it adopted natural boundaries as political borders. It also took into account the declared priorities of the two nations with respect to their territorial strategy. Both were primarily concerned with controlling the waterways, potential access routes to the Amazonian Basin.

As in the case of the border between Chile and Argentina (see chapter 2), however, the border agreement was signed before the reconnaissance of the lands had been completed. The U.S. Army's detailed cartography in the 1940s revealed the existence of a river, the Cenepa, that flowed across the designated border.[5] That discovery launched anew the two countries' conflict, which lasted until the mid-1990s. The peace treaty signed in 1998 handed over the Cenepa Basin to Peru and fixed the seventy-eight kilometers of disputed borderline along the ridge line of the Cordillera del Condor.[6] That put an end to one of the longest-running border conflicts in history.

The treaty, however, did not use the figure of the mountain merely as a principle for fixing the border. It also stipulated that the region be demilitarized and converted into a natural park. In so doing it was adopting the proposal of Yolanda Kakabadse, president at the time of the International Union for the Conservation of Nature and minister of environmental affairs in Ecuador.[7] The protection plan was motivated by the results of several international scientific expeditions that had just brought to light the tremendous botanical richness of the cordillera. A scientific work published in 1997 estimated that the Cordillera del Condor was home to the "richest flora of any similar-sized area anywhere in the New World."[8] Peru and Ecuador, assisted by several international foundations, agreed at the time to create "ecological protection zones," form a binational steering committee,[9] and, ultimately, establish a conservation corridor, named "Condor-Kutuku" (2004).[10] The International Tropical Timber Organization, which was responsible for proposing several scenarios for transnational protection, has regularly noted that the objectives of conservation and pacification are closely linked: "The conservation efforts in the Condor Mountain Range are not only contributing to the conservation of the extraordinary biological wealth shared by the two countries, but also to creating an environment of trust, an essential element for building a sound, imperturbable and lasting peace in the region."[11]

In the eighteenth century it seemed that peace was guaranteed by borders running along the mountain summits; but in the early twenty-first century, on the border between Peru and Ecuador, peace takes the form of transborder protected areas created to respect the extraordinary characteristics of the species and biotopes found there.

Cooperating in the Virunga in the Name of Mountain Gorillas

In the 1990s another mountain region came to emblematize the idea that transborder parks could promote peace in parts of the world torn apart by recurrent conflicts. The Virunga, a region of highlands and volcanoes, lay on the fringes of the British, Belgian, and German empires in the last decades of the nineteenth century, at a time when the three European powers were reshuffling their colonial claims in East Africa.

At that very moment, European explorers assigned to map the zone reported the presence of mountain gorillas. In 1921 the U.S. naturalist Carl Akeley, returning from an expedition, approached Belgium's King Albert I in an effort to persuade him to implement measures to protect the gorillas within the borders of the Belgian Congo and Rwanda (the League of Nations had entrusted the administration of that colony to Belgium after Germany's defeat in 1918). The king, impressed by his visit to Yellowstone

National Park a few years earlier, decided in 1925 to create the first "national" park of Africa; it would bear his name for several decades. In 1929
Albert National Park was expanded: it now straddled the border between
the Belgian Congo and Rwanda-Urundi.[12] In the 1930s the British colonial
administration created a hunting reserve just north of the national park
and then a forest reserve along its edge.

When the new nations in the region (Rwanda, Congo, and Uganda)
achieved independence, each turned the protection tools the colonial powers had created to its own advantage, using its own legal instruments. But
demographic pressure and food-production objectives led to a reduction
in the size of the protected area and to weakened monitoring of forest resources. The mountain gorillas were pushed farther up the mountain and
onto the steeper slopes.

Interest in transnational protection of the Virunga region acquired
new impetus in the 1980s as a result of two tragic events. In the 1970s
the mountain gorillas, whose habitats had been radically reduced in size,
found a convincing advocate in the mass media in the person of Dian Fos-
sey. Her gruesome murder at her research site in 1985 provoked a wave
of international protests. People once again began to think about protecting the region's species. Then, in the 1990s, the area became a civil and
international war zone. The wars claimed a large number of victims and
led to vast migrations of refugees. These conflicts complicated the task of
managers of the protected areas. At the time, the Virunga massif comprised
three national parks, governed by the laws of their respective countries. But
each of these parks remained fragile, facing pressures from farmers and
poachers in the region. Virunga National Park in Congo and Volcanoes National Park in Rwanda were the successors of Albert National Park, though
smaller in size; Mgahinga Gorilla National Park in Uganda replaced the
British colony's forest reserve in 1991.

In the 1990s many international organizations encouraged coordination of national protection policies. Several wanted to make the Virunga
an emblematic locale for promoting peace through protection of the natural mountain environment. There was no dearth of scientific arguments.
The massif as a whole displayed extraordinary signs of biodiversity, with
many species native to the region. The massif was part of a single ecoregion, called the Albertine Rift Afromontane Region Forests, that some environmentalist associations identified as one of the hot spots of biodiversity
on the planet.

Transnational initiatives took the form of a tourism development project to facilitate observation of the gorillas (the Mountain Gorilla Project),

a species protection program (the International Gorilla Conservation Program), and many programs for cooperation and exchange between the administrations and wardens in the three parks.[13] In the early 2000s an NGO, the Wildlife Conservation Society, proposed that this collaboration be extended to a larger entity: the Greater Virunga Landscape. As a result, in 2004 ministers from Rwanda, Uganda, and Congo signed a Trilateral Entente Protocol for the Conservation of Nature in the transborder protection regions of the central Albertine Rift.[14] The signatories expressed their desire "to coordinate and collaboratively manage these protected areas as one ecosystem." Their commitment thus extended well beyond the protection of the mountain gorillas.

As in the Cordillera del Condor, the initiatives in the Virunga pursued a dual objective: to protect nature and to promote peace. Supporters of the initiatives justified their actions by claiming that the Virunga was a unique natural region, an ecoregion that had to be conceived of as a totality if it was to be managed in an optimal manner. From that standpoint, they partially achieved their objectives, since the national parks administrations on both sides of the border have managed to collaborate on common objectives. But the idea that pacification of the region could result from such cooperation in nature management has not become a reality, as attested by the resumption of conflicts between the nations and interests present in the 1990s. The most optimistic console themselves by noting that the resumption of hostilities has not prevented those involved in local environmental protection from continuing to cooperate.[15]

Bioregions and Ecoregions: A New Order of Nature? A New Political Frame of Reference?

In the Cordillera del Condor, as in the Virunga Mountains, nature-protection initiatives have taken into account emerging paradigms in environmental management and new political conceptions of the mountain. In the first place, these initiatives, supported by international organizations and coordinated among bordering countries, became transborder. Nature protection—the protection of the mountains in particular—was until the mid-twentieth century a national prerogative. It was motivated as much by the national imaginary as by the nature imaginary (see chapter 4). Now nature protection tends to borrow a globalist rhetoric, toning down nationalist rhetoric if need be in the name of safeguarding species and promoting peace.

In the second place, in both the Cordillera del Condor and the Virunga

Mountains, the object to be protected has changed in nature. Animal and plant species are still at issue, foremost among them the mountain goril-las of Africa and the rare and native species whose presence was revealed by the botanical expeditions conducted in the Andes, campaign after cam-paign. But increasingly the advocates of nature protection have devoted their efforts to maintaining biotopes and ecosystems, and to preserving in-teractions between creatures and the free movement characteristic of these interactions. That evolution in the models of nature protection—from the protection of species, especially the most emblematic ones, to the protec-tion of environments—has granted increasing importance to the charac-terization of these biotopes and ecosystems. That gradual shift in the ob-jectives of nature protection explains why the identification of areas that possess combinations of species and specific dynamics now plays a deci-sive role. Among these areas, mountains occupy a key place.

The terms adopted to designate these places (ecoregion, bioregion, eco-logical region, biogeographical province, and so on) vary depending on the expert and the institution. So too do the definitions of these places and the ways of tallying them. The first attempts to divide up the earth's surface into large ecological regions date back to the work of Raymond F. Dasmann and Miklos D. F. Udvardy in the mid-1970s,[16] which identified nearly two hundred "biogeographic provinces." These first typologies were criticized for being too vast in scale and for having little usefulness in the promotion of management and protection measures. Later the WWF pro-moted the concept of the ecoregion, defining it as "a large area of land or water that contains a geographically distinct assemblage of natural commu-nities that share a large majority of their species and ecological dynamics; share similar environmental conditions; and interact ecologically in ways that are critical for their long-term persistence."[17] Conservation objectives thus clearly played a role in identifying the objects to be circumscribed. The criteria selected for identifying "assemblages," "environmental condi-tions," and "interactions" led the WWF to list 825 land ecoregions on the earth's surface, 426 river and lake ecoregions, and 229 marine ecoregions. To optimize its crusade to maintain biodiversity, the WWF also drew up a list of 238 high-priority ecoregions, called Global 200, on which it decided to focus its actions.[18]

The advocates of radical environmentalism, especially in the United States, are less concerned with protecting natural sites than with promoting alternative ways of life. They have adopted the term "bioregion" to refer "both to geographical terrain and a terrain of consciousness—to a place and the ideas that have developed about how to live in that place."[19] From that perspective, the identification of bioregions is both a scholarly assessment and an ethical and political method, since these bioregions are supposed to become the frameworks and referents for collective life, even collective identity. In fact, bioregionalism, as an ideology and as a movement propagating that way of thinking, encourages individuals to adopt "ecocentric identities": "An ecocentric approach to identity invites individuals to perceive themselves not simply as members of various human social groupings but as an integral part of a much larger whole, as components of a fundamentally interlinked, and interdependent, 'web of nature.' . . . This would be similar to but much greater than that experienced through identification with a people or nation."[20] It invites these same individuals to conceive of their actions and commitments within the framework of bioregions, that is, of "newly constituted territories."[21]

In that respect bioregionalism is a new utopianism: "These bioregions will be inhabited in a manner that respects ecological carrying capacity, engenders social justice, uses appropriate technology creatively, and allows for a rich interconnection between regionalized cultures."[22] Some have seen them rather as an anti-utopia that, in the name of a harmonious vision of the relationship between human beings and nature, promotes a social and moral order that shapes individual aspirations to conform to environmental objectives. The normative, even manipulative character of the reference to bioregions comes through in many publications, for example: "Once individuals begin to identify themselves in terms of their environmental resources, bioregional planning will be better accepted and implemented."[23]

Whatever interpretation is given to bioregionalism, it is certainly inspired by determinist thinking, according to which the characteristics and actions of living creatures—human beings in particular—are determined by the natural environments in which they live. But that determinism is not subject to the laws of nature painstakingly uncovered in the nineteenth century. This new determinism is deliberately promoted and freely assumed; it is a "'voluntary' environmental determinism,"[24] one that is therefore deeply paradoxical.

Forming Networks of Protected Areas: The Yellowstone to Yukon Conservation Initiative

Mountain ranges occupy a key place on the lists of ecoregions and bioregions drawn up since the 1980s. A comparison of the principal criteria used (geological, climatic, and botanical) casts into relief the individuality of many of these ranges, or at least vast portions of them. The establish-ment of management measures on the scale of entire mountain chains fol-lows from that preliminary objectification, which ignores political borders and administrative boundaries.

One of the most remarkable achievements in this area is the "Yellow-stone to Yukon" region. Harvey Locke, a Canadian attorney and envi-ronmental activist, is said to have devised it in 1993 in response to his travels in the mountains of British Columbia. Locke obviously knew that the range already had several protected areas, including a few famous na-tional parks (Yellowstone, Grand Teton, Glacier, Banff, Jasper, and others). He wondered, however, about the efficiency of managing discontinuous regions, a system that did not facilitate communication between the parks and the reserves or the free movement of species.

Therefore, on a topographical map, he drew a large zone that took in all the existing protected areas. Aware that it would be impossible to place the entire zone under strict protection regulations, he instead proposed the es-tablishment of a network of corridors, which would at least guarantee the free movement of protected species. The name he gave to that vast moun-tain region, "Yellowstone to Yukon," was adopted by a transnational orga-nization founded in 1997 that was responsible for implementing the vi-sion: the Yellowstone to Yukon Initiative (Y2Y).

Locke surely broke new ground in formulating that proposal. He was praised for "his ability to bridge the worlds of science, conservation, law and political activism."[25] In 1999 *Time Magazine Canada* named him one of Canada's leaders of the twenty-first century. Nevertheless, Locke was quick to downplay the import of his initiative. He claimed that he had simply formalized something that was already in the air: "But really what happened is that I labeled what people were thinking anyway. . . . So as the synthesizer and articulator of these yearnings and ideas and intellectual concepts, I was important. But I did not invent any of the things that were synthesized; I packaged them."[26]

The idea of the Yellowstone to Yukon as a vast ecoregion that ought to be managed in such a way as to optimize its natural dynamics actu-ally originated in conservation biology. Since the 1960s this joint branch of biology and ecology has been concerned with better understanding the mechanisms by which rare species vanish as well as the environmental con-ditions needed to maintain and restore populations.[27] This is why conser-vation biology, like biology generally, is as interested in species themselves as in their habitats, the purview of ecology. Research conducted at the time in the national parks of the Rocky Mountains and the vicinity showed that the parameters of the protected areas did not always offer optimal conditions for protecting emblematic species, especially the grizzly bear.

Starting in 1964 that observation led several agencies, including the U.S. National Park Service, the Forest Service, and the Fish and Wildlife Service, to conceive of joint modes of management coordinated across a vast region encompassing the largest national park in the United States, Greater Yellowstone.[28] The idea for a Yellowstone to Yukon region borrowed a great deal from that Greater Yellowstone Ecosystem. In the end it was only an extension of it considerably farther north. Like Yellowstone with its grizzly, it had its own emblematic animal species: the wolf.

In 1991 scientists equipped a she-wolf with a radio tracking collar to follow her movements. The animal, named Pluie (French for "rain," a name chosen because of the weather conditions prevailing during the operation), surprised scientists with the extent of her migrations. She covered an area of 100,000 square kilometers, crossing through the territories of British Columbia, Montana, Idaho, and Washington. A year and a half after beginning to track her, scientists lost trace of the wolf. They later recovered the battery from her radio collar; it had been pierced by a rifle bullet. In 1995 Pluie, a male, and three wolf cubs accompanying her were shot and killed by a hunter. The newspapers had been following her movements, and she acquired the status of a martyr. Her history became part of the legend of the Y2Y initiative.[29] In a sense, her tragic end provided an image for the idea that species protection had to be conceptualized within new frameworks.

At Yellowstone specialists in conservation biology demonstrated that protection policies were more effective when the protected areas were well adapted to the behaviors of species. They showed as well that threatened species greatly benefit from corridors that protect their movements.[30] By forming a network of protected areas in an ecoregion of considerable size, agencies were able to respond to the fragmentation of habitats and ecosystems, identified as one of the principal obstacles to the protection of biodiversity.

The Yellowstone to Yukon project adopted the idea of an ecological corridor. In December 1993 environmentalists, scientists, and others gathered for a workshop in Kananaskis, Alberta, Canada. Each participant drew on a map the boundaries of the region he or she wanted to demarcate based on the principles of his or her area of expertise and in accordance with personal criteria.[31] It quickly became clear from the resulting assemblage of maps that the migration zones for the large predators, foremost among them the grizzly bear and the wolf, were the prime focus of attention.

The region selected included the Mackenzie Mountains, the Canadian Rockies, and the Columbia Mountains. Along its outer edges it adjoined the protected regions of Selway-Bitteroot, Hells Canyon, and the Greater Yellowstone Ecosystem. Unlike for a national park or a nature reserve, the Yellowstone to Yukon Conservation Initiative also declined to fix strict boundaries for the chosen zone on the grounds that the notion of boundary could not be applied to the idea of an ecosystem without distorting it.

Finally, the Yellowstone to Yukon project declared one last ambition: that of establishing collaboration between the agencies and populations on either side of the border. In that respect it broadened the early experiment of Waterton National Park in Alberta and Glacier National Park in Montana. These two parks, separated only by the U.S.-Canada border, in 1931 became the first to institutionalize transborder cooperation. The term "peace park" originated there. A few decades later the Yellowstone to Yukon Conservation Initiative adopted the idea for its own use and systematized it, as if to show that ecoregions constitute the only pertinent entities for conceptualizing environmental management.

Within a few years Locke's idea had become an ambitious and wide-ranging project. It circumscribes a region that is sometimes called the "wild heart of North America."[32] Covering a vast mountainous region, it serves as a management model for this type of environment: "Of all the world's mountainous regions, the Yellowstone to Yukon landscape provides the best opportunity to preserve an intact, ecologically healthy mountain system where both human and wildlife communities can thrive and prosper."[33] It is not surprising, then, that the Yellowstone to Yukon Conservation Initiative has inspired operations in other mountain regions of the world.

Coordinating Policies across a Mountain Range: The Alpine Convention

The Afromontane forests of the Albertine Rift, and the Rockies between the Yellowstone region and the Yukon, among other examples, have been identified as ecoregions primarily to serve specific objectives of environmental and species protection. Neither the identification of these mountain regions nor the transnational cooperation that resulted has led to more general regional projects. In reality, few ecoregions and few transborder mountain regions have achieved that status. The Alps undoubtedly constitute the most remarkable exception.

The Alps range is one of the ecoregions (however that concept is defined) commonly identified in Europe. The idea that it constitutes a natural region meriting joint management across borders goes back a long time, at least to the 1950s. Around that time, advocates of protecting Alpine nature became concerned about the rate of tourist, urban, and industrial development in the region. At the general assembly of the International Union for the Conservation of Nature, held in Brussels in 1950, Renzo Videsott vehemently protested against the tourism projects threatening the Gran Paradiso National Park in Italy, where he served as director. He called for transborder coordination to defend protected areas and to protect threatened animal species. Shortly thereafter the Bavarian League for Nature Protection and the Society for Wildlife Protection in Germany formed an umbrella organization composed of representatives of associations from several Alpine countries. That was the beginning of the International Commission for the Protection of the Alps (CIPRA). Its founding documents announced its mission to "preserve habitats and flora and fauna in the Alps as well as to assess the influence of tourism on the landscape, plants, and animals." Although modest in size and ambitions, CIPRA became the first organization concerned with treating the Alps as a single entity and with monitoring the preservation of its natural qualities.

A second step was taken in the 1970s. After two decades of frenetic tourism development, many voices in associations, agencies, and political parties demanded a moratorium on the construction of buildings and infrastructure and also called for more protected areas. The idea that the Alps constituted a space apart, a unique environment in Europe, began to make inroads. The strong economic and demographic growth of the bordering regions, such as the Po Plain, Bavaria, the plains and corridors of the Rhône-Alps region, and the Swiss Plateau, made that mountain region look like a recreational area and a nature reserve for everyone's benefit. The CIPRA therefore revived the idea of cooperation among the Alpine countries to define a common vision for the massif as a whole. The idea led to an intergovernmental conference held in the Bavarian Alps in 1989, then to the signing of an International Convention for the Protection of the Alps, commonly called the Alpine Convention, in Salzburg in 1991. The signatories are the seven countries in which the range is located,[34] plus the European Union, which has thus demonstrated its desire to be a partner in those forms of transborder institutional cooperation that are consistent with its own regional policy (see chapter 10).

But the involvement of European countries and of the European Union in that unprecedented experiment in regional cooperation on the scale of a massif has been controversial because of the predominant orientation of the Alpine Convention. The first lines of the document single out

24. Alps made with faces. Cartographical rhetoric has a role to play in presentations of institutional projects having to do with mountain massifs. Maps displaying the region concerned may show how the project comprises a large number of border regions, which thereby establish closer ties to one another. But they may also suggest the fundamental, prepolitical unity of the range by using simplified representations of the terrain. This illustration, produced on behalf of the network of communes in the Alliance of the Alps, shows the diversity in that fundamental unity. Courtesy of "Alliance in the Alps," Design ID Connect.

the Alps from several standpoints at once: "The Alps are one of the largest continuous unspoilt natural areas in Europe, which, with their outstanding unique and diverse natural habitat, culture and history, constitute an economic, cultural, recreational and living environment in the heart of Europe." But article 2.1 clearly states that the convention is a "comprehensive policy for the preservation and protection of the Alps." The parts of the Alpine Convention that attest to an interest in supporting or promoting economic activities or improving the living conditions of the inhabitants seem like an afterthought. In fact, it did not escape anyone's attention that the convention, which is binding as an international treaty, was signed by ministers for environmental affairs in the countries concerned and not by heads of state. It was not until the late 1990s that references to sustainable development would take precedence over statements of strict protection objectives.

As a result, the Alpine Convention and the application protocols that allowed it to become operational have repeatedly encountered opposition in national parliaments, both from interest groups representing mountain population and from associations of elected representatives from mountain regions. These representatives of the mountain populations' interests argue that the convention pays too much attention to the objectives of environmental protection and that, when the convention was drawn up, the populations played only a marginal role. A gradual incorporation of development objectives has not appeased the principal opponents. That situation slowed and sometimes even blocked the protocols ratification process. In late 2009 specialized committees from both parliamentary chambers in Switzerland opposed ratification of the protocols on the grounds that they displayed an "imbalance between the protection and the exploitation of Alpine regions."[35]

The demarcation of the application zone of the Alpine Convention, once it became inevitable, gave rise to contradictory scenarios and discussions that reveal the positions of its promoters and opponents.[36] The advocates of a vast zone that would include the large cities located near the Alps, such as Munich, Vienna, and Milan, argue for the importance of involving political actors and populations as much as possible in actions that promote sustainable development. The proponents of a restricted application zone have often sought to limit as much as possible the dreaded constraints. For the associations representing the local populations, the primary issue at stake is to keep the mountains from becoming merely a reserve to satisfy the needs of the surrounding cities: "The Alps region must not become a complementary zone. That idea goes against good economic decentralization. The Alps must not become the Indian reservation of Europe."[37]

In response to the resistance of existing organizations, often created in an effort to define the sectoral policies of countries in the region (see chapter 5), the Alpine Convention has promoted a unified vision of the Alpine area (fig. 24). It has also encouraged the formation of other associations and groups, composed of advocates of protection or promoters of sustainable development in the region. The protected areas in the various countries now have a network of organizations, called ALPARC, that allows them to exchange information and practical knowledge and compare conceptions of their missions. That network promotes the establishment of ecological corridors on the model of the North American experiments so that the objectives of conserving animal and plant wildlife will not be limited to the protected areas. The Alliance in the Alps, an association of municipalities that strives to promote sustainable development initiatives at the local level, has nearly three hundred members.[38] Two other associations, one of Alpine towns and another of tourist resorts called the Pearls of the Alps (see chapter 10), are working on similar objectives in their respective contexts. Finally, the national and provincial alpine clubs that emerged in the last third of the nineteenth century have gradually been converted to the aims of protection and controlled development in the Alps. They federated into the Club Arc Alpin in 1995 "to represent their common interests, particularly in the areas of alpinism, nature conservation, alpine regional development and alpine culture, and to achieve the purposes documented in the Convention on the Protection of the Alps."[39] Nationalist rhetoric and the vaunting of athletic feats, which gave rise to these clubs in the nineteenth century, have given way to an environmentalist and internationalist vision oriented entirely toward the practices and management of the mountain massifs.

The Alpine Convention, the first international treaty designed to coordinate various public policies across a mountain massif, was subsequently promoted as a model for actions across Europe and in several other regions of the world as well.[40] At the instigation of the International Year of Mountains (see chapter 8), the agenda for action of the Alpine Conference in 2003–4 expressed a desire to conduct exchanges with other mountain regions of the world so that they could benefit from the Alpine experiment. A similar convention was thus signed in the Carpathians, and comparable projects are under discussion for the Balkans, the Caucasus, and the Altai.[41]

Diverging Forms of Political
Institutionalization in the Sierra Nevada

In 1991 the *Sacramento Bee*, a regional newspaper for central California, published a series of articles titled "Majesty and Tragedy: The Sierra in Peril" (fig. 25).[42] The journalist Tom Knudson presented the results of eight months of interviews, readings, and visits to the Sierra Nevada. The articles attracted attention both for the quality of the investigation and for the somber portrait that emerged from it: air pollution, death of the forests, poisoned rivers, threatened wildlife, soil erosion, forest fires. Shortly thereafter Knudson was awarded the Pulitzer Prize for journalism.

Nearly a century separates Knudson's portrait of the Sierra Nevada and the impassioned initiatives of John Muir (see chapter 4). In *The Mountains of California*, one of his most popular books, Muir, often described as one of the most influential environmentalists in the United States, describes the mountain range: "It still seems to me above all others the Range of Light, the most divinely beautiful of all the mountain-chains I have ever seen."[43] Muir was deeply moved by the splendor of the Sierra Nevada. Knudson was grief-stricken at the condition of its landscapes and its compromised biodiversity.

Both men had a clear idea of what the Sierra Nevada was and ought to be. The mountain range was clearly identified in the early phases of Spanish colonization. Geologists and botanists, from the late nineteenth century to the present, have supplied numerous scientific descriptions. Back in 1891 Andrew Lawson, professor of geology at the University of California,

25. Front page of the report on the Sierra Nevada in the *Sacramento Bee*. Tom Knudson, "Majesty and Tragedy: The Sierra in Peril," *Sacramento Bee*, June 9–13, 1991.

Berkeley, noted that the Sierra Nevada "constitutes a magnificent unit, one of the finest examples on the face of the globe of a single range, the type of its class."[44] But a century later, when Knudson was writing his articles, the Sierra, unlike the Appalachians (see chapter 5), was not managed as a single entity. Located almost wholly in the state of California, it is divided up among many counties—all of them straddling the mountain range and one or another of the basins alongside it—eight national forests, and three national parks. In short, the range is fragmented into countless political and administrative units, which makes joint management of the totality impossible.[45]

That is one of the problems Knudson highlighted: "There are no official estimates of overall environmental damage to the Sierra Nevada for one simple reason: No government agency, university or environmental group has taken an exhaustive look at the entire range."[46] And, he maintained, that was exactly what the mountain range needed: an overall vision that dealt with all the places and all the problems at once. His reporting would later be credited with reframing the issues on the scale of the entire mountain range: "These articles made the conservation community aware that there was no unified voice speaking out for the protection of the Sierra."[47]

In fact, at the time the report was published the construction of a unified vision and the establishment of the necessary institutions were already under way. The U.S. Forest Service, though one of the most controversial forces in the region, undertook the first form of regional coordination. The agency, which owns half the land in the Sierra, favors the commercialization of national forest timber located outside the national parks. It has regularly been blamed for the damage caused by commercial forest operations. In the late 1980s, to stave off such criticisms, the Forest Service sought to coordinate policies conducted in the Sierra's eight national forests with the aim of guaranteeing preservation of the habitats of the endangered spotted owl. The initiative grew in subsequent years with the adoption of the Sierra Nevada Forest Plan Amendment, usually known as the Sierra Nevada Framework, which set up discussions between the different parties in forest conflicts and promoted management modes acceptable to most of them.

The U.S. Forest Service's reframing of policies on the scale of the mountain range led to other such initiatives. Each attracted its own set of actors, and each came up with a specific plan of action. In 1991 the state of California, through the California Biodiversity Council (a committee made up of federal and state agencies as well as local collectivities), sought to promote its regional policies for ten "bioregions," all defined by biophysical criteria. One of these regions was the Sierra Nevada. The state of California also sponsored a scientific study intended to provide indisputable grounds for identifying a bioregion and for promoting appropriate public policies. The four-year (1992–96) study, titled "Sierra Nevada Ecosystem Project" (SNEP), emphasized the diversity of the ecosystems that constituted the Sierra Nevada and at the same time concluded that the promotion of environmental management on that scale was imperative. In November 1991 the Sierra Summit, also financed by the state of California, set out "to launch a consensus-building process for improving the management of natural resources in the Sierra Nevada."[48]

Unsatisfied with the modalities and results of that summit, environmental and nature-protection organizations joined with citizens groups promoting watershed management and held their own conference, Sierra Now, in August 1992. One outcome of this event was the launch of the Sierra Nevada Alliance. A little later a second environmental coordination group was created: the Sierra Nevada Forest Protection Campaign. In both cases these organizations assembled—or gave a new institutional form to—various volunteer groups that had previously had little contact with one another. And in both cases financial assistance from private foundations allowed these groups to come together.

Initially, then, the emergence of the Sierra Nevada as an object of institutional initiatives was motivated by a concern for environmental protection in the region. Although the various institutions and events expressed different points of view, the question of protection truly structured public debates. In light of that focus, a second series of initiatives arose in the mid-1990s, openly concerned with reframing the debate in terms of the living and working conditions of the populations and the interests of local businesses. In reaction to the Sierra Now conference and the creation of the Sierra Nevada Alliance, economic actors and regional developers held their own conference. Worried about the increasing power of radical proposals for environmental protection, they established the Sierra Economic Summit, at which they agreed to create the Sierra Communities Council.[49]

That polarization of regional coordination initiatives led in 1994 to efforts at reconciliation with the creation of the Sierra Business Council. The organization targeted businesses, hoping to raise their awareness of the benefits they could derive from a high-quality natural environment. This pragmatic attitude had the objective of preventing people in the Sierra from getting caught up in the usual posture of radical opposition, with protectors of the environment pitted against supporters of economic development.

In 2004 this approach, which favored integrating the objectives of environmental protection and economic development, acquired full institutional recognition. A state agency, the Sierra Nevada Conservancy, "initiates, encourages, and supports *efforts* that improve the *environmental, economic, and social well-being of the Sierra Nevada Region, its communities, and the citizens of California.*"[50] That "mission" rested on a vision: "The magnificent Sierra Nevada Region enjoys outstanding environmental, economic and social health with vibrant communities and landscapes sustained for future generations." The idea of a "Sierra Nevada region" took root in a lasting way, and the mountain environment as a frame of reference came to be central to its purpose. Environmentalists and promoters of economic development were invited to contribute jointly to sustainable development and to reasonable management of mountain resources. Promoters have thus had to familiarize political leaders with that new object. To that end, a Sierra Nevada Lobby Day is held every year.[51] Within the walls of the capitol in Sacramento, that day is dedicated to persuading California state legislators that "the Sierra Nevada exists as an entity and matters!" Lobbyists conduct a work session to remind legislators of the regional issues at stake, then go door to door to legislators' offices to ask them to support the lobbyists' actions. The idea of a new mountain region has taken root in the halls of the California legislature, the only entity that can decide to give it a long-term institutional form.

The Mountain as Emerging Regional Framework and as Referent for Collective Action

From the Virunga to the Alps, from the Rockies to the Andes, and from the Sierra Nevada to the Carpathians, mountain ranges and massifs, whether large or of modest size, are becoming political objects. Identified clearly by scientific protocols already proven for decades, these natural entities are apprehended increasingly often as management entities, even as fully functioning political entities. They are now part of the general rise in influence of transnational regions, which the imaginaries of globalization seem to encourage even as the exclusive prerogatives of nation-states weaken and transborder practices intensify: "The 'mythical' resurrection of the 'local' or 'regional' scale—both in theory and in practice—is an integral part of the 'myth' of globalization."[52]

Various factors may explain why the mountain regions appear well situated for that general reframing of the scales and objects of political and institutional life. References to their natural quality, their precedence in time over any territorial or political project, play a key role. That view is voiced in environmentalist organizations especially, both by those seeking to optimize policies for protecting threatened species and biotopes, such as the World Wildlife Fund or Conservation International, and by those who support a more radical or revolutionary conception of environmentalism, such as the staunchest bioregionalists. In addition, that environmentalist awareness is no longer the exclusive domain of such associations. It is seeping into public policies, particularly in specialized agencies and administrations in Europe and California, for example.

A second factor in the increasing influence of mountain ranges and massifs as political objects is that many of them lie on borders. The promotion of transborder cooperation across ridge lines may stem from a globalist or antinationalist rhetoric, from a pragmatic understanding of the limits of national public policies on the environment, or both. The image of the unifying mountain, of the mountain summit where different parties hold hands or signs accords, is replacing the image that prevailed for decades of a natural rampart between nation-states. But that rhetoric fools no one, especially not administrators or directors of political projects. Transborder cooperation is possible only if it engages preexisting legal political entities and recognizes their decisive role.[53]

The other great lesson of the preceding examples is that the objectification of mountain ranges and their irruption into international relations have not come about solely in the name of the environmentalist paradigm or in a single problematization—the mountain as a privileged environment for nature-protection policies. The organizations most wary or hostile with respect to radical environmentalism, and those that pursue radically different projects, tend to embrace these same mountain objects in order to champion their critical or alternative conceptions. The Alps and the Sierra Nevada illustrate that state of affairs. This is evidence of the triumph of the new political objectification of the mountain: it is being adopted not only by those who introduced it, but also by their opponents.

CONCLUSION

Normative Naturalism and Orocentric Identities

From the Mountain of Reims to the Sierra Nevada, by way of Fouta Djal-lon, the Andes of Patagonia, the Yukon, Mount Royal, the Naga Hills, and the center of Sulawesi, we have journeyed in this book to a great number of mountains, or at least to places designated as such. We might have been far more systematic had we stopped in Ethiopia and South Africa, in Australia and Greenland, or in Java. But it was never our aim to be exhaustive.

This panorama, though selective, has brought a number of different mountains into view. But the type of diversity that has interested us was not that of the mountains themselves but rather that of the figures through which the very notion of the mountain was invoked. The variety of these figures has served as a guide to this book: the mountain as territorial rampart for the modern state, as symbolic heart of the nation, as one motif among others of social and natural diversity, as the emblem of wilderness, as the place par excellence for implementing sustainable development policies, as a sphere of public policy, and so on.

In contrast to what has been done in most of the existing literature on the subject, we chose not to begin with what mountains are—not to treat them as objects already there on the surface of the earth. Our focus in this book has been on the mountain as figure, around and through which a set of conceptions of the natural, social, and political world has taken shape. Ultimately, however, there has been a return to the material reality of these mountains, since everything demonstrates that the figures through which the mountain is conceived are also motivations and guides for action: cols and passes are fortified because mountains along the border are attributed the role of rampart; protected areas come into being because wilderness and cultural heritage are valued; mountains are equipped with ski lifts or hydroelectric dams because of the place granted them more generally—

and more abstractly as well—within a social or territorial modernization plan. In short, the mountain as a category of cognition is in the first place a collection of ideas, before being a field for the implementation of operations, regulations, and public policies. It is therefore a category of collective action.

Let us conclude by noting that, beyond their variety, mountains have one characteristic in common: taken together, they remind us that modernity has maintained and still maintains a particular relationship to the idea of the mountain.

In fact, the idea of the mountain and the figures associated with it invariably refer to another, more general idea: that of nature. Modernity made nature one of the poles of its imaginary and made the mountain one of the emblematic illustrations of nature. As a result, modern thinkers identified mountains as being objectifiable, measurable, and mappable. They linked these landforms to a set of characteristics and elements: plants and animals, landscapes and ecosystems, inhabitants and users, moral values and codes of conduct. In other words, the primary identification, that of landforms, has been accompanied by a secondary identification: that of attributes, qualities, norms, and values.

It emerges quite clearly that this coupling of forms of identification is not without consequences. It naturalizes—in other words, inscribes in the order of nature—conventions and distinctions that actually lie within the social or political order. That process makes it possible to define nations "naturally" in terms of their mountain setting and to delimit nation-states by their "natural" borders. Mountain populations can be viewed as naturally different because they are lodged in the most natural of natural environments. The naturalistic identification of the mountain, therefore, has gone hand in hand with the naturalization of the mountaineers, who are viewed as being linked by necessity to the mountains. Through that eminently political operation, the mountain appears to be the most apolitical of geographical entities because it is the most natural.

Nevertheless, the ambiguity of the mountain's political status, though characteristic of modernity and in that respect at work throughout the period studied in this book, has not remained constant over the last three centuries. We have proposed the term "normative naturalism" to designate how the order of nature was invoked in the eighteenth century, when "natural borders" were being promoted. That concept of normative naturalism is undoubtedly valid for a number of other figures that arose in the following two centuries: the figure of the mountaineer as emblem of the nation; the figure that gave rise to the first forest policies in Europe and later the colonies; the figure of the mountain as a refuge of biodiversity. But this modern political philosophy is not exclusive: we have seen that economic and social policies targeting the development of mountain regions have sometimes adopted different frames of reference. Throughout the twentieth century, moreover, most radical bioregionalist militants invoked that philosophy to promote bioregions as the new frameworks for political and administrative organization and "ecocentric identities," but national and transnational organizations called for a different form of mountain identity to federate the populations most concerned. Thus, some participants in the ideological confrontations of the last century, in adopting the principal figures of the mountain, revived that "normative naturalism" and the "orocentric identities" that correspond to it, while others took an opposing view.

The political history of the mountain in modern times thus combines identification of a type of natural object and of spheres of action with identification of the individuals and collectives associated with that object and with various modes of self-identification adopted by such social groups. It is a history built of conventions, norms, and identities.

NOTES

FOREWORD

1. Jon Mathieu, *The Third Dimension: A Comparative History of Mountains in the Modern Era* (Cambridge: White Horse Press, 2011).

INTRODUCTION

1. Hans R. Stampfli, "Amanz Gressly, 1814–1865: Lebensbild eines außerordentlichen Menschen; Separatdruck," *Mitteilungen der Naturforschenden Gesellschaft des Kantons Solothurn* 32 (1986).

2. John Grand-Carteret, *La montagne à travers les âges* (Grenoble: H. Falque et F. Perrin, 1903); Margaret H. Nicolson, *Mountain Gloom and Mountain Glory: The Development of the Aesthetics of the Infinite* (Ithaca, NY: Cornell University Press, 1959).

3. Jean-Paul Bozonnet, *Des monts et des mythes: L'imaginaire social de la montagne* (Grenoble: Presses Universitaires de Grenoble, 1992); Edwin Bernbaum, *Sacred Mountains of the World* (Berkeley and Los Angeles: University of California Press, 1997); Samivel, *Hommes, cimes et dieux* (Paris: Arthaud, 1962); Jon Mathieu, *The Third Dimension: A Comparative History of Mountains in the Modern Era* (Cambridge: White Horse Press, 2011).

4. Jules Blache, *L'homme et la montagne* (Paris: Gallimard, 1933); Don Funnell and Romola Parish, *Mountain Environments and Communities* (London: Routledge, 2001); William A. B. Coolidge, *The Alps in Nature and History* (London: E. P. Dutton, 1908).

5. John R. Searle, *The Construction of Social Reality* (New York: Free Press, 1995); Ian Hacking, *The Social Construction of What?* (Cambridge, MA: Harvard University Press, 1999).

6. For Hacking, constructionism refers to "sociological, historical, and philosophical projects that aim at displaying or analyzing actual, historically situated, social interactions or causal routes that led to, or were involved in, the coming into being or establishing of some present entity of fact" (Hacking, *The Social Construction of What?*, 48)

7. Philippe Joutard, *L'invention du Mont Blanc* (Paris: Gallimard, 1986); Anne-Marie Martin, Jean-Claude Mermet, and Nadine Ribet, "L'invention du Mézenc," in *Campagnes de tous nos désirs: Patrimoine et nouveaux usages sociaux*, ed. Maurice Rautenberg, André Micoud, Laurence Bérard, and Philippe Marchenay (Paris: Maisons des sciences de l'homme, 2000), 45–57.

8. Henri Berarldi was no doubt one of the first authors to use the word "invention" in that very contemporary sense: "The Pyrenees have existed for only a hundred years. They are 'modern.' The Pyrenees were invented by Ramond." *Cent ans aux Pyrénées* (Paris: Danel, 1898), 1:1. See also Bernard Debarbieux, Marie-Claire Robic, and Céline Fuchs, eds., "Les géographes inventent les Alpes," *Revue de géographie alpine* 4 (2001); Karen Wigen, "Discovering the Japanese Alps: Meiji Mountaineering and the Quest for Geographical Enlightenment," *Journal of Japanese Studies* 31, no. 1 (2005): 1–26; Marina Frolova, *Les paysages du Caucase: Invention d'une montagne* (Paris: Comité des travaux historiques et scientifiques, 2006); and David Stradling, *Making Mountains: New York City and the Catskills* (Seattle: University of Washington Press, 2007).

9. Philippe Forêt, *La véritable histoire d'une montagne plus grande que l'Himalaya* (Paris: Bréal, 2004); Henri Chamussy, "Revisiter le concept de 'montagne méditerranéenne,'" *Montagnes Méditerranéennes* 12 (2000): 37–40.

10. Blache, introduction to *L'homme et la montagne*, 7.

11. Barry Smith and David Mark, "Do Mountains Exist? Towards an Ontology of Landforms," *Environment and Planning B* 30, no. 3 (2003): 411–27.

12. Bruno Messerli and Jack D. Ives, eds., *Mountains of the World* (London: Parthenon, 1997), 2–3.

13. Roderick Peattie, *Mountain Geography: A Critique and Field Study* (Cambridge, MA: Harvard University Press, 1936), 4.

14. Jean-Paul Guérin and Hervé Gumuchian, eds., *Les représentations en actes*, proceedings of the Lescheraines colloquium (Grenoble: Université de Grenoble, 1985).

15. The vocabulary and analytical tool kit introduced in what follows is greatly indebted to the many studies in social science that, over the last twenty years, have taken an interest in the cognitive dimensions of social action, public policies, and especially development. Although in this book we do not adopt all the concepts used in these approaches (*référentiel*, *cadrage*, "problem setting," "framing," and so on) and do not necessarily use terms in the sense that other authors have given them ("paradigm" and "figure," for example), that is not because we are unfamiliar with the literature. Far from it. But we do not make overly theoretical or conceptual claims in this book and therefore will not discuss these terminological issues here.

16. Marcel Roncayolo, *La ville et ses territoires* (Paris: Folio, 1990); Christian Topalov, "La ville, catégorie de l'action," *L'année sociologique* 58, no. 1 (2008): 9–17.

17. Vincent Berdoulay, ed., *Les Pyrénées: Lieux d'interaction des savoirs (XIXe–début XXe siècle)* (Paris: Comité des travaux historiques et scientifiques, 1995).

18. Pierre Bourdieu, *Language and Symbolic Power* (Cambridge, MA: Harvard University Press, 1991), 127.

19. Thomas L. Friedman, *The World Is Flat: A Brief History of the Twenty-first Century* (New York: Farrar, Straus and Giroux, 2005).

CHAPTER ONE

1. M. Clozier, "Sur la découverte d'une souche d'arbre pétrifiée trouvée dans une Montagne aux environs d'Etampes," in *Mémoires de mathématique et de physique présentés à l'académie royale des sciences* (Paris: Imprimerie Royale, 1755), 2:598–604.

2. Denis Diderot and Jean Le Rond d'Alembert, eds., *Encyclopédie ou dictionnaire raisonné des sciences, des arts et des métiers par une société de gens de lettres* (Paris, 1751–65), s.v. "montagne."

3. Bernard Debarbieux, "'The Mountain in the City: Social Uses and Transformations

of a Natural Landform in Urban Space," *Cultural Geographies* 5, no. 4 (1998): 399–431.

4. Florian Cajori, "History of Determinations of the Heights of Mountains," *Isis* 12, no. 3 (1929): 482–514; François de Dainville, *Le langage des géographes, 1500–1800* (Paris: Picard, 1964), 166.

5. Frederico Fernández Christlieb and Angel J. García-Zambrano, *Territorialidad y paisage en el altepetl del siglo XVI* (Mexico City: Fondo de Cultura Económica, 2006).

6. For a discussion of the scientific and popular terminology in use in the Himalayas of Nepal, see Joelle Smadja, ed., *Histoire et devenir des paysages en Himalaya* (Paris: CNRS Éditions, 2003), 52–89.

7. Yannick Lageat and Falmata Liman, "Le Lesotho: Enclavement et dépendance d'un royaume montagnard," in *Les montagnes tropicales: Identités, mutations, développement*, ed. François Bart, Serge Morin, and Jean-Noël Salomon (Pessac: Dynamiques des milieux et des sociétés dans les espaces tropicaux, 2001), 315–22.

8. These are not the only popular conceptions that existed before the eighteenth century, nor are they the only ones in relation to which Enlightenment naturalism positioned itself. One would have to mention as well the biblical and theological discourse on the mountain and earlier scientific works, for example, those of the Arab Muslim geographers between the ninth and eleventh centuries. But such discussions, necessary for a history of conceptions of the mountain, will be set aside here in light of our overall objective in this book.

9. Michel Foucault, *The Order of Things: An Archaeology of the Human Sciences* (New York: Vintage, 1973), translation of *Les mots et les choses* (Paris: Gallimard, 1966).

10. The expression, invented by René Descartes in *Règles pour la direction de l'esprit*, rule 4 (1628), was adopted by Foucault.

11. Bruno Latour, *We Have Never Been Modern* (London: Harvester Wheatsheaf, 1993), translation by Catherine Porter of *Nous n'avons jamais été modernes* (Paris: La Découverte, 1991).

12. The following section covers in great part material from earlier articles: Bernard Debarbieux, "The Mountains between Corporal Experience and Pure Rationality: The Contradictory Theories of Philippe Buache and Alexander von Humboldt," in *High Places*, ed. Denis Cosgrove and Veronica della Dora (London: IB Tauris, 2008), 87–104; Bernard Debarbieux, "The Various Figures of Mountains in Humboldt's Science and Rhetoric," *Cybergeo: European Journal of Geography* (online, 2013). On that journey, and more generally, on Humboldt's writings, see Charles Minguet, *A. de Humboldt, historien et géographe de l'Amérique espagnole, 1799–1804* (Paris: Maspero, 1969); Michael Dettelbach, "Global Physics and Aesthetic Empire: Humboldt's Physical Portraits of the Tropics," in *Visions of Empire: Voyages, Botany and Representations of Nature*, ed. David P. Miller and Peter H. Reill (Cambridge: Cambridge University Press, 1996), 258–92; Helmut de Terra, *Humboldt: The Life and Times of A. von Humboldt* (New York: Knopf, 1955); Jon Mathieu, *The Third Dimension: A Comparative History of Mountains in the Modern Era* (Cambridge: White Horse Press, 2011).

13. Alexander von Humboldt, *Personal Narratives of Travel to the Equinoctial Regions of the New Continent during the Years 1799–1804 by A. von Humboldt and Aimé Bonpland with Maps and Plans* (London: Longman, 1866), 5:461–62, translation by H. M. Williams of *Voyage aux régions équinoxiales . . .* (Paris: Maze, 1814–25).

14. Philippe Buache's biography has been thoroughly studied in Numa Broc, "Un géographe dans son siècle: Philippe Buache," *Dix-huitième siècle* 3 (1971): 223–35; and Lucie Lagarde, "Philippe Buache, cartographe ou géographe?" in *Terres à découvrir,*

terres à parcourir, ed. Danielle Lecocq and Antoine Chambard (Paris: L'Harmattan, 1998).

15. Philippe Buache, "Essai de géographie physique, où l'on propose des vues générales sur l'espèce de Charpente du Globe, composée des chaînes de montagnes qui traversent les mers comme les terres, avec quelques considérations particulières sur les différents bassins de la mer, et sur sa configuration intérieure," *Compte-rendus et mémoires de l'Académie Royale des Sciences* (Paris, 1752), 399–416.

16. Humboldt, *Personal Narratives of Travel*, 5:497.

17. Ibid., 450.

18. Ibid., 454.

19. Polycarpus Leyser, *Commentatio de vera geographiae methodo (Reflection on a True Method of Geography)* (1726). For a commentary, see Franco Farinelli, "Friedrich Ratzel and the Nature of (Political) Geography," *Political Geography* 19, no. 8 (2000): 943–55.

20. Comte de Buffon (Georges-Louis Leclerc), *Histoire et théorie de la terre* (Paris: Imprimerie Royale, 1749), 69.

21. Buache, "Essai de géographie physique," 402.

22. Isabelle Laboulais, "Le système de Buache, une 'nouvelle façon de considérer notre globe' et de combler les blancs de la carte," in *Combler les blancs de la carte: Modalités et enjeux de la construction des savoirs géographiques (XVIIe–XXe siècle)*, ed. I. Laboulais-Lesage (Strasbourg: Presses Universitaires de Strasbourg, 2004), 93–115.

23. Denis Diderot and Jean Le Rond d'Alembert, eds., *Encyclopédie ou dictionnaire raisonné des sciences, des arts et des métiers* (1757), vol. 7, s.v. "Géographie physique" (Nicolas Desmarets). The quotations that follow are taken from the same source.

24. Edwin Danson, *Weighing the World* (Oxford: Oxford University Press, 2006).

25. Alexander von Humboldt, *Asie Centrale: Recherches sur les chaînes de montagnes et la climatologie comparée* (Paris: Gide, 1843), 51, our translation.

26. Note that in the early nineteenth century, geography and geology (usually called "geognosy") were not, strictly speaking, clearly distinct disciplines but rather objects of curiosity and potentially redundant points of view. Humboldt sometimes speaks of "mineralogical geography," just as he speaks of the "geography of plants," to indicate that he thinks the analysis of the spatial distribution of rocks and plants contributes to an overall understanding of nature.

27. Humboldt reformulated the principal objective of that trip to Asia as being to study "the positional relationships between these plains and the great intumescences of the earth's crust." Humboldt, *Asie Centrale*, xxxii and xxxiii.

28. Peter Simon Pallas, *Observations sur la formation des montagnes et sur les changements arrivés au Globe, particulèrement à l'Empire de Russie* (Saint Petersburg: Acta Academiae Scientiarum Imperialis Petropolitanae, 1777).

29. Humboldt, *Asie Centrale*, 155, our translation.

30. Alexander von Humboldt, *Sur l'élévation des montagnes de l'Inde* (Paris: Feugueray, n.d.), 13, our translation.

31. Ferdinand Freiherr von Richthofen, *China: Ergebnisse eigener Reisen und darauf gegründeter Studien* (Berlin, 1877–1912).

32. Léonce Élie de Beaumont, *Notice sur le système des montagnes* (Paris: P. Bertrand, 1852).

33. Émile Argand, "Tectonique de l'Asie," in *Compte-rendu du 13e congrès Géologique International* (Brussels, 1924), 171–372.

34. On the role of instruments and measurements in Humboldt's work, see Sylvie Pro-

vost, "Les instruments de Humboldt et de Borda au volcan des Canaries," *Revue du Musée des arts et métiers* (2003): 112–14.

35. The most famous is Giovanni Riccioli's *Geographia Reformata* (1672).

36. Alexander von Humboldt, *Notice de deux tentatives d'ascension du Chimborazo* (Paris: Pihan de la Forest, 1838), 24, our translation.

37. Ibid., 25, our translation.

38. On this subject, see the analysis proposed by Sven Hedin and the controversies surrounding his proposals in Philippe Forêt, *La véritable histoire d'une montagne plus grande que l'Himalaya* (Paris: Bréal, 2004).

39. Jon Mathieu, "The Sacralization of Mountains in Europe during the Modern Age," *Mountain Research and Development* 26, no. 4 (2006): 343–49.

40. The most radical proposals question the ontological nature of the mountain and conclude that any objective definition is logically impossible. See Barry Smith and David Mark, "Do Mountains Exist? Towards an Ontology of Landforms," *Environment and Planning B* 30, no. 3 (2003): 411–27.

41. Paul Veyret and Germaine Veyret, "Essai de définition de la montagne," *Revue de géographie alpine* 50, no. 1 (1962): 5–35, quotation 5, our translation.

42. For example, Rhodes Whitmore Fairbridge, "Mountain and Hilly Terrain, Mountain Systems: Mountain Types," in *Encyclopedia of Geomorphology*, ed. R. W. Fairbridge (New York: Reinhold, 1968), 745–61; A. J. Gerrard, *Mountain Environments: An Examination of the Physical Geography of Mountains* (London: Belhaven, 1990); G. C. Miliaresis and D. P. Argialas, "Quantitative Representation of Mountain Objects Extracted from the Global Digital Elevation Model (GTOPO30)," *International Journal of Remote Sensing* 23, no. 5 (2002); Bruno Messerli and Jack Ives, *Mountains of the World* (London: Parthenon, 1997).

43. Veyret and Veyret, "Essai de définition de la montagne," 5.

44. Joseph Pitton de Tournefort, *Relation d'un voyage du Levant fait par ordre du roy* (Paris: Imprimerie Royale, 1717), 2 vols.

45. Carl von Troll, *Die tropischen Gebirge: Ihre dreidimensionale klimatische und Pflazengeographische Zonierung* (Bonn: Bonner geographische Abhandlungen, 1959), 25; and "High Mountain Belts between the Polar Caps and the Equator: Their Definition and Lower Limit," *Arctic and Alpine Research* 5, no. 3 (1973): 19–27.

46. Alexander von Humboldt, *Cosmos: A Sketch of the Physical Description of the Universe* (London: Harper & Brothers, 1877), 1:46, translation by E. C. Otté of *Kosmos* (1845).

47. This section has been excerpted and summarized from Bernard Debarbieux, "Le montagnard: Imaginaires de la territorialité et invention d'un type humain," *Annales de géographie* 660 (2008): 90–115.

48. Mentioned in Comte de Buffon (Georges-Louis Leclerc), *Supplément à l'Histoire naturelle*, vol. 4 (Paris: Imprimerie royale, 1778), 510.

49. Élie Bertrand, *Essai sur l'usage des Montagnes* (Zurich: Heidegger, 1754).

50. Jean-Louis Giraud-Soulavie, *Histoire naturelle de la France méridionale* (Paris: J. F. Quillau, 1780).

51. Comte de Buffon (Georges-Louis Leclerc), *De l'homme (Histoire naturelle de l'homme)* (1749; Paris: Maspero, 1971), 320.

52. A detailed analysis of these different ways of conceptualizing how the mountaineer is determined by the nature of the mountain can be found in François Walter, *Les figures paysagères de la nation* (Paris: Éditions de l'École des Hautes Études en Sciences Sociales , 2004).

53. Edme Mentelle, Conrad Malte-Brun, and P. Étienne Herbin De Halle, *Géographie mathématique, physique et politique de toutes les parties du monde*, vol. 7 (Paris: Chez Tardieu et Laporte, 1803), 589. The authors refer in a note to William Coxe, *Sketches of the Natural, Civil, and Political History of Switzerland* (1776).

54. Mentioned in Walter, *Les figures paysagères de la nation*.

55. Quoted ibid., 235.

56. François-Ignace d'Espiard de la Borde, *Essais sur le génie et le caractère des Nations* (Brussels, 1743).

57. Jean-André Deluc, *Lettres physiques et morales, sur les montagnes et sur l'histoire de la terre et de l'homme: Adressées à la Reine de la Grande Bretagne* (The Hague: Detune, 1778).

58. Onésime Reclus, *Le plus beau royaume sous le ciel* (Paris, 1899), 771.

59. Ernest Ingersoll, "Studies of Mountains," *Chatauquan* 7 (1886–87): 456–58.

60. Serge Briffaud, "Le temps du paysage: A. de Humboldt et la géohistoire du sentiment de nature," in *Géographies plurielles: Les sciences géographiques au moment de l'émergence des sciences humaines*, ed. H. Blais and I. Laboulais-Lesage (Paris: L'Harmattan, 2006), 275–99, quotation 278–79. The author adds: "The man of interest for the study of the cosmos is thus the man considered as a spectator of nature, that is, as a being who, inasmuch as he is 'impressed' by that spectacle, inasmuch as he feels it and sometimes thinks it, bears within himself the unity of the universe" (280).

61. Humboldt, *Cosmos*, 2:19.

62. Alexander von Humboldt, letter to Caroline von Wolzogen, May 14, 1806, in *Lettres américaines d'Alexandre de Humboldt, 1798–1807* (Paris: Librairie Orientale et Américaine, 1904), 211–12.

63. See John Ruskin, *Modern Painters* (London: Spottiswoode, 1843–60); and D. Cosgrove and J. E. Thorne, "Of Truth and Clouds: John Ruskin and the Moral Order in Landscape," in *Humanistic Geography and Literature*, ed. Douglas C. Pocock (London: Croom Helm, 1981).

64. Élisée Reclus, *The Earth and Its Inhabitants* (London: Virtue, 1876–94).

65. Élisée Reclus, *The Earth and Its Inhabitants: North America* (New York: Appleton, 1893), 80.

66. Fernand Braudel, *The Mediterranean and the Mediterranean World in the Age of Philip II* (Berkeley and Los Angeles: University of California Press, 1995), 1:25–51; translation of *La Méditerranée et le monde méditerranéen* (Paris: A. Colin, 1990).

67. That observation already held a prominent place in one of the first books on the history of scientific knowledge relating to the Pyrenees. See Cameida d'Almeina, *Les Pyrénées: Développement de la connaissance géographique de la chaîne* (Paris: A. Colin, 1893).

68. Robert Rhoades, "Integrating Local Voices and Visions into the Global Mountain Agenda," *Mountain Research and Development* 20, no. 1 (2000): 4–9, quotation 5.

69. See Humboldt, *Cosmos*, 30 and passim.

70. Élisée Reclus, *History of a Mountain* (New York: Harper & Brothers, 1881), translation by Bertha Ness and John Lillie of *Histoire d'une montagne* (Paris: Hetzei, 1880).

71. Jules Blache, *L'homme et la montagne* (Paris: Gallimard, 1933). See the analysis of Anne Sgard, "Un moment de la construction du savoir sur la montagne: Jules Blache dans 'L'homme et la montagne,'" in *L'effet géographique*, ed. B. Debarbieux and M. C. Fourny (Grenoble: Maison des Sciences de l'Homme, 2004).

72. Lucien Febvre, *A Geographical Introduction to History* (New York: Alfred Knopf, 1925),

translation by E. G. Mounford and J. H. Paxton of *La terre et l'évolution humaine: Introduction géographique à l'histoire* (Paris, 1922). This excerpt and those that follow are all taken from the French original, 211–20.

73. Albert Demangeon, "Introduction géographique à l'histoire," *Annales de géographie* 32, no. 176 (1923): 165–70, quotation 170.

74. *Die tropischen Gebirge*, the title of Troll's book, translates as "The Tropical Mountains"; Fernand Braudel uses the term "Mediterranean mountains" throughout *La Méditerranée: L'espace et les hommes* (Paris: Arts et métiers graphiques, 1977).

75. Guy Lasserre in 1988, quoted in Bart, Morin, and Salomon, eds., *Les montagnes tropicales*, 9.

76. Because of the considerable institutional weight of regional geography and geomorphology. That is less true in Germany, where these two branches of the discipline are less central.

77. For example, Martin Beniston, ed., *Mountain Envionments in Changing Climates* (London: Routledge, 1994).

78. That path was traced out a few decades ago by the U.S. geographer Roderick Peattie, who wrote, in a discussion on defining the mountain: "To a large extent then, a mountain is a mountain because of the part it plays in popular imagination. It may be hardly more than a hill but if it has distinct individuality, or plays a more or less symbolic role to the people, it is likely to be rated a mountain by those who live at its base." Roderick Peattie, *Mountain Geography: A Critique and Field Study* (Cambridge, MA: Harvard University Press, 1936), 4.

79. Smadja, *Histoire et devenir des paysages*.

80. Julien Dupuy, "Le 'modèle géographique éthiopien' à l'épreuve du temps," *Cahiers d'outre-mer* 235 (2006): 381–98.

81. Marie-Claire Robic, ed., *Du milieu à l'environnement: Pratiques et représentations du rapport homme-nature depuis la Renaissance* (Paris: Economica, 1992), 168.

INTRODUCTION TO PART ONE

1. Philippe Alliès, *L'invention du territoire* (Grenoble: Presses Universitaires de Grenoble, 1980), 10.

2. Jean Gottmann, *La politique des États et leur géographie* (Paris: Colin, 1952), 213.

CHAPTER TWO

1. Daniel Nordman, *Frontières de France, de l'espace au territoire, XVIe–XIXe siècle* (Paris: Gallimard, 1999). See also Peter Sahlins, *Boundaries: The Making of France and Spain in the Pyrenees* (Berkeley and Los Angeles: University of California Press, 1989).

2. Michel Foucault, *Security, Territory, Population: Lectures at the Collège de France, 1977–1978* (Basingstoke: Palgrave Macmillan, 2007), translation of *Sécurité, territoire, population: Cours au Collège de France, 1977–1978* (Paris: Gallimard/Seuil, 2004).

3. This expression is more satisfactory than "natural border," which seems contradictory (a border is necessarily political) and is not explicit enough about the underlying idea: the political mobilization of natural facts. That term, though unusual, is similar to those used long ago by J. Sölch (*naturgemarkten* or *natur-marken*) and R. Hartshorne ("naturally marked boundaries" or "boundaries marked in nature"). See Richard Hartshorne, "Geographic and Political Boundaries in Upper Silesia," *Annals of the Association of American Geographers* 23 (1933): 195–228, quotation 198.

4. John Pinkerton, *Modern Geography* (London, 1802); Johann C. Gatterer, *Abriss der Geographie* (Göttingen, 1775).

5. José Manuel Zavala, *Les Indiens mapuche du Chili, dynamiques interethniques et stratégies de résistance du XVIIIe siècle* (Paris: L'Harmattan, 2000).

6. Élisée Reclus, "Chili," in *Nouvelle géographie universelle* (Paris: Hachette, 1893), 18:695.

7. Lucien Gallois, "La frontière argentino-chilienne," *Annales de géographie* 12, no. 61 (1903): 47–53.

8. Michael Heffernan, "Historical Geographies of the Future: Three Perspectives from France," in *Geography and Enlightenment*, ed. David N. Livingstone and Charles W. J. Withers (Chicago: University of Chicago Press, 1999), 125–64.

9. Anne-Robert-Jacques Turgot, *Plan d'un ouvrage sur la géographie politique* (1751), reprinted in Gustav Schelle, *Oeuvres de Turgot et documents le concernant* (Glashütten im Taunus: Verlag Detlev Aumermann KG, 1972), 256–57.

10. Ibid., 257.

11. Anne-Robert-Jacques Turgot, *Fragment et pensées politiques pour servir à l'ouvrage sur la géographie politique* (1751), reprinted in Schelle, *Oeuvres de Turgot et documents le concernant*, 328.

12. Jean-Jacques Rousseau, "Judgment of Monsieur l'Abbé de Saint-Pierre's Plan for Perpetual Peace," in *The Collected Writings of Rousseau* (Hanover, NH: University Press of New England, 2005), translation of *Projet de paix perpétuelle de Mr. l'abbé de Saint Pierre* (1761), 21 (quoted passage from the French original).

13. Ibid., 24.

14. Nicolas Buache de la Neuville, *Essai d'une nouvelle division politique, ou moyen d'établir d'une manière fixe et invariable les bornes des possessions entre les différents royaumes*, Institut de France, ms. 2316, quoted in Daniel Nordman, "Buache de la Neuville et la frontière des Pyrénées," in Monique Pelletier, ed., *Images de la montagne* [exhibition catalog] (Paris: Bibliothèque Nationale de France, 1984), 107.

15. Buache de la Neuville, *Essai*, quoted ibid.

16. Ibid.

17. Johann Christoph Gatterer, *Abriss der Geographie* (1773), 22.

18. August Zeune, *Gea: Versuch einer wissenschaftlichen Erdbeschreibung* (Berlin, 1808), v.

19. Augusto Pinochet, *Geopolítica* (Santiago de Chile, 1968), translated into English as *Introduction to Geopolitics* (Santiago de Chile: Editorial Andrés Bello, 1981).

20. Rossitor Johnson, *The Clash of Nations* (New York: Thomas Nelson, 1914), 30.

21. Halford Mackinder, "The Geographical Pivot of History," *Geographical Journal* 23, no. 4 (1904): 421–37.

22. Eduard Maur, *Pamet hor* (Prague: Paseka, 2006).

23. That idea appears in several treaties (regarding Dauphiné) or drafts of treaties from the early seventeenth century on. The same arguments were also used to justify the transfer of Savoy to France, which was being considered in the early seventeenth century.

24. Alfred Dufrenoy and Léonce Élie de Beaumont, *Explication de la carte géologique de la France* (Paris: Librairie scientifique de F. Savy, 1841).

25. Georges Vigarello, "Le Tour de France," in *Les lieux de mémoire*, ed. Pierre Nora (Paris: Gallimard, 1992), vol. 2; Paul Boury, *Le Tour de France: Un espace sportif à géométrie variable* (Paris: L'Harmattan, 1997); Jean-Luc Boeuf and Yves Léonard, *La République du Tour de France, 1903/2003* (Paris: Seuil, 2003).

26. Carl von Clausewitz, *On War* (London : N. Trübner, 1873); translation of *Vom Kriege* (Berlin: Dümmlers Verlag, 1832).

27. Ibid., chapter 16.

28. Ibid., chapter 17.
29. Karl Ritter, "Über das historische Element in der geographischen Wissenschaft," in *Abhandlungen d. k. Akademie der Wissenschaft zu Berlin* (1833).
30. The concept is used in several of his publications and serves as the title of a book: Friedrich Ratzel, "Der Lebensraum: Eine biogeographische Studie," in K. Bücher, K. V. Fricker, F. X. Funk, G. von Mayr, and F. Ratzel, *Festgaben für Albert Schäffle zur siebenzigsten Wiederkehr seines Geburtstages* (Tübingen: H. Laupp, 1901), 101–89. For an analysis of Ratzel's use of the concept, see Woodruff D. Smith, "W. D. Friedrich Ratzel and the Origins of Lebensraum," *German Studies Review* 3, no. 1 (1980): 51–68.
31. Friedrich Ratzel, *Politische Geographie* (Munich: R. Oldenbourg, 1897). Moreover, he devotes the better part of chapter 15 of the first volume of his *Anthropogeographie* (Stuttgart: J. Engelhorn, 1882) to the mountain.
32. Élisée Reclus, *The Earth* (New York: Harper & Brothers, 1871), translation of *La terre* (Paris: L. Hachette, 1868), 186 (quoted passage from the French original).
33. Élisée Reclus, *L'homme et la terre* (Paris: Librairie universelle, 1905), 5:310.
34. Lucien Febvre, *A Geographical Introduction to History* (London: Routledge, 2009), translation of *La terre et l'évolution humaine: Introduction géographique à l'histoire* (Paris: Renaissance du Livre, 1922).
35. Jacques Ancel, *Géographie des frontières* (Paris: Gallimard, 1938).
36. Karl Haushofer, *De la géopolitique* (Paris: Fayard, 1986), 111–12. This is a collection of essays and lectures written during the interwar years. The excerpt quoted was written in 1931 for a radio broadcast and was never published in German.
37. It is therefore clear why, in 1941, an anthology of Ratzel's texts was published by Alfred Kroner Editions in Stuttgart, edited and with a preface by General Karl Haushofer (1869–1946), under the title *Erdenmacht und Völkerschicksal* (The Power of the Soil and the Destiny of the People).
38. Claude Raffestin, with Dario Lopreno and Yvan Pasteur, *Géopolitique et histoire* (Lausanne: Payot, 1995).
39. Quoted in *Le monde*, September 12, 2009.
40. Thomas Hallock, *From the Fallen Tree: Frontier Narratives, Environmental Politics, and the Root of a National Pastoral, 1749–1826* (Chapel Hill: University of North Carolina Press, 2003).
41. On this subject see Wilbur Jacobs, *The Appalachian Indian Frontier* (Lincoln: University of Nebraska Press, 1967), 97.
42. Hallock, *From the Fallen Tree*.
43. Jonathan Carver, *Travels through the Interior Parts of North America in the Years 1766, 1767 and 1768* (London, 1781).
44. Donald Jackson, *Thomas Jefferson and the Stony Mountains: Exploring the West from Monticello* (Urbana: University of Illinois Press, 1981).
45. According to H. A. Moulin, *Le litige chilo-argentin et la délimitation des frontières naturelles* (Paris: A. Rousseau, 1902), 48.
46. Quoted in Henry Nash-Smith, *The American West as Symbol and Myth* (New York: Vintage, 1959), 192. On this subject see also Roderick Nash, *Wilderness and the American Mind* (New Haven, CT: Yale University Press, 1967); and Simon Schama, *Landscape and Memory* (New York: Alfred A. Knopf, 1995).
47. Tom Chaffin, *Pathfinder: John Charles Fremont and the Course of American Empire* (New York: Hill & Wang, 2002).
48. Nash-Smith, *The American West*.

49. Ellen C. Semple, *American History and Its Geographic Conditions* (New York: Houghton Mifflin, 1933).

50. Frederick J. Turner, "The Significance of the Frontier in American History," *The Annual Report of the American Historical Association* (Washington, DC: U.S. Government Printing Office, 1893), 199–227.

51. Emmanuel de Martonne, "Le traité de Saint-Germain et le démembrement de l'Autriche," *Annales de géographie* 29, no. 157 (1920): 1–11, quotation 4.

52. Quoted in A. Asor Rosa, *Dall'unità a oggie: La Cultura*, vol. 4, pt. 2, of his *Storia d'Italia* (Turin: Einaudi, 1978), 953.

53. Especially Cesare Battisti, *Il Trentino: Saggio de geografia fisica et de antropogeografia* (Trento: G. Zippel, 1898); and *Il Trentino: Cenni geografici, storici, economici con un'appendice su l'Alto Adige* (Novara: Istituto Geographico de Agostini, 1915). Battisti changed his mind at the beginning of the war and was taken prisoner; he was executed as a traitor by the Austrians.

54. On this subject see François Walter, *Les figures paysagères de la nation* (Paris: École des Hautes Études en Sciences Sociales, 2004), especially chapter 6; Taline Ter Minassian, "Les géographes français et la délimitation des frontières balkaniques à la Conférence de la Paix en 1919," *Revue d'histoire moderne et contemporaine* 44, no. 2 (1997): 252–86; Emmanuelle Boulineau, "Les géographes et les frontières austro-slovènes des Alpes orientales en 1919–1920," *Revue de géographie alpine* 4 (2001): 173–84; and Wesley J. Reisser, "Self-Determination and the Difficulty of Creating Nation-States: The Case of Transylvania," *Geographical Review* 99, no. 2 (2009): 230–47.

55. Leon Dominian, *The Frontiers of Language and Nationality in Europe* (New York: Henry Holt, 1917). Nevertheless, Dominian's analysis of linguistic geography consisted in part of explaining the distribution of languages and nations by natural facts. He constantly makes reference to the role that topographical obstacles and corridors play in their diffusion. In particular, he cites the role of mountain chains in the French- and Polish-speaking regions and the "confinement of Czech to a plateau enclosed by mountains."

56. Quoted in Boulineau, "Les géographes et les frontières austro-slovènes," 175.

57. Emmanuel de Martonne, "La nouvelle Roumanie," *Annales de géographie* (1921): 1–31.

58. That was not the case, however, for all the proposals of experts enlisted in those years. For example, most validated the incorporation into the new Czechoslovakia of the German-speaking regions located on the near side of the ridge line that circles Bohemia. See Walter, *Les figures paysagères*, 414–15.

59. Jean Brunhes, "La géographie de l'histoire," in *Revue de géographie annuelle* 8, no. 1 (1914): 69.

60. Isaiah Bowman, "Constantinople and the Balkans," in *What Really Happened at Paris: The Story of the Peace Conference, 1918–1919*, ed. Edward M. House and Charles Seymour (New York: Charles Scribner's Sons, 1921), 142.

61. Skijastikn Sekliesco, *La Roumanie et la guerre* (Paris: Armand Colin, 1918), 81.

62. Lucian Boia, *History and Myth in Romanian Consciousness* (Budapest: Central European University Press, 2001), 132.

63. Patrice M. Dabrowski, "Constructing a Polish Landscape: The Example of the Carpathian Frontier," *Austrian History Yearbook* 39 (2008): 45–65.

64. As in Charles Wojatsek's *From Trianon to the First Vienna Arbitral Award: The Hungarian Minority in the First Czechoslovak Republic, 1918–1938* (Montreal: Institut des

Civilisations Comparées, 1981), which continues to accuse the Treaty of Trianon of "Czech expansionism" and "pan-Slavism."

65. Augustin Berque, *Japan: Nature, Artifice, and Japanese Culture* (Northamptonshire: Pilkington Press, 1997).

66. J. H. Jin, "Paektudaegan: Science and Colonialism, Memory and Mapping in Korean High Places," in *High Places: Cultural Geographies of Mountains, Ice and Science*, ed. Denis E. Cosgrove and Veronica Della Dora (London: IB Tauris, 2009), 196–216.

67. Yang Bo-Kyung, "Perceptions of Nature in the Choson Period," *Korea Journal* 37, no. 4 (1997): 134–55.

68. Jin, "Paektudaegan," 201.

69. On July 22, 1986, quoted ibid.

70. See, for example, Joydeep Sircar, "Oropolitics," *Alpine Journal* (1984).

71. Especially with the ascent of Mont Aiguille in Dauphiné in 1492, as the French armies passed through on their return from Italy; and, even earlier, the ascent of the Canigou by Peter III of Aragon in the thirteenth century.

72. Jin, "Paektudaegan."

73. Kenneth R. Olwig, *Landscape, Nature and the Body Politic* (Madison: University of Wisconsin Press, 2002).

74. Boia, *History and Myth*, 132.

75. Lucian Boia and James C. Brown, *Romania: Borderland of Europe* (London: Reaktion Books, 2001), 59.

CHAPTER THREE

1. Jim Gilchrist, "The Colour of Money," *The Scotsman*, April 1, 2000.

2. Stephen Goodwin, "Astride the Razor's Edge," *The Independent*, August 5, 2000.

3. Quoted in John Ross, "Macleod 'Gifts' Cuillin to Public," *The Scotsman*, July 10, 2003.

4. One example among many others: In *Le curieux antiquaire ou recueil géographique et historique des choses les plus remarquables qu'on trouve dans les quatres parties de l'univers* (Leiden, 1729; translation of the German edition of 1709), P. L. Berckenmeyer provides a classification of the parts of the world and of their populations. The Scotsman is chosen as the emblematic illustration of the mountaineer and Scotland as the prototype of the savage nation.

5. Quoted in T. C. Smout, *Exploring Environmental History: Selected Essays* (Edinburgh: Edinburgh University Press, 2009), 21.

6. H. Trevor-Roper, "The Highland Tradition of Scotland," in *The Invention of Tradition*, ed. E. Hobsbawn and T. Ranger (Cambridge: Cambridge University Press, 1983); Charles W. J. Withers, "The Historical Creation of the Scottish Highlands," in *The Manufacture of Scottish History*, ed. I. Donnachie and C. A. Whatley (Edinburgh: Polygon Press, 1992), 143–56; P. Womack, *Improvement and Romance: Constructing the Myth of the Highlands* (London: Macmillan, 1989); R. Clyde, *The Image of the Highlander, 1745–1822* (Edinburgh: Tuckwell Press, 1995).

7. English authors have also contributed to the construction of the cult of the wild and free Highlander. The poet William Wordsworth attests to it soberly but clearly: "A wilderness is rich with liberty"; quoted in Keith Thomas, *Man and the Natural World: A History of the Modern Sensibility* (New York: Pantheon, 1983), 267.

8. Charles W. J. Withers, "Place, Memory, Monument: Memorializing the Past in Contemporary Highland Scotland," *Ecumene* 3, no. 3 (1996): 325–44.

9. Quoted in Martina Frolova, *Paysages du Caucase: L'invention d'une montagne* (Paris: Comité des travaux historiques et scientifiques, 2006), 107.

10. The expression "nationalization of the mountains" is modeled on the formulation used by the historian Olivier Zimmer, who distinguishes between two periods and two complementary rhetorics within the modern history of Switzerland. During the long eighteenth century, narratives tended toward a "nationalization of nature," linking the image of the nation to that of the Alps, primarily through description and iconography. But with the apogee of European nationalism between 1870 and 1940, narratives (generally in the form of arguments) tended toward a "naturalization of the nation," explaining the singularity of Switzerland in terms of the nature of the Alps and their population. See Olivier Zimmer, "In Search of Natural Identity: Alpine Landscape and the Reconstruction of the Swiss Nation," *Comparative Studies of Society and History* 40 (1990): 637–65. On this subject, see also François Walter, "La montagne des Suisses: Invention et usages d'une représentation paysagère (XVIIIe–XXe)," *Études rurales* (1991): 121–24.

11. Claude Reichler and Roland Ruffieux, *Le voyage en Suisse* (Paris: Robert Laffont, 1998); Simon Schama, *Landscape and Memory* (New York: Alfred A. Knopf, 1995); François Walter, *Les figures paysagères de la nation* (Paris: École des Hautes Études en Sciences Sociales, 2004).

12. Philippe Bridel, *Le conservateur Suisse, ou Recueil complet des étrennes helvétiennes* (Lausanne: Louis Knab, 1815), 283.

13. On the general history of national Swiss identity, see George Kreis and Ulrich Imhof, *Mythos Schweiz: Identität—Nation—Geschichte* (Zurich: Verlag NZZ, 1991).

14. *États et délices de la Suisse, ou Description helvétique et géographique des XIII cantons suisses et de leurs alliés* (Basel, 1776). Let us note the importance given to cattle in this excerpt and even more in the rest of the work. Thanks to the emblematization of Switzerland through its mountains and by virtue of the same rhetorical figure, the cow has also acquired the status of emblematic animal of the Swiss mountains.

15. Quoted in Georg Kreis, *Helvetia—im Wandel der Zeiten: Die Geschichte einer nationalen Repräsentationsfigur* (1991, Zürich: NZZ Verlag, 1991); and Kreis and Imhof, *Mythos Schweiz*, 64.

16. Although ancient, that type of discourse became particularly widespread in the second half of the eighteenth century. For an anthology, see Reichler and Ruffieux, *Le voyage en Suisse*.

17. Albert Dupaigne, *Les montagnes* (Tours: Mame Éditions, 1872), 629–30.

18. François Walter, *Les Suisses et l'environnement* (Geneva: Éditions Zoé, 1990), 127.

19. Anselm Zurfluh, "L'Arc alpin, l'Europe, et l'Homo Alpinus," in Gérard-François Dumont et al., *L'arc alpin: Histoire et géopolitique d'un espace européen* (Paris: Economica, 1998), 106–30, especially 120.

20. Marco Cuaz, *Le Alpi* (Bologna: Il Mulino, 2005).

21. Marco Mondini, *Alpini: Parole e immagini di un mito guerriero* (Rome: Laterza, 2008), 111.

22. C. Corsi, "Della guerra in montagna," *Rivista Militare Italiana* 1 (1881): 225–26, quoted in Mondini, *Alpini*.

23. R. Boccardi, ed., *I Verdi: Cinquant'anni di storia alpina, 1872–1922* (Rome: Alfieri & Lacroix, 1922), quoted in Mondini, *Alpini*.

24. Cuaz, *Le Alpi*.

25. Aside from their literary quality, Mario Regoni Stern's novels owe part of their success to the way they recount how young Alpine dwellers found themselves embarking on an absurd war in distant Russia.

26. Some have spoken of "Alpine civilization," but no similar term is ever used for the Apennines, since parts of that chain belonged to totally different (and sometimes opposing) political entities until the late nineteenth century. The societies living in the Apennines, between Calabria, Latium, Tuscany, and Liguria, are only rarely described in terms of the points they have in common.

27. See, for example, Josip Roglic, "Die Gebirge als die Wiege des geschichtlichen Geschehens in Südosteuropa," in *Colloquium Geographicum*, ed. W. Lauer (Bonn: Ferdinand Dümmlers Verlag, 1970), 234ff. For a critical analysis of these discourses, see Ulf Brunnbauer and Robert Pichler, "Mountains as 'Lieux de Mémoire': Highland Values and Nation-Building in the Balkans," *Balkanologie* 6, nos. 1–2 (2002): 77–100; Karl Kaser, "Anthropology and the Balkanisation of the Balkans: Jovan Cvijic and Dinko Tomasic," *Ethnologia Balkanica* 2 (1998): 91; Karl Kaser, "Peoples of the Mountains, Peoples of the Plains: Space and Ethnographic Representation," in *Creating the Other: Ethnic Conflicts and Nationalism in Habsburg Central Europe*, ed. Nancy M. Wingfield (New York: Berghahn Books, 2003), 216–30; and David Norris, *In the Wake of the Balkan Myth: Questions of Identity and Modernity* (Basingstoke: Macmillan Press, 1999).

28. Jovan Cvijic, *La péninsule balkanique* (Paris: Armand Colin, 1928). See also, by the same author, "The Zones of Civilization of the Balkan Peninsula," and "The Geographical Distribution of the Balkan Peoples," *Geographical Review* 5 (1918): 470–82 and 345–61.

29. Irwin Sanders, *Rainbow in the Rocks: The People of Rural Greece* (Cambridge, MA: Harvard University Press, 1962), 80.

30. "Observations inédites sur l'état de la Grèce en 1829," *Revue des deux mondes* 40 (1830): 291.

31. Maria Todorova, *Imagining the Balkans* (Oxford: Oxford University Press, 1997); Norris, *In the Wake of the Balkans Myth*.

32. John Kifner, "The World; Through the Serbian Mind's Eyes," *New York Times*, April 10, 1994.

33. Sanders, *Rainbow in the Rocks*, 80.

34. In a different reading, mountain communities may be viewed through the prism of modernity. They are sometimes seen as the guardians of tradition in a changing world and sometimes as backward areas that need to catch up with the rest of the nation. That perspective will be discussed in chapter 5.

CHAPTER FOUR

1. Thomas Jefferson, *Notes on the State of Virginia* (London: John Stockdale, 1781); Alexander von Humboldt, *Researches Concerning the Institutions and Monuments of the Ancient Inhabitants of America with Descriptions and Views of Some of the Most Striking Scenes in the Cordilleras* (London: Longman, 1814), translation by H. M. Williams of *Vues des cordillères, et monuments des peuples indigènes de l'Amérique* (Paris: Schoell, 1810), 12.

2. Upon his return to Europe, he referred to Jefferson as a "magistrate whose name is dear to the true friends of humanity." Alexander von Humboldt, *Political Essay on the Kingdom of New Spain* (London: Longman, 1811), translation by John Black of *Essai politique sur le royaume de la Nouvelle Espagne*, 2 vols. (Paris: Schoell, 1881; reprint, Amsterdam: Theatrum Orbis Terrarum LTD, 1971), 10.

3. Donald Jackson, *Thomas Jefferson and the Stony Mountains: Exploring the West from Monticello* (Urbana: University of Illinois Press, 1981); John Logan Allen, *Passage*

through the Garden: Lewis and Clark and the Image of the American Northwest (Urbana: University of Illinois Press, 1975).

4. Letter from Thomas Jefferson to Caspar Wistar, June 7, 1804.

5. Alexandre Surell, *Étude sur les torrents des Hautes-Alpes* (Paris: Carilian-Goeuvry et Dalmont, 1841).

6. Félix Lenoble, "La légende du déboisement des Alpes," *Revue de géographie alpine* 11, no. 1 (1923): 5–116; Paul Veyret, "Un centenaire: L'étude sur les torrents des Hautes-Alpes de Surell," *Revue de géographie alpine* 31 (1943): 513–24.

7. Surell, *Étude sur les torrents*, 137.

8. Ibid., 139.

9. George Perkins Marsh, *Man and Nature* (New York: Charles Scribner's, 1864), 35. Marsh traveled to France and later painted a harsh portrait of the disastrous state of the French forests, in *The Earth as Modified by Human Action* (New York: Scribner & Armstrong, 1874), a revised version of *Man and Nature*.

10. Tamara L. Whited, "Extinguishing Disaster in Alpine France: The Fate of Reforestation as Technocratic Debacle," *GeoJournal* 51 (2000): 263–70.

11. Surell, *Étude sur les torrents*, 105.

12. Ibid., 144–45.

13. Ibid., 228.

14. On this subject see Jérôme Buridant, "De la découverte à l'action de terrain: Les forestiers français face à la montagne XVIIe–XIXe siècles," *Annales de ponts et chaussées* 103 (2002): 14–22; and Andrée Corvol, "La forêt de montagne à l'époque moderne," in *La montagne à l'époque moderne, acts du colloque de l'association des historiens modernistes* (Paris: Presses de Paris-Sorbonne, 1998), 99–133.

15. Quoted in Whited, "Extinguishing Disaster."

16. Prosper Demontzey, *Traité pratique du reboisement et du gazonnement des montagnes* (Paris: Rotschild, 1882).

17. Alexandre Surell, *Étude sur les torrents des Hautes-Alpes*, 2nd ed., with an afterword by Ernest Cézanne (Paris: Dunod, 1872); emphasis in original. The reformist Berger gave the same analysis: "Only the state, which does not perish, can think of the future of society and undertake to cultivate for itself those large plant forms that require centuries" (1865). Quoted in Bernard Kalaora and Antoine Savoye, *La forêt pacifiée: Les forestiers de l'École de Le Play, experts des sociétés pastorales* (Paris: L'Harmattan, 1986).

18. Anton Schuler, "La fondation de la société forestière suisse en l'an 1843 et son rôle dans la politique et la législation forestière helvétique," *Annales de ponts et chaussées* 103 (2002): 51–54, quotation 7.

19. Prosper Demontzey, *L'extinction des torrents en France par le reboisement* (Paris: Imprimerie Nationale, 1894); and Demontzey, "La correction des torrents et le reboisement des montagnes," *Revue des eaux et forêts* 29 (1890): 485–502.

20. Monsieur Noblemaire, "Notice biographique sur Alexandre Surell," *Annales de ponts et chaussées* 103 (2002): 4–13; originally published in 1888.

21. Frédéric Fresquet, "Un corps quasi-militaire dans l'aménagement du territoire: Le corps forestier et le reboisement des montagnes méditerranées en France et en Italie aux XIXe et XXe siècles" (Ph.D. diss., Université de Montpellier III, 1997).

22. E. Hanausek, S. Stauder, and J. Hopf, "Ouvrages de protection contre torrents et avalanches au Tyrol," *Revue forestière française* 5 (1982): 202–12.

23. Samuel P. Hays, *Conservation and the Gospel of Efficiency: The Progressive Conservation Movement, 1890–1920* (Cambridge, MA: Harvard University Press, 1959).

24. Roosevelt subordinated federal forest policy to a "patriotic duty of insuring the safety and continuance of the Nation"; quoted in Hays, *Conservation and the Gospel of Efficiency*, 125.

25. Gifford Pinchot, "The Relation of Forests to Stream Control," *Annals of the American Academy of Political and Social Science* 31 (1908): 219–27.

26. Quoted in Schuler, "La fondation de la société forestière suisse."

27. Anton Schuler, "Forêt: XIXe et XXe siècles," in *Dictionnaire historique de la Suisse* (2009), http://www.hls-dhs-dss.ch, accessed October 10, 2009.

28. Émile Cardot, *L'arbre, la forêt et les paturages de montagne: Manuel de l'arbre* (Paris: Touring-Club de France, 1907).

29. In Spain that is the generic name for the irrigated zones of a plain where fruits and vegetables are generally grown.

30. Henri Cavailhès, "La question forestière en Espagne," *Annales de géographie* 14 (1908): 318–31, quotations 331.

31. Alexandre Surell, quoted in René Favier, "Editorial," *Annales des ponts et chaussées* 103 (2002): 1.

32. Quoted in Benoît Boutefeu, "L'aménagement forestier en France: A la recherche d'une gestion durable à travers l'histoire," *VertigO—La revue électronique en sciences de l'environnement* 6, no. 2 (2005).

33. Quoted in Christian Küchli and Hansjakob Baumgartner, "La protection de l'environnement est née dans la forêt," *Magazine environnement* 2 (Bern: Office Fédérale de l'Environnement, 2001), 6–9.

34. *Le nouvelliste*, issue of April 16, 1919, quoted in Christophe Bolli, *Randonnaz, village disparu* (Sierre: Monographic, 1995), 140. The numerous protests and signs of unrest resulted in more lenient laws (in France between 1860 and 1882, for example) or in adjustments in their application (in Italy, for instance). Jérôme Buridant, "De la découverte à l'action de terrain: Les forestiers français face à la montagne XVIIe–XIXe siècles," *Annales des ponts et chaussées* 103 (2002): 14–22.

35. The historian Tamara Whited even went so far as to argue that "seventeen eighty-nine, and succeeding years, had left a dark image of the mountains and their disorderly people for lowland consumption: pillage, illegal pasturing, communal appropriations, and land clearance had seemingly 'created' the mountains visible in the mid-19th century." Whited, "Extinguishing Disaster."

36. Surell, "Étude sur les torrents" (1872), 192.

37. James Scott, *Seeing like a State: How Certain Schemes to Improve the Human Condition Have Failed* (New Haven, CT: Yale University Press, 1998).

38. Tania Murray Li, *Will to Improve: Governmentality, Development, and the Practice of Politics* (Durham, NC: Duke University Press, 2007).

39. John F. Sears, *Sacred Places: American Tourist Attractions in the Nineteenth Century* (New York: Oxford University Press, 1989), 138.

40. Josiah Whitney, *Yosemite Guide Book* (Cambridge, MA: University Press, Welch, Bigelow, 1870), 123.

41. Samuel Bowles, *Across the Continent* (Springfield, MA: S. Bowles, 1866), 224, quotations taken from Sears, *Sacred Places*, chapter 6.

42. Thomas Starr King, *A Vacation among the Sierras: Yosemite in 1860* (San Francisco: Book Club of California, 1962), 64.

43. In fact, the idea may date back to the 1830s and a proposal by the American painter George Catlin, who specialized in the representation of western landscapes and in portraits and everyday scenes featuring Native Americans.

44. The intrigues that led to that protection are not perfectly understood, but the historian Alfred Runte points out that the emphasis placed on the scarcity, or even the absence, of natural resources was probably a determining argument in making the case to protect Yosemite. Alfred Runte, *Yosemite: The Embattled Wilderness* (Lincoln: University of Nebraska Press, 1990), 19.

45. Frederick Law Olmsted, *Preliminary Report upon the Yosemite and Big Tree Grove*, August 1865, reprinted in *Landscape Architecture* 43 (1952).

46. "Olmsted looked upon his parks as symbols of nature in which the essence of nature is expressed in art." Irving D. Fisher, *Frederick Law Olmsted and the City Planning Movement in the United States* (Ann Arbor: University of Michigan Research Press, 1986), 63.

47. Olmsted, *Preliminary Report*, 258.

48. John Muir, *The Mountains of California* (New York: Century, 1894); Muir, *My First Summer in the Sierra* (Boston: Houghton Mifflin, 1911).

49. Robert Lewis, "Frontier and Civilization in the Thought of Olmsted," *American Quarterly* 29 (1977): 385–403.

50. Linda M. Graber, *Wilderness as Sacred Space* (Washington, DC: Association of American Geographers, 1976); Roderick F. Nash, *Wilderness and the American Mind* (New Haven, CT: Yale University Press, 1967).

51. Clarence Glacken, *Traces on the Rhodian Shore: Nature and Culture in Western Thought from Ancient Times to the End of the Eighteenth Century* (Berkeley and Los Angeles: University of California Press, 1976).

52. Barbara Nowak, *Nature and Culture: American Landscape and Painting* (New York: Oxford University Press, 1980).

53. John Muir, *The Yosemite* (New York: Century, 1912), 4.

54. General Grant National Park would become Kings Canyon National Park in 1940.

55. William Wyckoff, *Creating Colorado: The Making of a Western American Landscape* (New Haven, CT: Yale University Press, 1999).

56. Henry David Thoreau, *Journal*, entry of August 30, 1856.

57. Charles W. Webber, *Old Hicks: The Guide* (New York: Harper, 1855), 46.

58. Nowak, *Nature and Culture.*

59. Aldo Leopold, *A Sand County Almanac and Sketches Here and There* (Oxford: Oxford University Press, 1949).

60. As in Sid Marty, *Men for Mountains* (New York: Vanguard Press, 1978).

61. Jim Thorseel, "Protection of Nature in Mountain Regions," in *Mountains of the World: A Global Priority*, ed. B. Messerli and J. D. Ives (London: Parthenon, 1997), 237–48.

62. Wordworth's target was the Kendal and Windermere Railway, according to Margaret H. Nicholson, *Mountain Gloom and Mountain Glory: The Development of the Aesthetics of the Infinite* (Ithaca, NY: Cornell University Press, 1959), 17.

63. John Ruskin, *Sesame and Lilies* (London: Smith, Elder, 1865), 85.

64. John Ruskin, *Modern Painters* (New York: Merrill and Baker, 1843), 464.

65. Quoted in Jacques Gillerion, *Le parc national suisse* (Lausanne: Delachaux-Niestlé, 1996).

66. Mark D. Spence, *Dispossessing the Wilderness: Indian Removal and the Making of the National Parks* (New York: Oxford University Press, 1999), 4.

67. Colin Fisher, review of Mark D. Spence, *Dispossessing the Wilderness: Indian Removal and the Making of the National Parks, H-Environment, H-Net Reviews* (August 2000).

68. The term "purification process" is borrowed from Bruno Latour. See the introduction to his *We Have Never Been Modern* (London: Harvester, Wheatsheaf, 1993).
69. The term "climax" refers to the supposed original state of a plant formation or ecosystem before it is disturbed by human beings. That concept, widely used until the mid-twentieth century, has since been strongly criticized for of its inability to account for the existence of environmental transformations independent of human action.
70. Roderick F. Nash, *The Rights of Nature* (Madison: University of Wisconsin Press, 1989).
71. These are zones where access is strictly prohibited to human beings in order to allow the areas to evolve independently of any human intervention.
72. Ronald William Clark, *The Victorian Mountaineers* (London: Batsford, 1953); Dominique Lejeune, *Les alpinistes en France, 1875–1919: Étude d'histoire sociale, étude de mentalité* (Paris: Comité des travaux historiques et scientifiques, 1988); and Peter H. Hansen, *The Summits of Modern Man: Mountaineering after the Enlightenment* (Cambridge, MA: Harvard University Press, 2013).
73. Edward Whymper, *Scrambles amongst the Alps in the Years 1860–69* (London: John Murray, 1900).
74. In fact, the Deutsche Alpenverein was originally Bavarian, before aspiring to embrace all lovers of German culture: it was behind the merger in 1873 of the two Germanic clubs into the Deutsche und Österreichische Alpenverein. On the Club Alpin Français see Anneliese Gidl, *Alpenverein: Die Städter entdecken die Alpen; Der Deutsche und Österreichische Alpenverein von der Gründung bis zum Ende des Ersten Wetlkrieges* (Vienna: Böhlau, 2007).
75. Mountain climbing in Japan developed during the Meiji period. The modernization taking place in Japanese society at the time led it to look closely at the mountain imaginary in the West and to view its own mountains on the Alpine model; see Karen Wigen, "Discovering the Japanese Alps: Meiji Mountaineering and the Quest for Geographical Enlightenment," *Journal of Japanese Studies* 31, no. 1 (2005): 1–26.
76. Hence the Club Alpino Italiano was very proud to open its branch in L'Aquila in 1870. This city in the central Apennines is also at the foot of the main peak in that range, the Gran Sasso, which at the time was readily called the most Alpine of them all. Italian unification and the constitution of mountain institutions at the national level thus contributed to a reevaluation of the respective characteristics of different mountains. The Gran Sasso became a national park in 1991.
77. Lejeune, *Les alpinistes en France*; Ronald William Clark, *The Victorian Mountaineers*; Lee W. Holt, *Mountains, Mountaineering and Modernity: A Cultural History of German and Austrian Mountaineering, 1900–1945* (Ann Arbor: University of Michigan Press, 2008).
78. Ibid.
79. Ernest Cézanne, "La question des montagnes," *Annuaire du Club alpin français* (1874).
80. Yann Drouet, "The CAF at the Borders: Geopolitical and Military Stakes in the Creation of the French Alpine Club," *International Journal of the History of Sport* 22, no. 1 (2005): 59–69.
81. For example, in the well-received book by the Austrian writer Karl Heinrich Waggerl, *Brot* (Leipzig: Insel Verlag, 1930).
82. Website of Gran Paradiso National Park, http://www.pngp.it, accessed December 12, 2009.

CHAPTER FIVE

1. Joëlle Salomon Cavin, *La ville mal-aimée* (Lausanne: Presses Polytechniques et Universitaires Romandes, 2005).

2. Renzo Boccardi, ed., *I Verdi: Cinquant'anni di storia alpina, 1872–1922* (Rome: Alfieri & Lacroix, 1922), 90.

3. For analyses of Auguste Calvet's contribution, see Jean-Paul Métailié, "Auguste Calvet, pionnier du sylvopastoralisme dans les Pyrénées," in *Les Pyrénées: Lieux d'interaction des savoirs, 19e–début 20e siècles*, ed. Vincent Berdoulay (Paris: Comité des travaux historiques et scientifiques, 1994), 160–74; Bernard Kalaora and Antoine Savoye, *La forêt pacifiée: Les forestiers de l'École de Le Play, experts des sociétés pastorales* (Paris: L'Harmattan, 1986).

4. As a mining engineer, Le Play was primarily interested in forests as suppliers of wood for forges. He was the author of several monographs on rural and familial sociology. For an analysis, see the special issue of the review *Les études sociales* titled *Frédéric Le Play et la question forestière* 136 (1997); and Kalaora and Savoye, *La forêt pacifiée*.

5. Jean-Yves Puyo, "La science forestière vue par les géographes français, ou la confrontation de deux sciences 'diagonales' (1870–1914)," *Annales de géographie* 108, nos. 609–610 (1999): 615–34.

6. Charles Gardelle, "L'inalpage des vaches laitières: Réalités et images d'une spécificité alpine," *Revue de géographie alpine* 77, no. 1 (1989): 293–301.

7. Charles Flahaut, "Le devoir des botanistes en matière de géographie humaine," *9e Congrès international de géographie* 1 (1908).

8. Ernest Guinier, "La question des montagnes," *Annuaire de la Société des touristes du Dauphiné* (Grenoble, 1890).

9. On cadasters in Switzerland in the 1950s, see, for example, Gilles Rudaz, "Porter la voix de la montagne: Objectivation et différenciation du territoire par le Groupement de la population de montagne du Valais romand (1945–2004)" (Ph.D. diss., University of Geneva, 2005). For an example in France of the reunification of land parcels, see J.-L. Chaize, "Un exemple de remembrement rural en montagne (commune de Savournon Hautes-Alpes)," *Revue de géographie alpine* 46, no. 1 (1958): 81–96; and in the Swiss canton of Valais, see Jean Loup, *Pasteurs et agricultureurs valaisans: Contribution à l'étude des problèmes montagnards* (Grenoble: Imprimerie Allier, 1965).On stock-breeding initiatives in France, see Simone Bossy, "Associations foncières pastorales et groupements pastoraux: Bilan d'une décennie," *Revue de géographie alpine* 4 (1985): 439–63.

10. Office Fédéral de l'Industrie, des Arts et Métiers et du Travail (OFIAMT), *Mesures de la Confédération en faveur des populations montagnardes* (Bern, 1956), 5–6, 12.

11. Céline Broggio, "La politique de la montagne en France," *Hérodote* 107 (2002): 147–58, quotation 157.

12. Jacques Joly, "L'énergie électrique dans les Alpes slovènes," *Revue de géographie alpine* 56, no. 1 (1968): 141–52.

13. The Mediterranean mountains lag behind, primarily because of climate: seasonal differences in recorded rainfall make it difficult to turn a profit on costly investments.

14. Virginie Bodon, *La modernité au village: Tignes, Savines, Ubaye: La submersion de communes rurales au nom de l'intérêt général, 1920–1970* (Grenoble: Presses Universitaires de Grenoble, 2002).

15. André Guex, *Le demi-siècle de Maurice Troillet: Essai sur l'aventure d'une génération* (Martigny: Imprimerie Pillet, 1971), 279.

16. Paul Wagret, "L'équipement hydroélectrique de l'Autriche," *Revue de géographie alpine* 45, no. 1 (1957): 113–26.

17. Jacky Herbin, "Le tourisme au Tyrol: Une valeur d'exemple pour les Alpes Françaises?" *Revue de géographie alpine* 68, no. 1(1980): 83–96.

18. Hervé Gumuchian, *La neige dans les Alpes françaises du Nord: Les territoires de l'hiver ou la montagne française au quotiden* (Grenoble: Cahiers de l'Alpe, 1984); Rémy Knafou, *Les stations intégrées de sports d'hiver des Alpes françaises* (Paris: Masson, 1978); Jean-Claude Gallety, "Les sociétés montagnardes devant l'aménagement touristique: L'émergence du fait communal" (Ph.D. diss., Grenoble, Institut d'Urbanisme, 1983).

19. Bruno Cognat, *La montagne colonisée* (Paris: Éditions du Cerf, 1973).

20. James Briggs, "Ski Resorts and National Forests: Rethinking Forest Service Management Practices for Recreational Use," *Boston College Environmental Affairs Law Review* 28 (2000): 79–118; Hal Clifford, *Downhill Slide: Why the Corporate Ski Industry Is Bad for Skiing, Ski Towns, and the Environment* (San Francisco: Sierra Club Books, 2002).

21. Arnold Henry Guyot, "On the Appalachian Mountain System," *American Journal of Science and Arts*, 2nd ser., 31 (1861): 167–71; David Walls, "On the Naming of Appalachia," in *An Appalachian Symposium: Essays Written in Honor of Cratis D. Williams*, ed. J. W. Williamson (Boone, NC: Appalachian State University Press, 1977), 4; George Stewart, *Names on the Land: A Historical Account of Place-Naming in the United States* (Boston: Houghton Mifflin, 1945).

22. William Frost, "Our Contemporary Ancestors in the Southern Mountains," *Atlantic Monthly* (March 1899).

23. See Henry Shapiro, *Appalachia on Our Mind: The Southern Mountains and Mountaineers in the American Consciousness, 1870–1920* (Chapel Hill: University of North Carolina Press, 1978). On the "backwardness" of the region, see Walls, "On the Naming of Appalachia," 5.

24. Shapiro, *Appalachia on Our Mind*, 115.

25. Anthony Harkins, *Hillbilly: A Cultural History of an American Icon* (New York: Oxford University Press, 2005).

26. Ibid.

27. Shapiro, *Appalachia on Our Mind*, xiii.

28. Ibid., 32.

29. Ibid., 188.

30. United States Department of Agriculture, *Economic and Social Problems and Conditions of the Southern Appalachians*, Miscellaneous Report 205 (Washington, DC, 1935).

31. In 1960 income in that region of the United States was 23 percent lower than the average for the country as a whole. See http://www.arc.gov.

32. Ronald Eller, *Uneven Ground: Appalachia since 1945* (Lexington: University of Kentucky Press, 2008), 57. In his analysis of how the idea of underdevelopment in the Appalachians was constructed, Eller particularly emphasizes the role played by a photo report in *Look* magazine titled "Portrait of an Underdeveloped Country, Southern Appalachian Mountain Region" (65–66).

33. In 2013, 83 percent of the highway system was completed and 3 percent was under construction (http://www.arc.gov/adhs). See David E. Whisnant, *Modernizing the Mountaineer: People, Power, and Planning in Appalachia* (Knoxville: University of Tennessee Press, 1994); Eller, *Uneven Ground*.

34. Michael Bradshaw, *The Appalachian Regional Commission: Twenty-five Years of Government Policy* (Lexington: University of Kentucky Press, 1992).

35. Quoted in Eller, *Uneven Ground*, 102.

36. Eller has shown that these were usually cities with more than seven thousand residents (*Uneven Ground*, 182) and that the most isolated regions benefited proportionally less from investments (214).

37. Section 2 of the Appalachian Regional Development Act.

38. Whisnant, *Modernizing the Mountaineer*.

39. A "project focus" was promoted at the time in an effort to move away from the "handout focus" characteristic of previous policies, which encouraged individuals, businesses, and collectivities to position themselves as beneficiaries of public assistance.

40. Broggio, "La politique de la montagne"; François Gerbaux, *La montagne en politique* (Paris: L'Harmattan, 1994).

41. Broggio, "La politique de la montagne," 148.

42. Gioacchino Garofoli, "Local Development in Europe: Theoretical Models and International Comparisons," *European Urban & Regional Studies* 9, no. 3 (2002): 225–39.

43. Law of December 3, 1971, no. 1102, "Nuove norme per lo sviluppo della montagna."

44. Gilles Rudaz and Bernard Debarbieux, *La montagne suisse en politique* (Lausanne: Presses Polytechniques et Universitaires Romandes, 2013).

45. That strategy of choosing regional centers was also applied to development in the Appalachians.

46. In 1985 the program was expanded to the country as a whole. G. Hovorka, *Die Kulturlandschaft im Berggebiet in Österreich—OECD Fallstudie*, research report 43 (Vienna: Bundesanstalt für Bergbauernfragen, 1998).

47. www.bioalpin.at, accessed April 15, 2014.

48. Similar examples of promoting locally sourced products are found everywhere in Europe, indicating a desire to pair agricultural production and tourist consumption. The certified label "Product of Regional Natural Park," adopted in France and Spain, and even more locally the partnerships between hoteliers and farmers in response to visitors' requests to consume local products during their stay, attest to that tendency.

49. The mountain regions are particularly well represented on the certified labels granted for cheese: Gruyère, Etivaz, Emmental, Formaggio d'Alpe Ticinense, Berner Alpkäse (Switzerland); Beaufort, Reblochon, Cantal, Ossau-Iraty (France); Tiroler Bergkäse, Vorarlberger Alpkäse (Austria); Allgäuer Bergkäse (Germany); Serra da Estrela, Nisa (Portugal); Murcia, Cabrales, Alt Urgell (Spain); Bra d'alpeggio, Fontina (Italy).

50. Decree no. 2000-1231 of December 15, 2000, regarding the use of the term "mountain."

51. In fact, article 2 of the decree lists a series of exceptions, such as the absence of raw materials necessary for production or "technical conditions" that prevent part of the production or manufacture process from being carried out in a mountain zone.

52. Bernard Poche and Jean-Paul Zuanon, "Les collectivités de montagne: Image externe et représentation propre," in *Actes du XIe colloque franco-italien d'études alpines: Spécificité du milieu alpin?* (Grenoble: Université de Grenoble 2, 1986), 5–22, quotation 5; Bernard Crettaz, *La beauté du reste: Confession d'un conservateur de musée sur la perfection et l'enfermenent de la Suisse et des Alpes* (Carouge: Zoé, 1993).

53. [The English term "mountaineer" is generally not used in such contexts. Inhabitants of such regions prefer the English expressions "mountain people," "mountain dwellers," or "mountain populations."—Trans.]

54. Laurent Tissot, "Du touriste au guide de montagne: La question de l'identité alpine (1850–1920)," in *L'espace alpin de la modernité: Bilans et perspectives au tournant du*

siècle, ed. Daniel Grange (Grenoble: Presses Universitaires de Grenoble, 2004). On the celebration of a cattle-raising identity among farmers in the Swiss Alps, see Gérald Berthoud, Bernard Crettaz, and Yvonne Preiswerk, *Vache d'utopie* (Geneva: Slatkine, 1991).

55. Law no. 991 of July 15, 1952.

56. Called at the time the Swiss Union for Mountain Farmers.

57. Later, in the 1990s, organizations for the defense of mountain interests would appear in Eastern Europe: on the Swiss model in Bulgaria, Macedonia, and Romania; and on the French model in Poland and Slovakia.

58. Sometimes, in the name of that same principle of *solidarité*, associations for interregional cooperation, especially between the cities and the mountains, came into being. That was especially true for Switzerland in the 1940s with the creation of the Swiss Sponsorship of Mountain Communities (Schweizer Patenschaft für Berggemeinden) in 1940 and of Swiss Aid for Mountaineers (Aide Suisse aux Montagnards) in 1942.

59. Rudaz and Debarbieux, *La montagne suisse en politique*.

60. Peter Rieder, "L'agriculture de montagne dans le champ de tension entre le marché, la politique et la société," afterword to Didier Ruef and Ulrich Ladurner, *Paysans de nos montagnes* (Sierre: Monographic, 1998), 201–6.

61. More fine-grained analyses have shown that the evolution of public policies and the problematizations of the mountain question in Switzerland led local groups constantly to revise their own way of characterizing mountain questions and of presenting themselves to the outside. See Rudaz, "Porter la voix de la montagne."

62. For a broader frame of analysis consistent with the specific one adopted here, see Pierre Lascoumes and Jean-Pierre Le Bourhis, "Le bien commun comme construit territorial: Identités d'action et procédure," *Politix* 42 (1998): 37–66.

63. European Conference on Territorial Cooperation between European Mountains, Chambéry, June 8–9, 2006.

INTRODUCTION TO PART TWO

1. Miller proposes to call "globalism" the state of mind that is the very condition for globalization when "the framing of particular features of nature or society" is amenable to analysis "solely on a worldwide basis." Clark Miller, "Resisting Empire: Globalism, Relocalization, and the Politics of Knowledge," in *Earthly Politics: Local and Global in Environmental Governance*, ed. Sheila Jasanoff and Marybeth Long Martello (Cambridge, MA: MIT Press, 2004), 81–102, quotation 82.

2. David Delanay and Helga Leitner, "The Political Construction of Scale," *Political Geography* 16, no. 2 (1997): 93–97, quotation 94–95.

3. Roland Robertson, "Glocalization: Time-Space and Homogeneity-Heterogeneity," in *Global Modernities*, ed. Mike Featherstone, Scott Lash, and Roland Robertson (London: Sage, 1995), 25–44; Erik Swyngedow, "Globalisation or 'Glocalisation'? Networks, Territories, and Rescaling," *Cambridge Review of International Affairs* 17, no. 1 (2004): 25–48.

CHAPTER SIX

1. On the British presence in Kashmir between the late nineteenth century and World War I, see Martin Sökefeld, "Rumours and Politics on the Northern Frontier: The British, Pakhtun Wali and Yaghestan," *Modern Asian Studies* 36 (2002): 299–340; Robert A. Huttenback, "The 'Great Game' in the Pamirs and the Hindu-Kush: The

British Conquest of Hunza and Nagar," *Modern Asian Studies* 9 (1975): 1–29; and Chad Haines, "Colonial Routes: Reorienting the Northern Frontier of British India," *Ethnohistory* 51 (2004): 535–65.

2. Colonel Algernon Durand, *The Making of a Frontier: Five Years' Experience and Adventures in Gilgit, Hunza Nagar, Cjitral and the Eastern Hindu-Kush* (London: Thomas Nelson, 1889).

3. Ibid., 88.

4. Ibid.

5. Francis Younghusband, *The Heart of a Continent: A Narrative of Travels in Manchuria, across the Gobi Desert, through the Himalayas, the Pamirs and Chitral, 1884–94* (London: J. Murray, 1896).

6. Felix Driver, "Geography's Empire: Histories of Geographical Knowledge," *Environment and Planning D: Society and Space* 10 (1992): 23–40; Pierre Singarevelou, ed., *L'empire des géographes: Géographie, exploration et colonisation XIXe–XXe siècle* (Paris: Belin, 2008).

7. Tim Fulford, Debbie Lee, and Peter J. Kitson, *Literature, Science and Exploration in the Romantic Era* (Cambridge: Cambridge University Press, 2004); Richard Grove, *Green Imperialism: Colonial Expansion, Tropical Island Edens and the Origins of Environmentalism, 1600–1860* (Cambridge: Cambridge University Press, 1996); David Philip Miller and Peter Hanns Reill, eds., *Visions of Empire: Voyages, Botany, and Representations of Nature* (Cambridge: Cambridge University Press, 1996); David Livingstone, *The Geographical Tradition: Episodes in the History of a Contested Enterprise* (Oxford: Blackwell, 1992); Felix Driver, *Geography Militant: Cultures of Exploration and Empire* (Oxford: Blackwell, 2001); Anne Godlewska and Neil Smith, eds., *Geography and Empire* (Oxford: Blackwell, 1994).

8. Hélène Blais, *Voyages au Grand Océan: Géographies du Pacifique et colonisation, 1815–1845* (Paris: Comité des travaux historiques et scientifiques, 2005).

9. Anatole Leroy-Beaulieu, *La France, la Russie et l'Europe* (Paris: Calmann-Lévy, 1888), 212.

10. Timothy Mitchell, *Colonising Egypt* (Cambridge: Cambridge University Press, 1988).

11. David Livingstone, *Putting Science in Its Place: Geographies of Scientific Knowledge* (Chicago: University of Chicago Press).

12. Matthew H. Edney, *Mapping an Empire: The Geographical Construction of British India, 1765–1843* (Chicago: University of Chicago Press, 1997).

13. Clements Markham, address to the Royal Geographical Society, *Geographical Journal* 8, no. 1 (July 1896): 1–15, quotation 6.

14. Kapil Raj, "La construction de l'empire de la géographie, l'odyssée des arpenteurs de sa Très Gracieuse Majesté, la Reine Victoria en Asie Centrale," *Annales histoire, sciences sociales* 20 (1997): 1153–80.

15. Diderot and d'Alembert's *Encyclopédie* is generally laudatory of the geographer-czar, particularly the article that Chevalier Jaucourt devotes to the geography of Russia.

16. W. H. Parker, "Europe: How Far?" *Geographical Journal* 3 (1960): 278–97.

17. The question of whether Russia belonged to Europe or to Asia arose repeatedly between the seventeenth and eighteenth centuries. It is also essential for understanding how Westerners regarded Russia and how Russian leaders positioned themselves in that respect. See Svetlana Gorshenina, "De la Tartarie à l'Asie Centrale: Le coeur d'un continent dans l'histoire des idées entre la cartographie et la géopolitique" (Ph.D. diss., Universités de Paris I-Sorbonne et de Lausanne, 2007), 1:271–74, and especially 336ff. on the particular role conferred on the Urals.

18. Mark Bassin, "Inventing Siberia: Visions of the Russian East in the Early Nineteenth Century," *American Historical Review* 96 (1991): 763–94; Mark Bassin, "Russia between Europe and Asia: The Ideological Construction of Geographical Space," *Slavic Review* 50 (1991): 1–17.

19. Martina Frolova, *Paysages du Caucase: L'invention d'une Montagne* (Paris: Comité des travaux historiques et scientifiques, 2006); Gorshenina, "De la Tartarie à l'Asie Centrale."

20. Alexander von Humboldt, *Asie Centrale: Recherches sur les chaînes de montagnes et la climatologie comparée* (Paris: Gide, 1843).

21. Gorshenina, "De la Tartarie à l'Asie Centrale," 2:427.

22. Halford Mackinder, "The Geographical Pivot of History," *Geographical Journal* 24, no. 4 (1904): 421–37.

23. Younghusband, *The Heart of a Continent.*

24. Gorshenina, "De la Tartarie à l'Asie Centrale," 2:427.

25. Frederic Thomas, "L'invention des 'hauts plateaux' en Indochine: Conquête coloniale et production de savoirs," *Ethnologie française* 2 (2004): 639–49.

26. The term "kingdom" is used generically here, as it is by many authors, to designate very heterogeneous forms of government.

27. Sunait Chutintaranond, "Mandala, Segmentary State and Politics of Centralization of Medieval Ayudhya," *Journal of the Siam Society* 78, no. 1 (1990): 89–100.

28. Michel Bruneau, *L'Asie d'entre Inde et Chine: Logiques territoriales des États* (Paris: Belin, 2006), 47.

29. That conception was theorized after the fact by Lord Curzon and set out in George Curzon, *Frontiers* (Oxford: Clarendon, 1908). See also B. D. Hopkins, "The Bounds of Identity: The Goldsmid Mission and the Delineation of the Perso-Afghan Border in the Nineteenth Century," *Journal of Global History* 2 (2007): 233–54.

30. Nigel J. R. Allan, "Defining Places and People in Afghanistan," *Post-Soviet Geography and Economics* 42 (2001): 545–60.

31. Witt Raczka, "Xinjiang and Its Central Asian Borderlands," *Central Asian Survey* 17, no. 3 (1998): 373–407.

32. For part of its course, in fact, that line followed a boundary identified by the Indian princes, along which a road had been built to guarantee security. Manilal Bose, *History of Arunachal Pradesh* (New Delhi: Concept, 1997).

33. Lord Hardinge, quoted in Manilal Bose, *History of Arunachal Pradesh*, 121.

34. For example, in a book written in Kashmir under the Durand administration, Edward Knight argued: "The [British] influence ought at least to extend to the large chain of mountains [the Himalayas] that forms the natural border of India. For the protection of our empire, it is necessary that we master our side of the mountain gateways." Edward Frederick Knight, *Where Three Empires Meet: A Narrative of Recent Travel in Kashmir, Western Tibet, Gilgit, and the Adjoining Countries* (London: Longmans, Green, 1893), 289.

35. Thomas H. Holdich, *Tibet, the Mysterious* (London: Alston Rivers, 1906).

36. As Edward Said has shown in *Orientalism* (New York: Vintage, 1978).

37. Michel Bruneau, "Modèles spatiaux des États de l'Asie du Sud-Est continentale," *Cahiers de géographie du Québec* 35, no. 94 (1991): 89–116.

38. Quoted in Thomas, "L'invention des 'hauts plateaux.'"

39. The term was often understood by Westerners as a synonym for "slave," an interpretation contested by Jan Ovesen, "All Lao? Minorities in the Lao People's Democratic Republic," in *Civilizing the Margins: Southeast Asian Government Policies for the De-*

velopment of Minorities, ed. Christopher R. Duncan (Ithaca, NY: Cornell University Press, 2004), 214–40.

40. Pavie Mission, *Indochine, Atlas, notices et cartes* (Paris: A. Challamel, 1903), 10–11.

41. Quoted in Thomas, "L'invention des 'hauts plateaux,'" 416.

42. Jules Sion, "L'Asie des moussons," in *Géographie universelle,* ed. P. Vidal de la Blache and L. Gallois, vol. 9, pt. 2, *Inde, Indochine, Insulinde* (Paris: Librairie Armand Colin, 1929), 108.

43. Pierre Gourou, *L'Asie* (Paris: Hachette, 1953), 30–31. He attributes that divorce to the presence of malaria in the highlands, a malady to which the inhabitants of the plains had no immunity.

44. Wilhelm Credner, *Siam, das Land der Tai* (Stuttgart, 1935); Harald Uhlig, "Hill Tribes and Rice Farmers in the Himalayas and South-East Asia: Problems of the Social and Ecological Differentiation of Agricultural Landscape Types," *Transactions and Papers: The Institute of British Geographers* 47 (1969): 1–23.

45. John McKinnon and Jean Michaud, "Montagnard Domain in the South-East Asian Massif," in *Turbulent Times and Enduring Peoples: Mountain Minorities in the Southeast Asian Massif,* ed. Jean Michaud (London: Curzon, 2000), 1–25.

46. Especially Edmund R. Leach, *Political Systems of Highland Burma* (Cambridge, MA: Harvard University Press, 1954).

47. Oliver Wolters, *History, Culture and Religion in Southeast Asian Perspectives* (Singapore: Institute of Southeast Asian Studies, 1982), 32; Anthony Reid, *Southeast Asia in the Age of Commerce, 1450–1680,* vol. 1, *The Lands Below the Winds* (New Haven, CT: Yale University Press, 1988). These studies echo a few observations made by members of the Pavie Mission. Captain Cupet noted in 1900, for example, that the populations he called "Moï" expended considerable effort to elude the gaze of administrations of states on the plains and the coast: "The [Moï] population is feared. If you go among them in great numbers, they steal away and conceal what they possess. There are no longer any leaders with whom to reach an understanding and whose greed or ambition could be exploited. The bad reputation of the country in terms of salubrity makes you hesitate to go there, and its apparent poverty does not arouse any envy" (413).

48. James Scott, *The Art of Not Being Governed: An Anarchist History of Upland Southeast Asia* (New Haven, CT: Yale University Press, 2009). A rather similar theory had previously been defended for the Philippines: Felix Keesing, *The Ethnohistory of Northern Luzon* (Stanford, CA: Stanford University Press, 1962).

49. Scott, *The Art of Not Being Governed,* 5.

50. Bronevski, quoted in Frolova, *Paysages du Caucase,* 106.

51. Quoted in Jean Radvanyi, "Grand Caucase, la 'montagne des peuples' écartelée," *Hérodote* 107 (2002): 67. See also A. Grannes, "'Persons of Caucasian Nationality'— Russian Negative Stereotypes," in *Contrasts and Solutions in the Caucasus,* ed. Ole Hoiris and Sefa Martin Yürükel (Aarhus, Denmark: Aarhus University Press, 1998), 22.

52. Radvanyi, "Grand Caucase."

53. Captain Devaux, *Les Kebaïles du Djerdjera* (Marseilles, 1859). On the contribution of these study missions to knowledge of the peoples of the region, see especially Jacques Berque, "125 ans de sociologie maghrébine," *Annales ESC* 11, no. 3 (1956): 296–324.

54. Daniel Nordman, "Les sciences historiques et géographiques dans l'exploration scientifique de l'Algérie," in *Géographies plurielles: Les sciences géographiques au mo-*

ment de l'émergence des sciences humaines, 1750–1850, ed. Hélène Blais and Isabelle Laboulais (Paris: L'Harmattan, 2006), 235–53.

55. Hélène Claudot-Hawad, ed., *Berbères ou Arabes: Le tango des spécialistes* (Paris: Non Lieu, 2006).

56. On this subject see Florence Deprest's analysis in *Géographes en Algérie, 1880–1950* (Paris: Belin, 2009). A specialist in the history of colonial geography in Algeria, Deprest also examines other conceptions of Barbery and its relation to its physical surroundings, those of Augustin Bernard and Émile-Felix Gauthier especially.

57. Jean Célérier, "La montagne au Maroc: Essai de définition et de classification," *Hespéris* 25 (1938): 117–24. Earlier in the text he also writes: "The mountain is the *Bled es-Siba* par excellence, the land of the rebel tribes, where the sultan was unable to establish a minimum of order, where the splintering into enemy groups not long ago obliged peaceful travelers to enlist many costly protectors. . . . The most intelligent and most energetic sultans have exhausted themselves fighting against centrifugal force, which is the major product of the mountain" (117).

58. Said, *Orientalism*.

59. Hopkins, "The Bounds of Identity," 233.

60. Tania M. Li, *The Will to Improve: Governmentality, Development, and the Practice of Politics* (Durham, NC: Duke University Press, 2007).

61. As has been aptly shown for Sulawesi, for example, in Albert Schrauwers, *Colonial "Reformation" in the Highlands of Central Suawesi, Indonesia, 1892–1995* (Toronto: University of Toronto Press, 2000), 51.

62. Hopkins speaks of the "'indigenization' of Western norms of statehood." Hopkins, "The Bounds of Identity," 233.

63. Quoted in Benjamin Stora, *Algérie, Maroc: Histoires parallèles, destins croisés* (Paris: Zellige, 2002), 19.

64. Mohamed Othman Benjelloun, *Projet national et identité au Maroc: Essai d'anthropologie politique* (Paris: L'Harmattan, 2002). For the last decade or so the Moroccan state has proved more open to dealing explicitly with the Berber minority in its conception of the nation and in public policies. On this, see Mari Oiry-Varacca, "The Use of Amazigh Identities in Tourism Development Projects: Interconnectedness and Embeddedness Dynamics in the Moroccan Mountains," *Via@: International Interdisciplinary Review of Tourism* 2 (2012), http://www.viatourismreview.net/Article11_EN.php.

65. Jean Michaud, ed., *Turbulent Times and Enduring Peoples: The Mountain Minorities of the South-East Asian Massif* (London: Curzon, 2000); Jean Michaud and Tim Forsyth, eds., *Moving Mountains: Highland Identity and Livelihoods in Postsocialist China, Vietnam and Laos* (Stanford, CA: Stanford University Press, 2010).

66. Ovesen, "All Lao"; Vatthana Pholsena, *Post-War Laos: The Politics of Culture, History and Identity* (Ithaca, NY: Cornell University Press, 2006).

67. Jean Michaud, "Économie et identité chez les Hmong de la haute région du Vietnam septentrional," *Aséanie* 22 (2009): 48–69. The sociologist Ulrich Beck even uses the example of the Hmong network to illustrate his thesis on the organization of transnational social movements. See his *What Is Globalization?* (Cambridge: Polity, 1999).

68. Stephen P. Cohen, *The Indian Army: Its Contribution to the Development of a Nation* (Berkeley and Los Angeles: University of California Press, 1971).

69. Neville Maxwell, *India's China War* (London: Cape, 1970).

70. To be precise, two very different conceptions were in competition at the constituent assembly and the political debate in India in 1946–47. The political leaders of Assam strongly supported assimilation, whereas some of Nehru's public declarations were more open to recognizing cultural diversity and political autonomy.

71. Markus Franke, *War and Nationalism in South Asia: The Indian State and the Nagas* (London: Routledge, 2009).

72. J. Doresse, quoted in Alain Gascon, "Croissant 'aride' et Ethiopie 'heureuse': La montagne la plus peuplée du monde face aux crises," in *Les montagnes tropicales: Identités, mutations, développement*, ed. François Bart, Serge Morin, and Jean-Noël Salomon (Pessac: Dynamiques des milieux et des sociétés dans les espaces tropicaux, 2001), 193–204, quotation 200. See also Jean Gallais, "Perception et interprétation amhirique de la montagne," in *Ethnogéographies*, ed. Paul Claval and Singarevélou (Paris: L'Harmattan, 1995), 93–119.

73. By "second highest mountain" Lord Curzon meant Kangchenjunga, the third highest peak on earth, just after K2, as would later be discovered. Quoted in Walt Unsworth, *Everest: The Mountaineering History* (Seattle, WA: Mountaineers Books, 1998), 14.

74. Christopher Hale, *Himmler's Crusade: The Nazi Expedition to Find the Origin of the Aryan Race* (London: Wiley, 2003).

75. www.asiatimes.com, June 25, 2008.

76. Lt.-Gen. V. R. Raghavan, *Siachen: Conflict without End* (New Delhi: Viking, 2002), 2.

77. Thierry Lefebvre, "L'invention occidentale de la haute montagne andine," *Mappemonde* (2005): 79; Joy Logan, *Aconcagua: The Invention of Mountaineering on America's Highest Peak* (Phoenix: University of Arizona Press, 2011).

78. Annie Smith Peck, *High Mountain Climbing in Peru & Bolivia: A Search for the Apex of America Including the Conquest of Huascarán* (London, 1912).

CHAPTER SEVEN

1. Carl Troll, "Die geographischen Grundlagen der andinen Hochkulturen des Inkareiches," *Ibero-amerikanisches Archiv* 5 (1931); Troll, "Die Stellung der Indianer Hochkulturen im Landschaftsaufbau der tropischen Anden," *Zeitschrift der Gesellschaft für Erdkunde* 3 (1943): 93–128. For a discussion of these texts, see Olivier Dollfus, "Les Andes centrales tropicales vues par deux géographes: Isaiah Bowman et Carl Troll," *Bulletin de l'Institut français d'études andines* 7, nos. 1–2 (1978): 7–21.

2. Richard Pasquis and Pierre Usselmann, "Milieux, environnement et migrations dans les Andes centrales péruviennes," in *Les montagnes tropicales: Identités, mutations, développement*, ed. François Bart, Serge Morin, and Jean-Noël Salomon (Pessac: Dynamiques des milieux et des sociétés dans les espaces tropicaux, 2001), 249–57.

3. John Murra, "El 'control vertical' de un máximo de pisos ecológicos en la economía de las sociedades andinas," in *Visita de la provincia de León de Huánuco en 1562*, ed. John Murra, vol. 2 (Lima: Universidad Nacional Hermilio Valdizán, 1972), 427–68. See also Pierre Morlon, *Comprendre l'agriculture paysanne dans les Andes centrales* (Paris: Institut national de la recherche agronomique, 1992); and Benjamin Orlove, "Down to Earth: Race and Substance in the Andes," *Bulletin of Latin American Research* 17, no. 2 (1998): 207–22.

4. In his excellent analysis of Peruvian geography, Benjamin Orlove reports a true "obsession" on the part of Peruvian geographers with measurements and a "fascination" with determining elevations. According to him, the choice of an elevation threshold "gave Peru something new . . . : the highlands." Benjamin Orlove, "Put-

ting Race in Its Place: Order in Colonial and Postcolonial Peruvian Geography," *Social Research* 60, no. 2 (1993): 302–36, quotation 319.

5. Ibid.

6. Ibid., 325. See also Orlove, "Down to Earth."

7. Peter Blanchard, "Indian Unrest in the Peruvian Sierra in the Late Nineteenth Century," *The Americas* 38, no. 4 (1982): 449–62.

8. In Ecuador, by contrast, the Andean region lies at the heart the *sierra-selva-costa* triad. It is the most highly valued part of the national imaginary: the land of Inca civilization, of Hispanic centrality, the seat of the historic and present-day capital, and the principal center of autochthonous settlement. In this, the sierra concentrates the principal emblems of the nation. As a result, the indigenous populations are assimilated less to the Andes themselves than to the traditional rural areas. M. Crain, "The Social Construction of National Identity in Highland Ecuador," *Anthropological Quarterly* 15, no. 3 (1991): 43–59; Sarah A. Radcliffe, "Imaginative Geographies, Postcolonialism, and National Identities: Contemporary Discourses of the Nation in Ecuador," *Ecumene* 3, no. 1 (1996): 23–42; Jean-Paul Deler, *Genèse de l'espace équatorien: Essai sur le territoire et la formation de l'État national* (Paris: Association pour la diffusion de la pensée française, 1981). See also C. Bataillon, Jean-Paul Deler, and Hervé Théry, *Géographie universelle: Amérique Latine* (Paris: Hachette and Montpellier: Reclus, 1991), for a comparison of the concept of the Andes in the different countries of the region.

9. For example, Isaiah Bowman, *The Andes of Southern Peru: Geographical Reconnaissance along the Seventy-third Meridian* (New York: American Geographical Society, 1916).

10. Henri Senthiles, "Réforme agraire au Pérou," *Tiers-Monde* 11, no. 44 (1970): 759–66. See also Julio Cotler, "The Mechanics of Internal Domination and Social Change in Peru," *Studies in Comparative International Development* 3, no. 12 (1967): 229–46.

11. Jacques Malengreau, *Sociétés des Andes, des empires aux voisinages* (Paris: Karthala, 1995).

12. Senthiles, "Réforme agraire au Pérou."

13. Morlon, *Comprendre l'agriculture paysanne*, 139.

14. Sabin Berthelot, *Considérations sur l'acclimatement et la domestication, exposées dans le but de démontrer l'importance des jardins et des ménageries d'acclimatation pour la propagation des animaux et des plantes utiles* (Paris: Imprimerie Béthune, 1844), 2. At the time Berthelot was secretary general of the Société de Géographie de Paris and author of a monumental natural history of the Canary Islands.

15. On acclimatization societies see the works of Michael A. Osborne, especially *Nature, the Exotic, and the Science of French Colonialism* (Bloomington: Indiana University Press, 1994); Marie-Noëlle Bourguet and Christophe Bonneuil, "De l'inventaire du globe à la 'mise en valeur' du monde: Botanique et colonisation (fin 18e–début 20e siècle)," *Revue française d'histoire d'outre-mer* 322 (1999); Christophe Bonneuil, *Mettre en ordre et discipliner les tropiques: Les sciences du végétal dans l'empire français, 1870–1940* (Paris: Éditions des Archives contemporaines, 2002); and Christopher Lever, *They Dined on Eland: The Story of the Acclimatisation Societies* (London: Quiller, 1992). Shortly before the creation of the Société Zoologique d'Acclimatation, Isidore Geoffroy Saint-Hilaire published his account of the many experiments he conducted to create herds of llamas, alpacas, and Angora goats. The book allowed him to promote a scientific approach to acclimatization. See Isidore Geoffroy Saint-Hilaire, *Acclimatation et domestication des animaux utiles* (Paris: Imprimerie nationale, 1849).

16. The agronomist Loiseleur-Deslonchamps praised that choice in 1832 in the society's report. See Hildebert Isnard, "Vigne et colonisation en Algérie," *Annales de géographie* 31 (1949): 212–19.
17. Quoted in Osborne, *Nature, the Exotic*, 136.
18. James Scott, *Seeing like a State: How Certain Schemes to Improve the Human Condition Have Failed* (New Haven, CT: Yale University Press, 1998), 282. Scott later wrote that "the French desired to transform the fiscally sterile hills into a space that would be *rentable* and *utile*." Scott, *The Art of Not Being Governed: An Anarchist History of Upland Southeast Asia* (New Haven, CT: Yale University Press, 2009), 76.
19. That reading is not confined to colonial administrations. When the rice-growing societies of the coastal plains and of the interior basins encroached on the mountain—as did the Khin, whom the Vietnamese authorities settled there before and after French colonization—they too valorized sedentary agriculture. Scott, *The Art of Not Being Governed*, 77.
20. Julien, administrator in chief of the colonies (1911), quoted in Anne Bergeret, "Les forestiers coloniaux français," in *Les sciences hors d'Occident au XXe siècle*, ed. Roland Waast (Paris: Office de la recherche scientifique et technique outre-mer, 1995), 59–74, quotation 63. On Madagascar, see also the analysis of Jean Fremigacci, "La forêt en situation coloniale à Madagascar (1900–1940), une économie de la délinquance," *Cahiers du Centre de Recherches Africaines* (1995).
21. Scott, *The Art of Not Being Governed*.
22. Tania Li, *The Will to Improve: Governmentality, Development, and the Practice of Politics* (Durham, NC: Duke University Press, 2007); Li, ed., *Transforming the Indonesian Uplands: Marginality, Power and Production* (London: Routledge, 1999); Albert Schrauwers, *Colonial "Reformation" in the Highlands of Central Sulawesi, Indonesia, 1892–1995* (Toronto: University of Toronto Press, 2000).
23. Carel Sirardus Willem Hogendorp, *Coup d'oeil sur l'île de Java et les autres possessions néerlandaises dans l'archipel des Indes* (Brussels: Imprimerie de Mat, 1830), 343.
24. Arun Agrawal, *Environmentality: Technologies of Government and the Making of Subjects* (Durham, NC: Duke University Press, 2005), 67.
25. Jean-Yves Puyo, "Sur le mythe colonial de l'inépuisabilité des ressources forestières (Afrique occidentale et orientale française, 1900–1940)," *Cahiers de géographie du Québec* 45, no. 126 (2001): 479–96.
26. Quoted ibid.
27. Jack D. Ives, *Himalayan Perceptions: Environmental Change and the Well-being of Mountain Peoples* (London: Routledge, 2004), 2; Berthold Ribbentrop, *Forestry in British India* (Calcutta: Government Printing Office, 1900).
28. Agrawal, *Environmentality*, 240 n. 10.
29. Agrawal speaks of the "making of forests"; ibid., 34.
30. Bernhard Eduard Fernow, *A Brief History of Forestry in Europe, the United States and Other Countries* (Toronto: University Press, 1907), 322. Later analyses have shown the role of scientists and scientific institutions, especially the Royal Geographical Society, in the production and circulation of a form of knowledge adapted to the natural conditions of the colonies. See Richard Grove, "Imperialism and the Discourse of Desiccation: Global Environmental Concerns and the Role of the Royal Geographic Society, 1860–1880," in *Geography and Imperialism, 1820–1940*, ed. Morag Bell, Robin Butlin, and Michael Heffernan (Manchester: Manchester University Press, 1995), 36–52.

31. Louise Fortman and Sally Fairfax, "American Forestry Professionalism in the Third World," *Economic and Political Weekly* 24, no. 32 (1989): 1839–44.

32. Anne Bergeret, "Discours et politiques forestières coloniales en Afrique et à Madagascar," in *Colonisations et environnement*, ed. Jacques Pouchepadass (Paris: L'Harmattan, 1993), 69.

33. Ibid., 23–47. See also Maurice Benchetrit, "Le problème de l'érosion des sols en montagne et le cas du Tell algérien," *Revue de géographie alpine* 43, no. 3 (1955): 605–40.

34. Quoted in Bergeret, "Discours et politiques forestières," 26.

35. J. P. Challot, "La forêt et la montagne marocaine," *Hesperis* (1938): 233–46.

36. Jean-François Troin, *Maroc: Régions, pays, territoires* (Paris: Maisonneuve et Larose, 2002).

37. Augustin Chevalier, "Les hauts plateaux du Fouta Djalon," *Annales de géographie* 18, no. 99 (1909): 253–61; Charles Robequain, "À travers le Fouta Djallon," *Revue de géographie alpine* 25, no. 4 (1937): 545–81.

38. Quoted at a conference held in Conakry in 1995, organized by the Guinean Ministry of Agriculture and the European Union. See Véronique André, Gilles Pestaña, and Georges Rossi, "Foreign Representations and Local Realities: Agropastoralism and Environmental Issues in the Fouta Djalon Tablelands, Republic of Guinea," *Mountain Research and Development* 23, no. 2 (2003): 149–55. See also Véronique André and Gilles Pestaña, "Les visages de Fouta-Djalon," *Les cahiers d'outre-mer* 217 (2002); and the more general critical analysis by Georges Rossi, "Une relecture de l'érosion en milieu tropical," *Annales de géographie* 107, no. 601 (1998): 318–29.

39. Jacques Richard-Molard, "Essai sur la vie paysanne au Fouta-Djalon," *Revue de géographie alpine* 32, no. 2 (1944): 135–239.

40. Christopher Allan Conte, *Highland Sanctuary: Environmental History in Tanzania's Usambara Mountains* (Athens: Ohio University Press, 2004), 3.

41. Ibid., 11.

42. Ibid., 12. It has been noted that similar descriptive procedures, initially guided by a European view of things, occurred in the neighboring region of Kilimanjaro: Pascal Mazurier, "Les représentations du Kilimandjaro de l'Antiquité à nos jours," in *Les montagnes tropicales: Identités, mutations, développement*, ed. François Bart, Serge Morin, and Jean-Noël Salomon (Pessac: Dynamiques des milieux et des sociétés dans les espaces tropicaux, 2001), 271–82.

43. Hans G. Schabel, "Tanganyika Forestry under German Colonial Administration, 1891–1919," *Forest and Conservation History* 34, no. 3 (1990): 130–41.

44. Roderick Neumann, *Imposing Wilderness: Struggles over Livelihood and Nature Preservation in Africa* (Berkeley and Los Angeles: University of California Press, 1998), 99.

45. Conte, *Highland Sanctuary*, 10.

46. Thomas Spear, *Mountain Farmers: Moral Economies of Land and Agricultural Development in Arusha and Meru* (Berkeley and Los Angeles: University of California Press, 1997), 90.

47. Neumann, *Imposing Wilderness*, 55.

48. W. A. Rodgers, R. Nabanyumya, E. Mupada, and L. Persha, "Community Conservation of Closed Forest Biodiversity in East Africa: Can It Work?" *Unasylva* 209 (2002): 41–49.

49. Tim Forsyth and Andrew Walker, *Forest Guardians, Forest Destroyers: The Politics of Environmental Knowledge in Northern Thailand* (Seattle: University of Washington Press, 2008), 40.

50. Reiner Buerguin, "'Hill Tribes' and Forests," Socio-Economics of Forest Use in the Tropics and Subtropics (SEFUT) Working Paper no. 7, Universität Freiburg, 2000.

51. Pierre-Yves Le Meur, "Les hautes terres du nord de la Thaïlande en transition: Développement, courtage et construction nationale," Tiers-Monde 41, no. 162 (2000): 370.

52. Forsyth and Walker, Forest Guardians, Forest Destroyers, 59.

53. Ibid., 29.

54. The historian Willem van Schendel introduced the term "Zomia" in his "Geographies of Knowing, Geographies of Ignorance: Jumping Scale in Southeast Asia," Development and Planning D: Society and Space 20 (2002): 647–68. Several researchers adopted the term, including Scott, who popularized it in The Art of Not Being Governed.

55. Julie Guthman, "Representing Crisis: The Theory of Himalayan Environmental Degradation and the Project of Development in Post-Rana Nepal," Development and Change 28, no. 1 (1997): 45.

56. A. P. Gautam, G. P. Shivakoti, and E. L. Webb, "A Review of Forest Policies, Institutions, and Changes in the Resource Condition in Nepal," International Forestry Review 6, no. 2 (2004): 140.

57. Guthman, "Representing Crisis," 45–69; Gautam, Shivakoti, and Webb, "A Review of Forest Policies," 136–48.

58. Mary Hobley, Participatory Forestry: The Process of Change in India and Nepal (London: Overseas Development Institute, 1996).

59. Analyses that contradicted these conclusions were available but were disregarded. Jack Ives, "Himalayan Misconceptions and Distortions: What Are the Facts?" Himalayan Journal of Sciences 3, no. 5 (2005): 15–24.

60. Erik Eckholm, Losing Ground: Environmental Stress and World Food Prospects (New York: Norton, 1976).

61. The geographers Jack Ives and Bruno Messerli proposed the formulation. They also analyzed the normative discourse implicit in that theory. See Jack D. Ives and Bruno Messerli, The Himalayan Dilemma: Reconciling Development and Conservation (London: Routledge, 1989).

62. Tim J. Forsyth, "Mountain Myths Revisited: Integrating Natural and Social Environmental Science," Mountain Research and Development 18, no. 2 (1998): 107–16.

63. World Bank, Nepal: Development Performance and Prospects (Washington, DC: South Asia Regional Office, World Bank, 1979).

64. Especially Thomas Hofer and Bruno Messerli, Floods in Bangladesh: History, Dynamics and Rethinking the Role of the Himalayas (Tokyo: United Nations University Press, 2006); Monique Fort, "Des milieux à risque en Himalaya central: Le cas du Népal," in Bart, Morin, and Salomon, eds., Les montagnes tropicales, 43–59; Hans Kienholz, Heinrich Hafner, and Guy Schneider, "Stability, Instability, and Conditional Instability: Mountain Ecosystem Concepts Based on a Field Study of the Kakani Area in the Middle Hills of Nepal," Mountain Research and Development 4, no. 1 (1984): 55–62; Ives, Himalayan Perceptions, 43.

65. Michael Thompson, Michael Warburton, and Tom Hatley, Uncertainty on a Himalayan Scale: An Institutional Theory of Environmental Perception and a Strategic Framework for the Sustainable Development of the Himalaya (London: Ethnographica, 1986).

66. Ives, Himalyan Perceptions, 5–6.

67. Thompson, Warburton, and Hatley, Uncertainty on a Himalayan Scale; Ives, Himalayan Perceptions.

68. Tim Forsyth, "Science, Myth and Knowledge: Testing Himalayan Environmental Degradation in Thailand," *Geoforum* 27, no. 3 (1996): 375–92; Li, *The Will to Improve*; Rossi, "Une relecture de l'érosion."

69. It has been estimated that more than a third of the protected areas globally are mountain regions, that is, 260 million out of a total of 785 million hectares. Bruno Messerli and Jack Ives, *Mountains of the World: A Global Priority* (New York: Parthenon, 1997), 217. For protected areas of more than a thousand hectares, that share surpasses 50 percent in many tropical regions: Bernard Debarbieux, Jean-Jacques Delannoy, Jean-François Dobremez, and François Vigny, *Les pays du monde et leurs montagnes* (Grenoble: Revue de géographie alpine, 2000).

70. Neumann, *Imposing Wilderness*, 13.

71. Roderick Neumann, "Nature-State-Territory: Toward a Critical Theorization of Conservation Enclosures," in *Liberation Ecologies*, ed. R. Peet and M. Watts (London: Routledge, 2004), 195–217; Mark Dowie, *Conservation Refugees: The Hundred-Year Conflict between Global Conservation and Native Peoples* (Cambridge, MA: MIT Press, 2009).

72. Terence Ranger, *Voices from the Rocks: Nature, Culture, and History in the Matopos Hills of Zimbabwe* (Oxford: James Currey, 1999).

73. Ibid., 39.

74. The academic literature abounds in case studies of that type. For Mexico, see José Luis Castilla Vallejo, *Naturaleza y postdesarrollo: Estudio sobre la sierra Gorda de Querétaro* (Tenerife, Spain: Universidad de la Laguna, 2009).

75. Gouvernement Général de l'Algérie, Service des Eaux et Forêts, *Les parcs nationaux en Algérie* (Algiers, 1930).

76. A botanist at the Muséum National d'Histoire Naturelle in Paris served as coordinator of the book that resulted from that colloquium. See A. Aubreville, A. Barbey, E. N. Barcaly, C. Bressou, P. Chouard, J. Dufrénoy, M.-L. Dufrénoy, F. Evrard, et al., *Contribution à l'étude des réserves naturelles et des parcs nationaux* (Paris: P. Lechevalier, 1937). On this subject see Caroline Ford, "Nature, Culture and Conservation in France and Her Colonies, 1840–1940," *Past and Present* 183 (2004): 173–98.

77. Mark Carey, "Disasters, Development, and Glacial Lake Control in Twentieth-Century Peru," in *Mountains: Sources of Water, Sources of Knowledge*, ed. Ellen Wiegandt (Dordrecht, The Netherlands: Springer, 2008), 181–96, quotation 188.

78. César Morales Arnao, "El parque national Huascarán," *Boletín de la Sociedad Geográfica de Lima* 78, nos. 1–2 (1961): 27–29. On the heightened awareness of high mountain scenery in the Andes resulting from the adoption of the Western aesthetic, see also Deborah Poole, "Landscape and the Imperial Subject: U.S. Images of the Andes, 1859–1930," in *Close Encounters of Empire: Writing the Cultural History of U.S.-Latin Relations*, ed. Gilbert Joseph (Durham, NC: Duke University Press, 1998).

79. Mary Baker, "National Parks, Conservation, and Agrarian Reform in Peru," *Geographical Review* 70, no. 1 (1980): 1–18.

80. Thierry Lefebvre, "L'invention occidentale de la haute montagne andine," *Mappemonde* 79, no. 3 (2005): 11.

81. Baker, "National Parks."

82. Introductory statement on Sagarmatha National Park at the World Heritage Website, http://whc.unesco.org/en/list/120, accessed April 15, 2014.

83. Information sheet on Sagarmatha National Park at the United Nations Environmental Program, World Conservation Monitoring Center Website, http://www.unep-wcmc.org/sites/wh/pdf/Sagarmatha.pdf, accessed April 15, 2014, 3.

84. Barbara Brower and Ann Dennis, "Grazing the Forest, Shaping the Landscape? Continuing the Debate about Forest Dynamics in Sagarmatha National Park, Nepal," in *Nature's Geography: New Lessons for Conservation in Developing Countries*, ed. K. S. Zimmerer and K. R. Young (Madison: University of Wisconsin Press, 1998), 184–208; Alton Beyers, "Contemporary Human Impacts on Alpine Ecosystems in the Sagarmatha (Mt. Everest) National Park, Khumbu, Nepal," *Annals of the Association of American Geographers* 95, no. 1 (2005): 112–40.

85. Stanley F. Stevens, *Claiming the High Ground: Sherpas, Subsistence, and Environmental Change in the Highest Himalaya* (Berkeley and Los Angeles: University of California Press, 1993).

86. "Political forests are a critical part of colonial-era state-making both in terms of the territorialization and legal framing of forests and the institutionalization of forest management as a technology of state power." Nancy Lee Peluso and Peter Vandergeest, "Genealogies of the Political Forest and Customary Rights in Indonesia, Malaysia, and Thailand," *Journal of Asian Studies* 60, no. 3 (2001): 761–812, quotation 762. See also Scott, *Seeing like a State*.

87. Catherine Aubertin, "La montagne, un produit du développment durable," *Revue de géographie alpine* 89, no. 2 (2001): 51–58.

CHAPTER EIGHT

1. Fausto Sarmiento, "Mount Chimborazo: In the Steps of Alexander von Humboldt," *Mountain Research and Development* 19, no. 2 (1999): 77–78.

2. Marie-Claude Smouts, *Tropical Forests, International Jungle: The Underside of Global Geopolitics* (New York: Palgrave Macmillan, 2003), 26.

3. Peter Haas, "Introduction: Epistemic Communities and International Policy Coordination," *International Organisation* 46, no. 1 (1992): 1–35.

4. Jack Ives and Bruno Messerli, "AD 2002 Declared by United Nations as 'International Year of the Mountains,'" *Arctic, Antarctic, and Alpine Research* 31, no. 3 (1999): 211–13, especially 211.

5. Simon Dalby, "Critical Geopolitics: Discourse, Difference, and Dissent," *Environment and Planning D: Society and Space* 9 (1991): 261–83; Daniel Momtaz, "The United Nations and the Protection of the Environment: From Stockholm to Rio de Janeiro," *Political Geography* 15, nos. 3–4 (1996): 261–71.

6. Quoted in Jack Ives, "Mountain Agenda: A Progress Report," *Mountain Research and Development* 15, no. 4 (1995): 349–54, quotation 350. See also Gilles Rudaz, "The Cause of Mountains: The Politics of Promoting a Global Agenda," *Global Environmental Politics* 11, no. 4 (2011): 43–65; Jack Ives, *Sustainable Mountain Development—Getting the Facts Right* (Lalitpur, Nepal: Himalayan Association for the Advancement of Science, 2013).

7. Resolution 53/24, United Nations General Assembly, November 10, 1998.

8. Pratap Chatterjee and Mathias Finger, *The Earth Brokers: Power, Politics and World Development* (London: Routledge, 1994), 82.

9. http://www.mountain-portal.co.uk/test/MtnAgenda.html, accessed July 14, 2007.

10. Bernard Debarbieux and Martin Price, "Representing Mountains: From Local and National to Global Common Good," *Geopolitics* 113, no. 1 (2008): 148–68, quotation 154.

11. Jack Ives and Bruno Messerli, *The Himalayan Dilemma: Reconciling Development and Conservation* (London: Routledge, 1989).

12. http://www.mountain-portal.co.uk/test/MtnAgenda.html, accessed July 14, 2007.

In view of its initial goal, the group was initially called "Mountain Agenda-UNCED 1992." Then, when its mission continued beyond the conference, it was renamed "Mountain Agenda." Peter Stone, ed., *The State of the World's Mountains: A Global Report* (London: Zed Books, 1992), xiv.

13. Stone, ed., *State of the World's Mountains*, xiv.
14. Ibid., xvi.
15. Jack Ives, Bruno Messerli, and Ernst Spiess, "Mountains of the World—A Global Priority," in *Mountains of the World: A Global Priority*, ed. B. Messerli and J. Ives (New York: Parthenon, 1997), 1–15, quotation 2.
16. Available at http://www.mountain-portal.co.uk/text/MtnAgenda.html, accessed July 14, 2007.
17. Bruno Messerli and Erwin Bernbaum, "The Role of Culture, Education, and Science for Sustainable Mountain Development," in *Key Issues for Mountain Areas*, ed. Martin F. Price, Libor Jansky, and Andrei Iatsenia (Tokyo: United Nations University Press, 2004), 210–33, quotation 230–31.
18. Martin Price, "Mountains: Globally Important Ecosystems," *Unasylva* 195 (1998); Jack Ives, "Along a Steep Pathway," *Our Planet* 2 (2002).
19. Ives and Messerli, "AD 2002 Declared," 211.
20. Jayanta Bandyopadhyay and Shama Perveen, "Emergence of and Future Steps for Sustainable Mountain Development in the Global Environmental Agenda," in *Protection of Mountain Areas in International Law: Rio, Johannesburg and Beyond*, ed. T. Treves, L. Pineschi, and A. Fodella (Milan: Giuffrè, 2004).
21. For commentary on these statistics, see Martin Price and Bruno Messerli, "Fostering Sustainable Mountain Development: From Rio to the International Year of Mountains, and Beyond," *Unasylva* 208 (2002): 6.
22. Ives, Messerli, and Spiess, "Mountains of the World," 2.
23. V. Kapos, J. Rhind, M. Edwards, C. Ravilious, and M. F. Price, "Developing a Map of the World's Mountain Forests," in *Forests in Sustainable Mountain Development: A Report for 2000*, ed. M. F. Price and N. Butt (Wallingford, UK: CAB International, 2000), 4–9.
24. Barbara Huddleston, Ergin Ataman, and Luca Fé d'Ostiani, *Towards a GIS-Based Analysis of Mountain Environments and Populations* (Rome: Food and Agriculture Organization of the United Nations, 2003).
25. Michel Meybeck, Pamela Green, and Charles Vörösmarty, "A New Typology for Mountains and Other Relief Classes: An Application to Global Continental Water Resources and Population Distribution," *Mountain Research and Development* 21 (2001): 34–45. The second estimate relies on a coarser grid, which explains why the results differ.
26. United Nations, Agenda 21 (1992).
27. Thomas Kohler and Daniel Maselli, eds., *Mountains and Climate Change: From Understanding to Action* (Bern: Geographica Bernensia, 2009).
28. Bruno Messerli, "From the Earth Summit 1992 to the IYM 2002: The Role and Responsibility of Switzerland," in *Mountains and Peoples: An Account of Mountain Development Programmes Supported by the Swiss Agency for Development and Cooperation* (Bern: SDC, 2001), 12–14.
29. Stone, ed., *State of the World's Mountains*.
30. Mountain Agenda, 1992.
31. Stone, ed., *State of the World's Mountains*, xiv.
32. Ibid., xv.
33. Ibid., xvi.

34. Interview by the authors with Bruno Messerli, Bern, Switzerland, July 2006.
35. Stone, ed., *State of the World's Mountains*, xv.
36. Messerli and Ives, eds., *Mountains of the World*.
37. Jack Ives, "Mountain Agenda: A Progress Report," *Mountain Research and Development* 15, no. 4 (1995): 349–54, especially 352.
38. Ives, Messerli, and Spiess, "Mountains of the World," 1.
39. Interview with Messerli, July 2006.
40. Jack Ives, Bruno Messerli, and Robert Rhoades, "Agenda for Sustainable Mountain Development," in Messerli and Ives, eds., *Mountains of the World*, 455–66.
41. Lorenza Mondada, "La 'montagne' comme objet de savoir co-construit dans le débat scientifique," *Revue de géographie alpine* 89, no. 2 (2001): 79–92. In any event, although some scientists still strive to promote the term (with clear success, since it has entered *The Oxford English Dictionary*), it is rarely used, and few people are familiar with it.
42. Global Mountain Biodiversity Assessment, http://gmba.unibas.ch/index/index/htm, accessed March 16, 2010.
43. http://mri.scnatweb.ch, accessed April 16, 2014.
44. Alfred Becker and Harald Bugmann, eds., *Global Change in Mountain Regions: The Mountain Research Initiative* (Stockholm: Royal Swedish Academy of Sciences, 2001); Astrid Björnsen Gurung, *Global Change and Mountain Regions (GLOCHAMORE) Research Strategy* (Zurich: Mountain Research Initiative, 2006).
45. Bernard Debarbieux, Jörg Balsiger, Dusan Djordjevic, Simon Gaberell, and Gilles Rudaz, "Scientific Collectives in Region-Building Processes," *Environmental Science and Policy* 42 (October 2014): 149–59.
46. Hector Cisneros, Elías Mujica, and Ana María Ponce: "Condesan: Watershed Management and Rural Development in the Andes," *Mountain Research and Development* 24, no. 3 (2004): 258–59.
47. Price and Messerli, "Fostering Sustainable Mountain Development," 7.
48. Mountain Institute, *International NGO Consultation on the Mountain Agenda, Summary Report and Recommendations to the United Nations Commission on Sustainable Development* (Franklin, WV: Mountain Institute, 1995); Ives, "Mountain Agenda: A Progress Report," 349.
49. Mountain Institute, *International NGO Consultation*, 2.
50. Jack Ives, "Mountain Environments," in *A Modern Approach to the Protection of the Environment*, ed. Giovanni Battista Marini-Bettolo (Oxford: Pergamon, 1989), 289–345, quotation 291.
51. Initial Organizing Committee of the Mountain Forum, *Report of the Initial Organizing Committee of the Mountain Forum, 21–25 September 1995* (Spruce Knob, WV: Mountain Institute, 1995), 5, 11.
52. http://www.mountainpartnership.org/members/en, accessed February 19, 2014.
53. Mountain Partnership, 2003, 5.
54. Jane Ross, "The Mountain Partnership at the CSD Partnerships Fair," *Mountain Research Development* 26, no. 4 (2006): 373–77, quotation 375.
55. Interview with Messerli, July 2006.
56. Price, Jansky, and Iatsenia, eds., *Key Issues for Mountain Areas*, 7.

CHAPTER NINE

1. That idea, however commonplace, must be applied with caution. Although mountain communities are often depicted as being cut off from the rest of the

world, historical analyses have shown that some do not lack for relations with the outside world, and a few have even occupied pivotal positions between economic and political spaces. For the Alps, that is what the historian Pier Paolo Viazzo calls the "Alpine paradox" (*Upland Communities: Environment, Population, and Social Structure in the Alps since the Sixteenth Century* [New York: Cambridge University Press, 1989], 142). As with many questions regarding the mountain populations, it is important to respect nuances and distinguish between conventional images and documented practices.

2. Valentine Moghadam, *Globalizing Women: Transnational Feminist Networks* (Baltimore: Johns Hopkins University Press, 2005), 30; Manuel Castells, *The Rise of the Network Society* (Malden, MA: Blackwell, 1996).

3. See as an example Jack Ives, *Sustainable Mountain Development—Getting the Facts Right* (Lalitpur, Nepal: Himalayan Association for the Advancement of Science, 2013).

4. Catherine Aubertin, "La montagne, un produit du développement durable," *Revue de géographie alpine* 89, no. 2 (2001): 51–58.

5. www.mtnforum.org, accessed March 12, 2010.

6. Jack Ives, "Along a Steep Pathway," *Our Planet* 131 (2002).

7. Initial Organizing Committee of the Mountain Forum (1995).

8. Ibid., 5.

9. United Nations, *Johannesburg Plan of Implementation* (Johannesburg: United Nations, 2002).

10. Initial Organizing Committee of the Mountain Forum, annex 2.

11. Bruno Messerli and Edwin Bernbaum, "The Role of Culture, Education, and Science for Sustainable Mountain Development," in *Key Issues for Mountain Areas*, ed. Martin F. Price, Libor Jansky, and Andrei Iastenia (Tokyo: United Nations University Press, 2004), 212. See also Robert Rhoades, "Integrating Local Voices and Visions into the Global Mountain Agenda," *Mountain Research and Development* 20, no. 1 (2000): 4–9.

12. Sheila Jasanoff and Marybeth Long Martello, eds., *Earthly Politics: Local and Global in Environmental Governance* (Cambridge, MA: MIT Press, 2004).; emphasis in original.

13. Neelendra K. Joshi and Vir Singh, eds., *Traditional Ecological Knowledge of Mountain People: Foundation for Sustainable Development in the Hindu Kush–Himalayan Region* (Delhi: Daya, 2009); Bernadette Montanari, "The Future of Agriculture in the High Atlas Mountains of Morocco: The Need to Integrate Traditional Ecological Knowledge," in *The Future of Mountain Agriculture*, ed. Stefan Mann (Berlin Heidelberg: Springer-Verlag, 2013).

14. Edwin Bernbaum, *Sacred Mountains of the World* (Berkeley and Los Angeles: University of California Press, 1998). Among NGOs, the Panos Institute has shown particular interest in these spiritual connections.

15. P. S. Ramakrishan, K. G. Saxena, and U. Chandrasekhara, *Conserving the Sacred: For Biodiversity Management* (New Delhi: Oxford and IBH, 1998); P. S. Ramkarishnan, K. G. Saxena, and K. S. Rao, *Shifting Agriculture and Sustainable Development of North-Eastern India: Tradition in Transition* (New Delhi: Oxford and IBH, 2006).

16. The Mountain Institute, *Sacred Mountains and Environmental Conservation: A Practitioner's Workshop* (Franklin, WV: Mountain Institute, 1998).

17. Ives, "Mountain Environments," 339.

18. Charlotte Bretherton, "Movements, Networks, Hierarchies: A Gender Perspective on Global Environmental Governance?" *Global Environmental Politics* 3, no. 2 (2003): 103–18, quotation 105.

19. Ives, "Along a Steep Pathway," 3–5.
20. Commission on Sustainable Development, *Report on the Third Session*, quoted in Bernard Debarbieux and Martin F. Price, "Representing Mountains: From Local and National to Global Common Good," *Geopolitics* 13, no. 1 (2008): 159.
21. Bretherton, "Movements, Networks, Hierarchies," 105.
22. Rhoades, "Integrating Local Voices and Visions," 6.
23. Ibid., 5.
24. David Barkin and Michèle Dominy, "Les régions montagneuses: Terres de refuge ou écosystème pour l'humanité?" in *La montagne: Un objet de recherches?*, ed. Bernard Debarbieux (Grenoble: Institut de géographie alpine, 2001), 67–72.
25. Hector Cisneros, Elías Mujica, and Ana María Ponce, "Condesan: Watershed Management and Rural Development in the Andes," *Mountain Research and Development* 24, no. 3 (2004): 258–59.
26. http://www.underutilized-species.org/Documents/PUBLICATIONS/session1-kbgp-english.pdf, accessed September 20, 2009.
27. Bernard Debarbieux and Gilles Rudaz, "Linking Mountain Identities throughout the World: The Experience of Swiss Communities," *Cultural Geographies* 15, no. 4 (2008): 497–517.
28. Jack Ives, Bruno Messerli, and Robert Rhoades, "Agenda for Sustainable Mountain Development," in *Mountains of the World: A Global Priority*, ed. Jack Ives and Bruno Messerli (New York: Parthenon, 1997), 456.
29. Derek Denniston, *High Priorities: Conserving Mountain Ecosystems and Cultures* (Washington, DC: Worldwatch Institute, 1995), 54.
30. Rhoades, "Integrating Local Voices and Visions."
31. Katrina Payne, Siobhan Warrington, and Olivia Bennett, *High Stakes: The Future for Mountain Societies* (London: Panos Institute, 2002).
32. http://www.vista360.org/flash/program1.swf, accessed December 6, 2012.
33. http://www.vista360.org/, accessed December 6, 2012.
34. Association Nationale des Élus de la Montagne, *Pour la montagne* 93 (April 2000): 4.
35. Ibid., 4.
36. Pierre Remy, "Joint Efforts for Mountain Development," in *Bishkek Global Mountain Summit: A Look into the Future*, ed. Asylbek Aidaraliev (Bishkek: Network of United Nations in the Kyrgyz Republic, 2003), 62–63, quotation 63.
37. Charter of the WMPA.
38. www.mountainpeople.org, accessed on October 10, 2009.
39. Benedict Anderson, *Imagined Communities* (London: Verso, 1983).
40. Rosi Braidotti, Ewa Charkiewica, Sabine Häusler, and Saskia Wieringa, *Women, the Environment and Sustainable Development: Towards a Theoretical Synthesis* (Atlantic Highlands, NJ: Zed Books, 1994).
41. Elizabeth Byers and Meeta Sainju, "Mountain Ecosystems and Women: Opportunities for Sustainable Development and Conservation," *Mountain Research and Development* 14, no. 3 (1994): 213–28, quotation 220.
42. Mountain Institute, annex 1.
43. United Nations General Assembly, resolution 60/198, "Sustainable Mountain Development," adopted March 8, 2006 (New York: United Nations, 2006).
44. http://www.mtnforum.org/rs/dl.cfm, accessed September 20, 2009.
45. http://www.mountainpartnership.org/; Douglas McGuire, "The Role of the Mountain Partnership in Promoting Gender Equality in Mountain Areas," in *Proceedings of*

    the International Conference Women of the Mountains, March 7–10, 2007, Orem, Utah, ed. Rusty Butler (Orem: Utah Valley State College, 2008), CD-ROM.

46. http://www.mtnforum.org/rs/dl.cfm, accessed September 20, 2009.

47. International Center for Integrated Mountain Development, *International Workshop on Women, Development and Mountain Resources: Approaches to Internalizing Gender Perspectives* (Katmandu: International Center for Integrated Mountain Development, 1988); *Women of the Mountains*, Orem, Utah, March 2007. One hundred and ten participants from twenty countries attended the Orem conference.

48. Phuntshok Tshering and Ojaswi Josse, *Advancing the Mountain Women's Agenda: A Report on a Global Gathering "Celebrating Mountain Women" in Bhutan, October 2002* (Katmandu: International Center for Integrated Mountain Development, 2003), 4.

49. http://www.mtnforum.org/calendar/events/0205mwaa.htm, accessed 2002.

50. Tshering and Thapa, *Celebrating Mountain Women*, 8.

51. In Tshering and Josse, *Advancing the Mountain Women's Agenda*, annex 5.

52. United Nations General Assembly, *Report A/62/292 of the Secretary-General "Sustainable Mountain Development,"* August 23, 2007.

53. Phuntshok Tshering and Rosemary Thapa, *Celebrating Mountain Women* (Katmandu: International Center for Integrated Mountain Development, 2003), v, 2.

54. Tshering and Josse, *Advancing the Mountain Women's Agenda*, 3.

55. Manjari Mehta, "Our Lives Are No Different from That of Our Buffaloes," in *Feminist Political Ecology, Global Issues and Local Experiences*, ed. D. Rocheleau, B. Thomas-Slayter, and E. Wangari (London: Routledge, 1996), 180–210.

56. Gurung and Rana, quoted in Mountain Forum, "Challenges Facing Mountain Women: General Discussion on Income Generation," *Legal Rights and Empowerment Bulletin* 3 (1999): 3–5.

57. Anita Anand and Ojaswi Josse, "Celebrating Mountain Women: Moving Mountains, Moving Women," *Mountain Research and Development* 22, no. 3 (2002): 233–35.

58. Susanne Wymann von Dach, "Integrated Mountain Development: A Question of Gender Mainstreaming," *Mountain Research and Development* 22, no. 3 (2002): 236–39.

59. Using the concept of intersectionality; Gill Valentine, "Theorizing and Researching Intersectionality: A Challenge for Feminist Geography," *Professional Geographer* 59, no. 1 (2007): 10–21.

60. Nira Yuval-Davis, "Beyond Difference: Women and Coalition Politics," in *Connections: Women's Studies, Women's Movements, Women's Lives*, ed. Mary Kennedy, Cathy Lubelska, and Val Walsh (London: Taylor and Francis, 1993), 3–10; Valentine Moghadam, *Globalizing Women: Transnational Feminist Networks* (Baltimore, MD: Johns Hopkins University Press, 2005).

61. Chandra Mohanty, "Under Western Eyes: Feminist Scholarship and Colonial Discourses," *Feminist Review* 30 (1988): 61–88; Carolyn Sachs, "Gendered Fields: Rural Women, Agriculture, and Environment" (Boulder, CO: Westview Press, 1966).

62. Quoted in Gilles Rudaz and Bernard Debarbieux, "'Mountain Women': Silent Contributors to the Global Agenda for Sustainable Mountain Development," *Gender, Place and Culture* 19, no. 5 (2012): 615–34, quotation 628.

63. Marie Bennigsen Broxup, ed., *The North Caucasus Barrier: The Russian Advance towards the Muslim World* (London: Hurst, 1992), 149–50.

64. Alexandre Grigoriantz, *La montagne du sang: Histoire, rites et coutumes des peuples montagnards du Caucase* (Geneva: Georg, 1998), 64–65.

65. Cem Oguz, "From the Idea of Caucasian Unity to Regional Fragmentation: The North Caucasus, 1990–1999," in *The Caspian Region*, ed. Moshe Gammer (London: Routledge, 2004), 42.

66. Mari Oiry-Varacca, "The Use of Amazigh Identities in Tourism Development Projects: Interconnectedness and Embeddedness Dynamics in the Moroccan Mountains," *Via@: International Interdisciplinary Review of Tourism* 2 (2012), http://www.viatourismreview.net/Article11_EN.php.

67. Camille Lacoste-Dujardin, "Un effet du 'postcolonial': Le renouveau de la culture berbère," *Hérodote* 120, no. 1 (2006): 96–117.

68. Otero Gerardo, "Global Economy, Local Politics: Indigenous Struggles, Civil Society and Democracy," *Canadian Journal of Political Science* 37, no. 2 (2004): 325–46.

69. Robert Andolina, Sarah Radcliffe, and Nina Laurie, "Gobernabilidad e identidad: Indigeneidades transnacionales en Bolivia," in *Pueblos indígenas, Estado y Democracia*, ed. Pablo Dávalos (Buenos Aires: Consejo Latinoamericano de ciencas sociales, 2005), 133–70.

70. Jacques Malengreau, *Sociétés des Andes, des empires aux voisinages* (Paris: Karthala, 1995), 230.

71. Sarah Hilbert, "For Whom the Nation? Internationalization, Zapatismo, and the Struggle over Mexican Modernity," *Antipode* 29, no. 2 (1997): 115ff; Manuel Castells, *The Power of Identity*, vol. 2 of *The Information Age* (Malden, MA: Blackwell, 1997); Jérôme Baschet, *L'étincelle zapatiste: Insurrection indienne et résistance planétaire* (Paris: Denoël, 2002); Rosado Marquez, *Colonisation, crise paysanne et développement durable dans les montagnes de la forêt Lacandon, Chiapas, Mexique*, in *Les montagnes tropicales: Identités, mutations, développement*, ed. François Bart, Serge Morin, and Jean-Noël Salomon (Pessac: Dynamiques des milieux et des sociétés dans les espaces tropicaux, 2001), 641–54.

72. Anne-Laure Amilhat-Szary, "Are Borders More Easily Crossed Today? The Paradox of Contemporary Trans-Border Mobilities in the Andes," *Geopolitics* 1 (2007): 1–18; Amilhat-Szary, "Ruralité, ethnicité et montagne," *Revue de géographie alpine* 97, no. 2 (2009); James Anderson, ed., *Transnational Democracy: Political Spaces and Border Crossings* (London: Routledge, 2002); Andolina, Radcliffe, and Laurie, "Gobernabilidaad e identidad."

CHAPTER TEN

1. Claire Waterton, "From Field to Fantasy: Classifying Nature, Constructing Europe," *Social Studies of Science* 32 (2002): 177–204.

2. Erik Swyngedouw, "Authoritarian Governance, Power and the Politics of Rescaling," *Environment and Planning D: Society and Space* 18 (2000): 63–76.

3. Lisa Sotto, "L'affirmation d'un nouvel espace géopolitique en Europe: Émergence et construction d'un espace alpin des transports," paper delivered at the 45th colloquium of the Association des Sciences Régionales de Langue Française, 2008, http://asrdlf2008.uqar.qc.ca/Papiers%20en%20ligne/SUTTO.pdf, accessed March 18, 2010.

4. Neil Brenner, "Globalization and Reterritorialization: The Re-scaling of Urban Governance in the European Union," *Urban Studies* 36 (1999): 431–51.

5. Council Directive 75/268/EEC of April 28, 1975, on mountain and hill farming and farming in certain less-favored areas.

6. Habitats Directive, 2007.

7. Data taken from Martin Price, ed., *Integrated Assessment of Europe's Mountain Areas* (Copenhagen: European Environmental Agency, 2010).

8. Céline Broggio, "Les enjeux d'une politique montagne pour l'Europe," *Revue de géographie alpine* 80, no. 4 (1992): 26–39.

9. Erik Gløersen, Kaisa Lähteenmäki-Smith, and Alexandre Dubois, "Polycentricity in Transnational Planning Initiatives: ESDP Applied or ESDP Reinvented?," in *European Territorial Governance*, ed. Wil Zonneveld, Jochem de Vries, and Leonie Janssen-Jansen (Delft: IOS Press, 2012): 237–62.

10. On the genesis of that concept, see Andreas Faludi, "Territorial Cohesion: An Unidentified Political Objective," *Town Planning Review* 76, no. 1 (2005): 1–13; Simin Davoudi, "Territorial Cohesion: An Agenda That Is Gaining Momentum," *Town and Country Planning* 73, nos. 7–8 (2004): 224–27; and Marjorie Jouen, *La cohésion territoriale, de la théorie à la pratique* (Paris: Notre Europe, 2008).

11. Assembly of European Regions, *Regions and Territories in Europe—The Regions' View of the Territorial Effects of European Policies* (Strasbourg: Assembly of European Nations, 1995). On the genesis of the idea of a Europe of regions, see Roderick Rhodes, "Regional Policy and a 'Europe of Regions': A Critical Assessment," *Regional Studies* 8, no. 2 (1974): 105–14.

12. In article 16 on services of general economic interest, which states that they must "work for social and territorial cohesion."

13. The Community Strategic Guidelines on Cohesion, 2007–13.

14. In keeping with André Sapir's report *An Agenda for a Growing Europe* (2003), which openly criticizes the agricultural and cohesion policies for their lack of cost-effectiveness.

15. European Commission, *Green Paper on Territorial Cohesion: Turning Territorial Diversity into Strength* (Brussels: Office for Official Publications of the European Communities, 2008).

16. Françoise Gerbaux, "Pressure Groups and the Defense of European Mountain Areas," *Journal of Alpine Research/Revue de géographie alpine* 92, no. 2 (2004): 27–37.

17. Luciano Caveri, "Cinq arguments pour une reconnaissance de la singularité des montagnes d'Europe," in *La montagne entre sciences et politiques*, ed. Bernard Debarbieux and Pierre-Antoine Landel (Grenoble: Dossier de la Revue de géographie alpine, 2002).

18. Frank Gaskell, "Interview of the President of Euromontana," *Inforegio panorama* 8 (2002): 2–3.

19. Opinion of the Economic and Social Committee, *The Future of Upland Areas in the EU* (2003/C 61/19), Brussels.

20. European Economic and Social Committee, *The Future of Upland Areas* (September 18, 2002), Brussels, article 5.4.

21. European Parliament: resolution of May 27, 1987; European Economic and Social Committee: opinion of the ESC, 461/88; Committee of the Regions: opinions of April 21, 1995, and September 18, 1997.

22. Committee of the Regions, "Report of the Committee of the Regions on 'Community Action for Mountain Areas'" (2003/C/128/05), *Official Journal of the European Union* 128 (2003): 25–40.

23. Opinion of the ESC, 461/88.

24. CEMAT (Conférence Européenne des Ministres Responsables de l'Aménagement du Territoire), *Guiding Principles for Sustainable Spatial Development of the European Continent* (Brussels: Conseil de l'Europe, 2002).

25. European Commission, *Reports on Economic, Social and Territorial Cohesion*, 2001, 2004, and 2007, Brussels.

26. European Commission, *Unity, Solidarity, Diversity for Europe, Its People and Its Territory: Second Report on Economic and Social Cohesion* (Brussels: European Communities, 2001).

27. Committee of the Regions, "Report."

28. European Commission, *Mountain Areas in Europe: Analysis of Mountain Areas in EU Member States, Acceding and Other European Countries* (Stockholm: Nordregio, 2004).

29. V. Kapos, J. Rhind, M. Edwards, C. Ravilious, and M. F. Price , "Developing a Map of the World's Mountain Forests," in *Forests in Sustainable Mountain Development: A State of Knowledge Report for 2000*, ed. M. F. Price and N. Butt (Wallingford, UK: CAB International, 2000), 4–9.

30. European Commission, *Mountain Areas in Europe*, 19.

31. Martin F. Price, Igor Lysenko, and Erik Gløersen, "Delineating Europe's Mountains," *Journal of Alpine Research/Revue de géographie alpine* 92, no. 2 (2004): 61–86, quotation 68. See also Erik Gløersen, "Renewing the Theory and Practice of European Applied Territorial Research on Mountains, Islands and Sparsely Populated Areas," *Regional Studies* 46, no. 4 (2012): 443–57.

32. The Swiss Confederation, though not a member of the European Union, had asked to be covered by that study, as had Norway. These two countries appear to be the most mountainous on the continent.

33. Philippe De Boe, Thérèse Hanquet, and Luc Maréchal, "Zones de montagne d'Europe . . . et de Wallonie," *Les cahiers de l'urbanisme* 57 (2005): 8–15.

34. *Panorama inforegio* 28 (December 2008): 6.

35. Committee of the Regions, Own-Initiative Opinion of the Committee of the Regions, "For a Green Paper—Towards a European Union Policy for Upland Regions: A European Vision for Upland Regions" (2008/C 257/07), *Official Journal of the European Union* 128 (2008): 36–40.

36. Stacy VanDeveer, "Ordering Environments: Regions in European International Environment Cooperation," in *Earthly Politics: Local and Global in Environmental Governance*, ed. Sheila Jasanoff and Marybeth Long-Martello (Cambridge, MA: MIT Press, 2004), 309–34.

37. Bernard Debarbieux, ed., "Mountain Regions as Referents for Collective Action," *Journal of Alpine Research/Revue de géographie alpine* 97, no. 2 (2009).

CHAPTER ELEVEN

1. Quoted in Trevor Sandwith, Clare Shine, Lawrence Hamilton, and David Sheppard, *Transboundary Protected Areas for Peace and Co-operation* (Gland: International Union for the Conservation of Nature—The World Conservation Union, 2001), vii.

2. Ibid. The International Union for the Conservation of Nature (IUCN) played a decisive role in the emergence and spread of the notion of peace parks.

3. The full official name is the "Protocol of Peace, Friendship, and Borders between Peru and Ecuador."

4. Ronald Bruce St. John, Rachel Bradley, and Clive H. Schofield , *The Ecuador-Peru Boundary Dispute: The Road to Settlement* (Durham: University of Durham Press, 1999), 23.

5. Alejandra Ruiz-Dana, "Peru and Ecuador: A Case Study of Latin American Integration and Conflict," in *Regional Trade Integration and Conflict Resolution*, ed. Shaheen Rafi Khan (New York: Routledge/IDRC, 2008); Gabriel Marcella, *War and Peace in the Amazon: Strategic Implications for the United States and Latin America of the 1995 Peru-Ecuador War* (Carlisle, PA: Army War College, 1995).

6. St. John, Bradley, and Schofield, *The Ecuado-Peru Boundary Dispute*, 43.

7. Sandwith, Shine, Hamilton, and Sheppard, *Transboundary Protected Areas*, 9. Kakabadse become president of the WWF in 2010.

8. Thomas Schulenberg and Kim Awbrey, *The Cordillera del Condor Region of Ecuador and Peru: A Biological Assessment*, RAP Working Papers 7 (Washington, DC: Conservation International, 1997).

9. Ken Conca, Alexander Carius, and Geoffrey Dabelko, "Building Peace through Environmental Cooperation," in *State of the World 2005: Redefining Global Security*, ed. Linda Starke (New York: Norton, 2005), 144–55, quotation 144; Russell A. Mittermeier, Cyril F. Kormos, Cristina Goettsch Mittermeier, and Patricio Robles Gil, *Transboundary Conservation: A New Vision for Protected Areas* (Mexico City: Cemex, 2005), 45. Arnaud Cuisinier-Raynal defended the critical view that this conservation policy was still the best means of cleaning up a border region and constituting a no-man's-land. Arnaud Cuisinier-Raynal, "La frontière au Pérou entre fronts et synapses," *L'espace géographique* 3 (2001): 213–30.

10. Ali Saleem Hassan, *Peace Parks: Conservation and Conflict Resolution* (Cambridge, MA: MIT Press, 2007).

11. Carlos F. Ponce and Martin Alcalde, "The Condor Corridor," *Tropical Forest Update* 13, no. 2 (2003): 13–14, quotation 14.

12. Charles Chester, "Transboundary Protected Areas," in *Encyclopaedia of Earth*, ed. Cutler J. Cleveland (Washington, DC: Environmental Information Coalition, National Council for Science and the Environment, 2008).

13. Helga Rainer, Stephen Asuma, Maryke Gray, Jose Kalpers, Anecto Kayitare, Eugene Rutagarama, Mbake Sivha, and Annette Lanjouw, "Regional Conservation in the Virunga-Bwindi Region: The Impact of Transfrontier Collaboration through the Experiences of the International Gorilla Conservation Programme," in *Transboundary Protected Areas: The Viability of Regional Conservation Strategies*, ed. Uromi Manage Goodale, Marc J. Stern, Cheryl Margoluis, Ashley G. Lanfer, and Matthew Fladeland (New York: Food Products Press, 2003), 189–204.

14. Trilateral Memorandum of Understanding between the Office Rwandais du Tourisme et des Parcs Nationaux, the Uganda Wildlife Authority, and the Institut Congolais pour la Conservation de la Nature.

15. Eric van Giessen, *Peace Park amid Violence? A Report on Environmental Security in the Virunga-Bwindi Region* (The Hague: Institute for Environmental Security, 2005).

16. Raymond F. Dasmann, *A System for Defining and Classifying Natural Regions for Purposes of Conservation*, Occasional Paper 9 (Morges: The International Union for the Conservation of Nature, 1973); Miklos D. F. Udvardy, *A Classification of the Biogeographical Provinces of the World*, Occasional Paper 18 (Morges: International Union for the Conservation of Nature, 1975).

17. www.worldwildlife.org.

18. David M. Olson and Eric Dinerstein, "The Global 200: A Representation Approach to Conserving the Earth's Most Biologically Valuable Ecoregions," *Conservation Biology* 12, no. 3 (1998): 502–15.

19. Peter Berg and Raymond Dasmann, "Reinhabiting California," *Ecologist* (1977), reprinted in *A Separate Country: A Bioregional Anthology of Northern California*, ed. Peter Berg (San Francisco: Planet Drum, 1978), 217–20. For an overview of the theoretical propositions of bioregionalism, see Kirkpatrick Sale, *Dwellers in the Land: The Bioregional Vision* (Philadelphia: New Society, 1985).

20. Charlotte Bretherton, "Ecocentric Identity and Transformatory Politics," *International Journal of Peace Studies* 6, no. 2 (2001).

21. Doug Aberley, *Boundaries of Home: Mapping for Local Empowerment* (Philadelphia: New Society Publishers, 1993), 71.

22. Ibid., 3.

23. Callahan. quoted in Daniel Press, "Environmental Regionalism and the Struggle for California," *Society and Natural Resources* 8 (1995): 289–306, quotation 303.

24. Stephen Frenkel, "Old Theories in New Places? Environmental Determinism and Bioregionalism," *Professional Geographer* 46, no. 3 (1994): 289–95, quotation 291.

25. http://www.cpawsyukon.org/conservation/y2y-excerpt.html, accessed December 15, 2009.

26. Chester Charles, "Landscape Vision and the Yellowstone to Yukon Conservation Initiative," in *Conservation across Borders: Biodiversity in an Interdependent World* (Washington, DC: Island Press, 2006), 134–216, quotation 145.

27. Michael Soule, "What Is Conservation Biology?" *BioScience* 35, no. 11 (1986): 727–34.

28. Heather J. Lynch, Stephanie Hodge, Christian Albert, and Molly Dunham, "The Greater Yellowstone Ecosystem: Challenges for Regional Ecosystem Management," *Environmental Management* 41, no. 6 (2008): 820–33.

29. Paul Paquet, the world specialist on wolves who conducted the Pluie tracking operation, says that "this was the founding story of Y2Y. . . . Really, the whole idea evolved out of it." Quoted in Cornelia Dean, "Wandering Wolf Inspires Project," *New York Times*, May 23, 2006.

30. Jodi A. Hilty, William Z. Lidicker, and Adina M. Merinlender, *Corridor Ecology: The Science and Practice of Linking Landscapes for Biodiversity Conservation* (Washington, DC: Island Press, 2006); Michel Soulé and John Terborgh, eds., *Continental Conservation* (Washington, DC: Island Press, 1999). The slogan adopted by the Yellowstone to Yukon Initiative—"making connections, naturally"—attests to its adherence to that corridor ecology.

31. Interview with David Mattson (U.S. Geological Survey), Cambridge, MA, June 13, 2008.

32. Harvey Locke, "Preserving the Wild Heart of North America: The Wildlands Project and the Yellowstone to Yukon Biodiversity Strategy," *Borealis: The Magazine for Canadian Parks and Wilderness* 15 (1994).

33. http://www.y2y.net, accessed November 15, 2009.

34. The principality of Monaco later joined the initiative.

35. www.parlament.ch/e/mm/2009/Pages/mm-urek-n-2009-11-10.aspx, accessed November 10, 2009.

36. Werner Bätzing, "La region alpine dans le sens de la délimitation de la Convention Alpine," in *La convention alpine*, ed. W. Danz and S. Ortner (Vaduz: Commission Internationale pour la Protection des Alpes, 1993), 236–47.

37. Note from René Gex-Fabry to M. R. Deffer for the Committee of the Swiss Center for Mountain Regions (SAB) of June 13, 1991, quoted in Gilles Rudaz, "Porter la voix de la montagne: Objectivation et différentiation du territoire par le Groupement de la population de la montagne du Valais romand (1945–2004)," Ph.D. diss., Université de Genève, 2005, 214.

38. Christina Del Biaggio, "The Institutionalization of the Alpine Region: An Analysis Based on a Study of Two Pan-Alpine Networks (Alliance in the Alps and Alparc)," *Journal of Alpine Research/Revue de géographie alpine* 97, no. 2 (2009).

39. http://club-arc-alpin.org, accessed February 12, 2014.

40. The ministers of the member states of the Alpine Convention promptly became ad-

vocates of transferring their model. After one of their meetings (in Merano, 2002), for example, they declared: "In the international context, we see the Alpine Convention as a model of sustainable development for transboundary mountain regions." Declaration of Merano, part 4, November 19, 2002.

41. Juliet Fall and Harald Egerer, "Constructing the Carpathians: The Carpathian Convention and the Search for a Spatial Ideal," *Journal of Alpine Research/Revue de géographie alpine* 92, no. 2 (2004): 98–106; Bernard Debarbieux, Martin Price and Jörg Balsiger, "The Institutionalization of Mountain Regions in Europe." *Regional Studies* (2013), http://dx.doi.org/10.1080/00343404.2013.812784.

42. Tom Knudson, "Majesty and Tragedy: The Sierra in Peril," *Sacramento Bee*, June 9–13, 1991.

43. John Muir, *The Mountains of California* (New York: Century, 1894), 3.

44. Quoted in Francis P. Farquhar, *History of the Sierra Nevada* (Berkeley and Los Angeles: University of California Press, 1995), 1. See also David Beesley, *Crow's Range: An Environmental History of the Sierra Nevada* (Reno: University of Nevada Press, 2004).

45. Jörg Balsiger, *Uphill Struggles: The Politics of Sustainable Mountain Development in Switzerland and California* (Cologne: Lambert, 2009); Jörg Balsiger, "The Impact of Ecoregional Mobilization on Mountain Policies in the Swiss Alps and California's Sierra Nevada," *Journal of Alpine Research/Revue de géographie alpine* 97, no. 2 (2009).

46. Knudson, "Majesty and Tragedy," 1.

47. http://sierranevadaalliance.org.

48. Sierra Summit Steering Committee, *The Sierra Nevada: Report of the Sierra Summit Steering Committee* (1992), 2.

49. Timothy Duane, *Shaping the Sierra: Nature, Culture, and Conflict in the Changing West* (Berkeley and Los Angeles: University of California Press, 1999). The inventors of that alternative form of activism thus embrace the "wise use movement," which in the United States designates the promotion of the environment and of nature for the benefit of society. That movement is often perceived as an alternative to radical environmentalism, even as an anti-environmentalism.

50. Leslie Laird, *Sierra Nevada Conservancy Act*, California Public Resources, 2004, code 33302; emphasis in original.

51. Gilles Rudaz, "The Sierra Nevada Lobby Day: Putting the 'Range of Light' on the Map," *Mountain Research and Development* 27, no. 4 (2007): 375–76.

52. Erik Swyngedouw, "Authoritarian Governance, Power and the Politics of Rescaling," *Environment and Planning D: Society and Space* 18 (2000): 63–76, quotation 63–64.

53. Juliet Fall, "Beyond Handshakes: Rethinking Cooperation in Transboundary Protected Areas as a Process of Individual and Collective Identity Construction," *Journal of Alpine Research/Revue de géographie alpine* 2 (2009): 61–73.

INDEX

The letter *f* following a page number denotes a figure.